W9-BCO-106

# AN ASSESSMENT OF THE SBIR PROGRAM AT THE DEPARTMENT OF DEFENSE

Committee for
Capitalizing on Science, Technology, and Innovation:
An Assessment of the Small Business Innovation Research Program

Policy and Global Affairs

Charles W. Wessner, *Editor*

NATIONAL RESEARCH COUNCIL
*OF THE NATIONAL ACADEMIES*

THE NATIONAL ACADEMIES PRESS
Washington, D.C.
**www.nap.edu**

**THE NATIONAL ACADEMIES PRESS 500 Fifth Street, N.W. Washington, DC 20001**

NOTICE: The project that is the subject of this report was approved by the Governing Board of the National Research Council, whose members are drawn from the Councils of the National Academy of Sciences, the National Academy of Engineering, and the Institute of Medicine. The members of the committee responsible for the report were chosen for their special competences and with regard for appropriate balance.

This study was supported by Contract/Grant No. DASW01-02-C-0039 between the National Academy of Sciences and U.S. Department of Defense, NASW-03003 between the National Academy of Sciences and the National Aeronautics and Space Administration, DE-AC02-02ER12259 between the National Academy of Sciences and the U.S. Department of Energy, NSFDMI-0221736 between the National Academy of Sciences and the National Science Foundation, and N01-OD-4-2139 (Task Order #99) between the National Academy of Sciences and the U.S. Department of Health and Human Services. The content of this publication does not necessarily reflect the views or policies of the Department of Health and Human Services, nor does mention of trade names, commercial products, or organizations imply endorsement by the U.S. Government. Any opinions, findings, and conclusions or recommendations expressed in this material are those of the author(s) and do not necessarily reflect the views of the National Aeronautics and Space Administration. Any opinions, findings, conclusions, or recommendations expressed in this publication are those of the author(s) and do not necessarily reflect the views of the organizations or agencies that provided support for the project.

International Standard Book Number-13: 978-0-309-10947-5
International Standard Book Number-10: 0-309-10947-7

Limited copies are available from the Policy and Global Affairs Division, National Research Council, 500 Fifth Street, N.W., Washington, D.C. 20001; 202-334-1529.

Additional copies of this report are available from the National Academies Press, 500 Fifth Street, N.W., Lockbox 285, Washington, DC 20055; (800) 624-6242 or (202) 334-3313 (in the Washington metropolitan area); Internet, http://www.nap.edu.

Copyright 2009 by the National Academy of Sciences. All rights reserved.

Printed in the United States of America

# THE NATIONAL ACADEMIES

*Advisers to the Nation on Science, Engineering, and Medicine*

The **National Academy of Sciences** is a private, nonprofit, self-perpetuating society of distinguished scholars engaged in scientific and engineering research, dedicated to the furtherance of science and technology and to their use for the general welfare. Upon the authority of the charter granted to it by the Congress in 1863, the Academy has a mandate that requires it to advise the federal government on scientific and technical matters. Dr. Ralph J. Cicerone is president of the National Academy of Sciences.

The **National Academy of Engineering** was established in 1964, under the charter of the National Academy of Sciences, as a parallel organization of outstanding engineers. It is autonomous in its administration and in the selection of its members, sharing with the National Academy of Sciences the responsibility for advising the federal government. The National Academy of Engineering also sponsors engineering programs aimed at meeting national needs, encourages education and research, and recognizes the superior achievements of engineers. Dr. Charles M. Vest is president of the National Academy of Engineering.

The **Institute of Medicine** was established in 1970 by the National Academy of Sciences to secure the services of eminent members of appropriate professions in the examination of policy matters pertaining to the health of the public. The Institute acts under the responsibility given to the National Academy of Sciences by its congressional charter to be an adviser to the federal government and, upon its own initiative, to identify issues of medical care, research, and education. Dr. Harvey V. Fineberg is president of the Institute of Medicine.

The **National Research Council** was organized by the National Academy of Sciences in 1916 to associate the broad community of science and technology with the Academy's purposes of furthering knowledge and advising the federal government. Functioning in accordance with general policies determined by the Academy, the Council has become the principal operating agency of both the National Academy of Sciences and the National Academy of Engineering in providing services to the government, the public, and the scientific and engineering communities. The Council is administered jointly by both Academies and the Institute of Medicine. Dr. Ralph J. Cicerone and Dr. Charles M Vest are chair and vice chair, respectively, of the National Research Council.

**www.national-academies.org**

## Committee for
## Capitalizing on Science, Technology, and Innovation: An Assessment of the Small Business Innovation Research Program

**Chair**
**Jacques S. Gansler (NAE)**
Roger C. Lipitz Chair in Public Policy and Private Enterprise
and Director of the Center for Public Policy and Private Enterprise
School of Public Policy
University of Maryland

**David B. Audretsch**
Distinguished Professor and Ameritech Chair of Economic Development
Director, Institute for Development Strategies
Indiana University

**Gene Banucci**
Executive Chairman
ATMI, Inc.

**Jon Baron**
Executive Director
Coalition for Evidence-Based Policy

**Michael Borrus**
Founding General Partner
X/Seed Capital

**Gail Cassell (IOM)**
Vice President, Scientific Affairs and
Distinguished Lilly Research Scholar for Infectious Diseases
Eli Lilly and Company

**Elizabeth Downing**
CEO
3D Technology Laboratories

**M. Christina Gabriel**
Director, Innovation Economy
The Heinz Endowments

**Trevor O. Jones (NAE)**
Founder and Chairman
Electrosonics Medical, Inc.

**Charles E. Kolb**
President
Aerodyne Research, Inc.

**Henry Linsert, Jr.**
CEO
Columbia Biosciences Corporation

**W. Clark McFadden**
Partner
Dewey & LeBoeuf, LLP

**Duncan T. Moore (NAE)**
Kingslake Professor of Optical Engineering
University of Rochester

**Kent Murphy**
President and CEO
Luna Innovations

**Linda F. Powers**
Managing Director
Toucan Capital Corporation

**Tyrone Taylor**
President
Capitol Advisors on Technology, LLC

**Charles Trimble (NAE)**
CEO, *retired*
Trimble Navigation

**Patrick Windham**
President
Windham Consulting

## PROJECT STAFF

**Charles W. Wessner**
Study Director

**McAlister T. Clabaugh**
Program Associate

**David E. Dierksheide**
Program Officer

**Sujai J. Shivakumar**
Senior Program Officer

**Adam H. Gertz**
Program Associate

**Jeffrey C. McCullough**
Program Associate

## RESEARCH TEAM

**Zoltan Acs**
University of Baltimore

**Alan Anderson**
Consultant

**Philip A. Auerswald**
George Mason University

**Robert-Allen Baker**
Vital Strategies, LLC

**Robert Berger**
Robert Berger Consulting, LLC

**Grant Black**
University of Indiana South Bend

**Peter Cahill**
BRTRC, Inc.

**Dirk Czarnitzki**
University of Leuven

**Julie Ann Elston**
Oregon State University

**Irwin Feller**
American Association for the Advancement of Science

**David H. Finifter**
The College of William and Mary

**Michael Fogarty**
University of Portland

**Robin Gaster**
North Atlantic Research

**Albert N. Link**
University of North Carolina

**Rosalie Ruegg**
TIA Consulting

**Donald Siegel**
University of California at Riverside

**Paula E. Stephan**
Georgia State University

**Andrew Toole**
Rutgers University

**Nicholas Vonortas**
George Washington University

**POLICY AND GLOBAL AFFAIRS**

Ad hoc Oversight Board for
Capitalizing on Science, Technology, and Innovation:
An Assessment of the Small Business Innovation Research Program


**Robert M. White (NAE), Chair**
University Professor Emeritus
Electrical and Computer Engineering
Carnegie Mellon University

**Anita K. Jones (NAE)**
Lawrence R. Quarles Professor of
Engineering and Applied Science
School of Engineering and Applied
Science
University of Virginia

**Mark B. Myers**
Senior Vice President, *retired*
Xerox Corporation

# Contents

## APPENDIXES

# Preface

Today's knowledge economy is driven in large part by the nation's capacity to innovate. One of the defining features of the U.S. economy is a high level of entrepreneurial activity. Entrepreneurs in the United States see opportunities and are willing and able to take on risk to bring new welfare-enhancing, wealth-generating technologies to the market. Yet, while innovation in areas such as genomics, bioinformatics, and nanotechnology present new opportunities, converting these ideas into innovations for the market involves substantial challenges.[1] The American capacity for innovation can be strengthened by addressing the challenges faced by entrepreneurs. Public-private partnerships are one means to help entrepreneurs bring new ideas to market.[2]

The Small Business Innovation Research (SBIR) program is one of the largest examples of U.S. public-private partnerships. An underlying thesis of the program is that small businesses are a strong source of new ideas and economic growth, but that it will be difficult to find financial support for these ideas in the early stages, thus the desirability for public-private partnerships in the small business, high-technology arena to encourage innovation and to help the government achieve its missions. Founded in 1982, the SBIR program was designed to encourage small business to develop new processes and products and to provide quality research in support of the many missions of the U.S. government. By

[1]See Lewis M. Branscomb, Kenneth P. Morse, Michael J. Roberts, Darin Boville, *Managing Technical Risk: Understanding Private Sector Decision Making on Early Stage Technology Based Projects*, Gaithersburg, MD: National Institute of Standards and Technology, 2000.

[2]For a summary analysis of best practice among U.S. public-private partnerships, see National Research Council, *Government-Industry Partnerships for the Development of New Technologies: Summary Report*, Charles W. Wessner, ed., Washington, DC: The National Academies Press, 2002.

including qualified small businesses in the nation's R&D (research and development) effort, SBIR grants are intended to stimulate innovative new technologies to help agencies meet the specific research and development needs of the nation in many areas, including health, the environment, and national defense.

As the SBIR program approached its twentieth year of operation, the U.S. Congress asked the National Research Council to conduct a "comprehensive study of how the SBIR program has stimulated technological innovation and used small businesses to meet federal research and development needs" and to make recommendations on still further improvements to the program.[3] To guide this study, the National Research Council (NRC) drew together an expert Committee that includes eminent economists, small businessmen and women, and venture capitalists, led by Dr. Jacques Gansler of the University of Maryland (formerly Undersecretary of Defense for Acquisition and Technology.) The membership of this Committee is listed in the front matter of this volume. Given the extent of "green-field research" required for this study, the Committee in turn drew on a distinguished team of researchers to, among other tasks, administer surveys and conduct case studies, and develop and analyze statistical information about the program. The membership of this research team is also listed in the front matter of this volume.

This report is one of a series published by the National Academies in response to the congressional request. The series includes reports on the Small Business Innovation Research Program at the Department of Defense, the National Institutes of Health, the National Aeronautics and Space Administration, the Department of Energy, and the National Science Foundation—the 5 agencies responsible for 96 percent of the program's operations. It includes, as well, an Overview Report that provides assessment of the program's operations across the federal government. Other reports in the series include a summary of the 2002 conference that launched the study, and a summary of the 2005 conference on *SBIR and the Phase III Challenge of Commercialization* that focused on the transition issues face by program participants at the Department of Defense and NASA.[4]

## PROJECT ANTECEDENTS

The current assessment of the SBIR program follows directly from an earlier analysis of public-private partnerships by the National Research Council's Board on Science, Technology, and Economic Policy (STEP). Under the direction of Gordon Moore, Chairman Emeritus of Intel, the NRC Committee on Government-Industry Partnerships prepared eleven volumes reviewing the driv-

[3]See the SBIR Reauthorization Act of 2000 (H.R. 5667, Section 108).

[4]National Research Council, *SBIR: Program Diversity and Assessment Challenges*, Charles W. Wessner, ed., Washington, DC: The National Academies Press, 2004. National Research Council, *SBIR and the Phase III Challenge of Commercialization*, Charles W. Wessner, ed., Washington, DC: The National Academies Press, 2007.

ers of cooperation among industry, universities, and government; operational assessments of current programs; emerging needs at the intersection of biotechnology and information technology; the current experience of foreign government partnerships and opportunities for international cooperation; and the changing roles of government laboratories, universities, and other research organizations in the national innovation system.[5]

This analysis of public-private partnerships included two published studies of the SBIR program. Drawing from expert knowledge at a 1998 workshop held at the National Academy of Sciences, the first report, *The Small Business Innovation Research Program: Challenges and Opportunities*, examined the origins of the program and identified operational challenges critical to the program's future effectiveness.[6] The report also highlighted the relative paucity of research on the SBIR program.

Following this initial report, the Department of Defense (DoD) asked the NRC to assess the Department's Fast Track Initiative in comparison with the operation of its regular SBIR program. The resulting report, *The Small Business Innovation Research Program: An Assessment of the Department of Defense Fast Track Initiative*, was the first comprehensive, external assessment of the Department of Defense's program. The study, which involved substantial case study and survey research, found that the SBIR program was achieving its legislated goals. It also found that DoD's Fast Track Initiative was achieving its objective of greater commercialization and recommended that the program be continued and expanded where appropriate.[7] The report also recommended that the SBIR program overall would benefit from further research and analysis, a recommendation subsequently adopted by the U.S. Congress.

## SBIR REAUTHORIZATION AND CONGRESSIONAL REQUEST FOR REVIEW

As a part of the 2000 reauthorization of the SBIR program, Congress called for a review of the SBIR programs of the agencies that account collectively for 96 percent of program funding. As noted, the five agencies meeting this criterion, by size of program, are the Departments of Defense, the National Institutes of

[5]For a summary of the topics covered and main lessons learned from this extensive study, see National Research Council, *Government-Industry Partnerships for the Development of New Technologies: Summary Report*, op. cit.

[6]See National Research Council, *The Small Business Innovation Research Program: Challenges and Opportunities*, Charles W. Wessner, ed., Washington, DC: National Academy Press, 1999.

[7]See National Research Council, *The Small Business Innovation Research Program: An Assessment of the Department of Defense Fast Track Initiative*, Charles W. Wessner, ed., Washington, DC: National Academy Press, 2000. Given that virtually no published analytical literature existed on SBIR, this Fast Track study pioneered research in this area, developing extensive case studies and newly developed surveys.

Health, the National Aeronautics and Space Administration, the Department of Energy, and the National Science Foundation.

HR 5667 directed the NRC to evaluate the quality of SBIR research and evaluate the SBIR program's value to the agency mission. It called for an assessment of the extent to which SBIR projects achieve some measure of commercialization, as well as an evaluation of the program's overall economic and noneconomic benefits. It also called for additional analysis as required to support specific recommendations on areas such as measuring outcomes for agency strategy and performance, increasing federal procurement of technologies produced by small business, and overall improvements to the SBIR program.

## ACKNOWLEDGMENTS

On behalf of the National Academies, we express our appreciation and recognition for the insights, experiences, and perspectives made available by the participants of the conferences and meetings, as well as by survey respondents and case study interviewees who participated over the course of this study. We are also very much in debt to officials from the leading departments and agencies. Among the many who provided assistance to this complex study, we are especially in debt to Kesh Narayanan, Joseph Hennessey, and Ritchie Coryell of the National Science Foundation; Michael Caccuitto, Victor Ciardello, and John Williams of the Department of Defense; Robert Berger and later Larry James of the Department of Energy; Carl Ray and Paul Mexcur of NASA; and Jo Anne Goodnight and Kathleen Shino of the National Institutes of Health.

The Committee's research team deserves major recognition for their role in the preparation of this report. Special thanks are due to Dr. Robin Gaster who stepped in to lead the DoD research team. Without his enormous energy, persistence, and productivity, this report would not have been completed. The DoD report and project as a whole are in debt to Peter Cahill, who made available his unparalleled knowledge of the program and its data. The important contributions made by Dr. Irwin Feller, who provided early, insightful, draft of the DoD study and conducted a large number of case studies are gratefully acknowledged. Paul Fowler also provided a valuable empirical perspective. Dr. Zoltan Acs carried out a number of case studies and contributed his valuable insights on the challenge of early-stage finance for innovative small businesses. Sujai Shivakumar also merits thanks for his careful review, edits, analysis, and written contributions, which were essential for the preparation of this report. Without collective efforts of these individuals, amidst many other competing priorities, it would not have been possible to prepare this report.

## NATIONAL RESEARCH COUNCIL REVIEW

This report has been reviewed in draft form by individuals chosen for their diverse perspectives and technical expertise, in accordance with procedures approved by the National Academies' Report Review Committee. The purpose of this independent review is to provide candid and critical comments that will assist the institution in making its published report as sound as possible and to ensure that the report meets institutional standards for objectivity, evidence, and responsiveness to the study charge. The review comments and draft manuscript remain confidential to protect the integrity of the process.

We wish to thank the following individuals for their review of this report: Robert Barnhill, Arizona State University; William Bonvillian, Massachusetts Institute of Technology; Bronwyn Hall, University of California, Berkeley; and Heidi Jacobus, Cybernet Systems Corporation.

Although the reviewers listed above have provided many constructive comments and suggestions, they were not asked to endorse the conclusions or recommendations, nor did they see the final draft of the report before its release. The review of this report was overseen by Robert Frosch, Harvard University, and Robert White, Carnegie Mellon University. Appointed by the National Academies, they were responsible for making certain that an independent examination of this report was carried out in accordance with institutional procedures and that all review comments were carefully considered. Responsibility for the final content of this report rests entirely with the authoring Committee and the institution.

Jacques S. Gansler Charles W. Wessner

# Summary

## I. INTRODUCTION

The Small Business Innovation Research (SBIR) program was created in 1982 through the Small Business Innovation Development Act. The program was designated as having four distinct purposes: "(1) to stimulate technological innovation; (2) to use small business to meet federal research and development needs; (3) to foster and encourage participation by minority and disadvantaged persons in technological innovation; and (4) to increase private sector commercialization innovations derived from federal research and development."[1]

As the SBIR program approached its 20th year of operation, the U.S. Congress requested the National Research Council (NRC) of the National Academies to conduct a "comprehensive study of how the SBIR program has stimulated technological innovation and used small businesses to meet federal research and development needs," and to make recommendations on improvements to the pro-

[1]Small Business Innovation Development Act (PL 97-219). In reauthorizing the program in 1992 (PL 102-564) Congress expanded the purposes "to emphasize the program's goal of increasing private sector commercialization developed through federal research and development and to improve the federal government's dissemination of information concerning the small business innovation research program, particularly with regard to program participation by woman-owned small business concerns and by socially and economically disadvantaged small business concerns." The evolution of the SBIR legislation was influenced by an accumulation of evidence beginning with David Birch in the late 1970s suggesting that small businesses were assuming an increasingly important role in both innovation and job creation. This trend gained greater credibility in the 1980s and was confirmed by empirical analysis, notably by Zoltan Acs and David Audretsch of the U.S. Small Business Innovation Data Base, which confirmed the increased importance of small firms in generating technological innovations and their growing contribution to the U.S. economy. See Zoltan Acs and David Audretsch, *Innovation and Small Firms*, Cambridge, MA: MIT Press, 1990.

gram.[2] Mandated as a part of SBIR's reauthorization in late 2000, the NRC study has assessed the SBIR program as administered at the five federal agencies that together make up some 96 percent of SBIR program expenditures. The agencies, in order of program size, are the Department of Defense (DoD), the National Institutes of Health (NIH)[3], the National Aeronautics and Space Administration (NASA), the Department of Energy (DoE), and the National Science Foundation (NSF).

Based on that legislation, and after extensive consultations with both Congress and agency officials, the NRC focused its study on two overarching questions.[4] First, how well do the agency SBIR programs meet four societal objectives of interest to Congress: (1) to stimulate technological innovation; (2) to increase private sector commercialization of innovations (3) to use small business to meet federal research and development needs; and (4) to foster and encourage participation by minority and disadvantaged persons in technological innovation.[5] Second, can the management of agency SBIR programs be made more effective? Are there best practices in agency SBIR programs that may be extended to other agencies' SBIR programs?

To satisfy the congressional request for an external assessment of the program, the NRC analysis of the operations of SBIR program involved multiple sources and methodologies. Extensive NRC commissioned surveys and case studies were carried out by a large team of expert researchers. In addition, agency-compiled program data, program documents, and the existing literature were reviewed. These were complemented by extensive interviews and discussions

---

[2]See Public Law 106-554, Appendix I—H.R. 5667, Section 108.

[3]The legislation designates the Department of Health and Human Services (DHHS) as the agency responsible for the SBIR program, and some components of DHHS, other than NIH, have SBIR programs. The DHHS program is dominated by NIH awards and the study's focus remains the NIH, in this case taken to represent the entire department.

[4]Three primary documents condition and define the objectives for this study: These are the Legislation—H.R. 5667, the NRC-Agencies *Memorandum of Understanding*, and the NRC contracts accepted by the five agencies. These are reflected in the Statement of Task addressed to the Committee by the Academies leadership. Based on these three documents, the NRC Committee developed a comprehensive and agreed set of practical objectives to be reviewed. These are outlined in the Committee's formal Methodology Report, particularly Chapter 3, "Clarifying Study Objectives." National Research Council, *An Assessment of the Small Business Innovation Research Program—Project Methodology*, Washington, DC: The National Academies Press, 2004, accessed at *<http://books.nap.edu/catalog.php?record_id=11097#toc>*.

[5]These congressional objectives are found in the Small Business Innovation Development Act (PL 97-219). In reauthorizing the program in 1992, (PL 102-564) Congress expanded the purposes to "emphasize the program's goal of increasing private sector commercialization developed through federal research and development and to improve the federal government's dissemination of information concerning small business innovation, particularly with regard to woman-owned business concerns and by socially and economically disadvantaged small business concerns."

**BOX S-1**
**Special Features of the Department of Defense SBIR Program**

**Scale.** The SBIR program at DoD is the largest of all the SBIR programs. At $943 million in 2005, DoD accounts for over half the program's funding.

**Diversity of Operation.** The program is spread across the three services and seven agencies involving widely different missions, ranging from missile defense to Navy submarines to Army support for special forces to the special needs of DARPA.

**The Acquisition Objective.** Unlike some major agency participants in the program (e.g., NIH & NSF), DoD seeks to acquire and use many of the technologies and products developed through the program. For many DoD officials, this is the primary objective of the program.

**Testing and Certification.** Given the demands inherent in the Defense mission, the services have stringent requirements for testing and certification that typically require substantial additional investments before commercialization is realized through DoD acquisition. Similarly, the long lead times involved in the procurement process means that careful attention must be paid to identify and integrate relevant SBIR projects if they are eventually to find a place in a weapons system.

**Innovation and Experimentation.** The department has launched a series of management initiatives and experiments in an effort to enhance the program's return, especially through greater commercialization, over the last decade. The department has also led the way in commissioning external evaluations of its program.

with program managers, program participants, agency 'users' of the program, as well as program stakeholders.[6]

The study as a whole sought to understand operational challenges and measure program effectiveness, including the quality of the research projects being conducted under the SBIR program, the challenges and achievements in commercialization of the research, and the program's contribution to accomplishing agency missions. To the extent possible, the evaluation included estimates of the benefits (both economic and noneconomic) achieved by the SBIR program, as well as broader policy issues associated with public-private collaborations for technology development and government support for high-technology innovation.

Taken together, this study is the most comprehensive assessment of SBIR to

[6]The Committee's methodological approach is described in National Research Council, *An Assessment of the Small Business Innovation Research Program—Project Methodology*, op. cit. For a summary of potential biases in innovation survey responses, see Box 4-1.

date. Its empirical, multifaceted approach to evaluation sheds new light on the operation of the SBIR program in the challenging area of early-stage finance. As with any assessment, particularly one across five quite different agencies and departments, there are methodological challenges. These are identified and discussed at several points in the text. This important caveat notwithstanding, the scope and diversity of the report's research should contribute significantly to the understanding of the SBIR program's multiple objectives, measurement issues, operational challenges, and achievements.

## II. SUMMARY OF KEY PROGRAM FINDINGS

The SBIR program at the Department of Defense is meeting the legislative and mission-related objectives of the program. The program is contributing directly to enhanced capabilities for the Department of Defense and the needs of those charged with defending the country. With regard to the specific legislated objectives of the program, the DoD SBIR program is:

- **Achieving significant levels of commercialization.**[7] Within DoD, commercialization can take multiple forms, sometimes involving insertion in the acquisition process, and/or direct sales to the government of through private commercial markets, licensing of technologies, and the acquisition of SBIR firms by larger Defense suppliers in the private sector. It is also true that the potential for private commercialization is sometimes inhibited by the very nature of the defense mission.
    - Commercialization and the potential for commercialization can be measured in a variety of fashions, ranging from sales and licensing to additional DoD or private investment to acquisition of the technology or the firm by other companies.
    - A significant proportion of the SBIR awards achieve commercialization, although other factors naturally also contribute to this process. For example, 46 percent of projects responding to the NRC Phase II Survey reported some sales or licensing revenues; a further 18 percent anticipated such revenues in the future.[8]
    - Commercial success tends to be concentrated. As is true with private sector early-stage projects (e.g., those funded by venture capital), a small number

[7]See Finding A in Chapter 2. A more detailed discussion of tools and metrics for assessing commercialization can be found in Chapter 4. It should be noted that the complex character of commercialization means that it cannot be captured in any individual metric.

[8]See Figure 4-1.

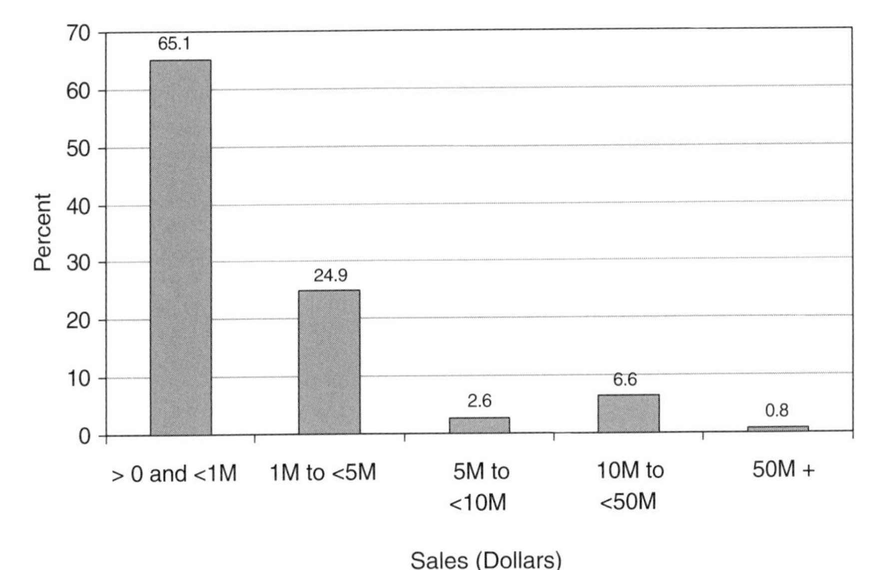


**FIGURE S-1** Distribution of sales, by size (percent of projects with >$0 in sales).
SOURCE: NRC Phase II Survey.

of firms account for a large proportion of commercial success.[9] Other SBIR firms often provide valuable services and products but do not reach the highest levels of commercial returns.[10]

---

[9]SBIR awards often occur earlier in the technology development cycle than where venture funds normally invest. Nonetheless, returns on venture funding tend to show the same high skew that characterizes commercial returns on the SBIR awards. See John H. Cochrane, "The Risk and Return of Venture Capital," *Journal of Financial Economics* 75(1):3-52, 2005. Drawing on the VentureOne database Cochrane plots a histogram of net venture capital returns on investments that "shows an extraordinary skewness of returns. Most returns are modest, but there is a long right tail of extraordinary good returns. 15 percent of the firms that go public or are acquired give a return greater than 1,000 percent! It is also interesting how many modest returns there are. About 15 percent of returns are less than 0, and 35 percent are less than 100 percent. An IPO or acquisition is not a guarantee of a huge return. In fact, the modal or 'most probable' outcome is about a 25 percent return." See also Paul A. Gompers and Josh Lerner, "Risk and Reward in Private Equity Investments: The Challenge of Performance Assessment," *Journal of Private Equity* 1(Winter 1977):5-12. Steven D. Carden and Olive Darragh, "A Halo for Angel Investors," *The McKinsey Quarterly* 1, 2004, also show a similar skew in the distribution of returns for venture capital portfolios.

[10]Unlike venture or angel investors, DoD SBIR funds research projects, not R&D companies as a whole. Angel investors or venture capitalists are an appropriate referent group, though not an appropriate group for direct comparison. Venture capital investors normally provide significant equity in exchange for ownership of a significant portion of the firm. Venture investors participate in firm governance and firm strategy and advise and recruit the firm's management team. SBIR awards essentially support projects.

- **Meeting the agency mission.** The DoD SBIR program is contributing significant enhancements to the department's mission capabilities.[11]
    - The innovations developed by SBIR companies are contributing to U.S. technological dominance, reducing the cost of operation of support systems, providing new capabilities, and providing increased responsiveness to new challenges (e.g., improvised explosive devices[12]).
    - Case studies indicate that SBIR projects have provided valuable mission technologies, some within a very short space of time, and others with very large impacts.[13]
    - The highly structured topic development process is designed to ensure that SBIR projects are aligned with specific Defense needs, and the department has made increased efforts to ensure that this is the case.
    - SBIR offers significant advantages for DoD with respect to its mission by providing:
        - Shorter planning horizons, which provide what agency staff see as unique flexibility within the execution year;
        - A low risk "technological probe" or research tool for finding new solutions, new technologies, and new suppliers, contributing to enhanced quality and capacity for systems and operations;
        - Access to technologies and providers otherwise largely excluded from the prime-dominated R&D process at DoD;[14]
        - Sole source contracts for successful technologies permitting more rapid acquisition with fewer constraints than normal procurements conducted under the Federal Acquisition Regulation (FAR).
    - In some cases, projects meet agency mission objectives without generating substantial commercial outcomes.

- **Supporting small business and competition.**[15] The DoD program provides substantial benefits for small business participants in terms of market access, funding, and recognition. The program supports a diverse array of small businesses contributing to the vitality of the defense industrial base while providing greater competition and new options and opportunities for DoD managers.
    - **New entrants.** The program attracts a substantial number (37 percent) of small business participants who are new to the program each year.[16]
    - **Formation of new, innovative companies.** A significant portion of re-

[11]See Finding B in Chapter 2.
[12]See Box 4-3.
[13]See Section 4.3.2 for references to SBIR companies.
[14]See evidence in Chapter 4, especially Section 4.3.
[15]See Finding C in Chapter 2.
[16]See Figure 3-3.

spondents to the NRC Firm Survey (25 percent) reported that they were founded entirely or partly because of an SBIR award.[17]

- **Support for academic researchers to transition ideas to the defense market.** About 25 percent of projects had some significant relationship to a university.[18]
- **Encouragement for university-industry and other partnerships.** DoD SBIR funding provides the resources for small firms to engage academic consultants and other private sector partners, as reflected in case studies.
- **Substantial impact on project initiation.** A large percentage of the surveyed firms (about 70 percent) reported that their project would definitely or probably not have gone ahead without the SBIR funding; many of the remainder indicated that they would have anticipated substantial delays without the SBIR award.[19]
- **Market recognition.** An SBIR award provides markets with additional information concerning the technical and commercial potential, generating a certification effect with regard to potential investors and customers.[20]

- **Supporting woman-owned small business concerns and by socially and economically disadvantaged small business concerns.**[21] The SBIR provides important support, notably in light of the contributions noted just above, to minority- and woman-owned firms.
  - **Trends in awards to woman-owned firms are positive.** Awards to woman-owned firms have continued to increase both in absolute numbers (303 Phase I awards in 2005) and as a percentage of the overall awards (12.9 percent of Phase I awards in 2005).[22]
  - **Trends in awards to minority-owned firms are more problematic.** The share of Phase I awards to minority-owned firms has declined substantially since the mid-1990s, falling below 10 percent for the first time in 2004, where it has remained. (See Figure S-2.) This trend in awards for minority

[17]See Table 4-15 (NRC Firm Survey, Question 1). Data reported in Table 4-15 are for firms with at least one DoD award. NRC Firm Survey results reported in Appendix B are for all agencies (DoD, NIH, NSF, DoE, and NASA).

[18]NRC Phase II Survey, Question 31. See also Table 4-19.

[19]See Figure 4-10 (NRC Phase II Survey, Question 13).

[20]Innovation awards "may be a signal to non-government sources of funding, such as banks, venture capital firms, and other potential investors, that the firm has a potential future stream of revenue from a reliable customer (the U.S. government.)" See Joshua Lerner, "'Public Venture Capital': Rationales and Evaluation," in National Research Council, *The Small Business Innovation Research Program: Challenges and Opportunities*, Charles W. Wessner, ed., Washington, DC: National Academy Press, 1999. See also Maryann P. Feldman and Maryellen R. Kelly, "Leveraging Research and Development: The Impact of the Advanced Technology Program," in National Research Council, *The Advanced Technology Program: Assessing Outcomes*, Charles W. Wessner, ed., Washington, DC: National Academy Press, 2001, p. 204.

[21]See Finding E in Chapter 2.

[22]See Figure 3-12 based on the DoD awards database.

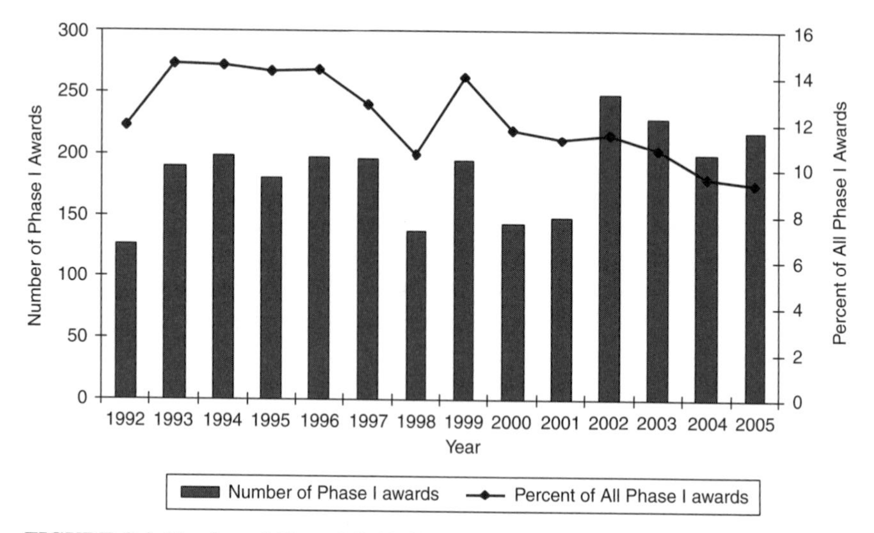


**FIGURE S-2** Number of Phase I DoD SBIR awards to minority-owned firms and percentage of all DoD Phase I awards, 1992–2005.
SOURCE: DoD awards database.

firms suggests the need for further empirical examination. There may be a need for enhanced outreach by the department.[23]

- **Knowledge generation.** The DoD SBIR program is generating significant intellectual capital. The program contributes to new scientific and technological knowledge, as SBIR companies have generated numerous publications and patents. This knowledge takes many forms, not all of which are easily measured.[24]
    - **Publications.** Thirty-two percent of DoD survey respondents published at least one peer-reviewed article based on the SBIR project surveyed by the NRC, and about 15 percent of them generated more than five articles.[25]
    - **Patents.** Nearly 35 percent of the DoD projects surveyed generated at least one patent application and just over 25 percent had received a patent related to the project.[26]
    - **Indirect effects.** Interviews by the research team offer strong anecdotal evidence with regard to indirect effects of projects that provide investigators, research staff, and DoD management with knowledge that can become useful in a different context. In some cases, knowledge is transferred from

[23]Based on DoD awards database.
[24]See Finding F in Chapter 2.
[25]See Figure 4-18 (NRC Phase II Survey, Question 18).
[26]See Table 4-18.

one company to another; in others, the knowledge becomes relevant in a different technical context.[27]

- ◦ **Nonlinear innovation paths.** By its three-phase structure, the SBIR program suffers from an implicit linear myth, namely that a single grant for a single project is sufficient to fully develop a technology and drive long-term growth of the company. In practice, a single grant is often not sufficient to commercialize a product. Often multiple related projects, complementary technologies, and varied funding sources are needed, in addition to effective management, to bring a product to market. With regard to additional awards, NRC survey data indicate that returns are maximized where firms have received 10-25 previous Phase II awards.[28]

## III. SUMMARY OF KEY PROGRAM RECOMMENDATIONS

**As noted in the Findings section above, the Department of Defense has an effective SBIR program. The recommended improvements listed below should enable the DoD SBIR managers to address the four mandated congressional goals in a more efficient and effective manner.**

- **Improve the Phase III transition.** DoD should continue and expand its work on improving the Phase III transition (the transition from SBIR-funded Phase I and Phase II research to further research along the road toward commercialization—especially testing and evaluation funded by other DoD sources).[29] Areas for possible action include:
  - ◦ **Provide incentives.** Expansion of positive incentives for program officers to utilize the SBIR program for their own research needs, beyond the current requirements for their involvement in topic development.[30]
  - ◦ **Adopt roadmaps.** Inclusion of SBIR in the technology and program development roadmaps that guide programs through the TRL stages, and eventually through acquisitions.[31]
  - ◦ **Identify and apply best practices across services and between services.** Some components have been particularly successful at Phase III transition. Techniques used by these components may be transferable.[32]

---

[27]See Section 4.4.5.4. See also Box 4-6 in Chapter 4 for a case study of Advanced Ceramics Research (ACR). The value of ACR's low-cost, small unmanned aerial vehicle (UAV), initially developed for whale watching around Hawaii, became apparent in Iraq, where it is used as a highly flexible, general-purpose battlefield surveillance tool.

[28]See Section 4.4.3.2.

[29]See Recommendation A in Chapter 2.

[30]See Recommendation A-1 in Chapter 2.

[31]See Recommendation A-5 in Chapter 2.

[32]See Recommendation D-3 and Box 2-3.

    - **Expand linkages.** Improved linkages with other programs (e.g., ManTech) might be enhanced to facilitate the Phase III transition.[33]
    - **Interest primes.** Encouragement for the growing interest of the prime contractors on the SBIR Program's outputs and opportunities for partnering with SBIR companies. Consideration should be given to performance incentives for further encouraging development of SBIR supported technologies.[34]
    - **Expand commercialization assistance.** Expanded commercialization programs that provide training, counseling, and networking opportunities.[35]
- **Develop program profile.** DoD leadership should take steps to improve the perception of the SBIR program's potential and accomplishments within the department, address obstacles to fulfilling the program's potential, and seek resources to enhance program operation and outcomes as a means of generating the attention and partnering needed for the program to operate most effectively.[36]
- **Encourage pilot programs.** Making changes initially through pilot programs allows DoD to alter selected areas on a provisional basis; a uniform approach is unlikely to work well for all components of a program that funds highly diverse projects with very different capital requirements and very different product development cycles.[37] DoD should also identify best practices across the SBIR program, and implement them as appropriate, across the Department.[38]
- **Improve management and assessment.** Additional funding should be provided for program management and assessment in order to encourage and support the development of a results-oriented SBIR program. Effective management requires additional staff and funding. To manage the program effectively requires better monitoring of awardees, enhanced efforts to facilitate commercialization, the regular collection of higher-quality data and its more systematic assessment. Currently, sufficient resources are not available for these functions.
- **Expand evaluation.** DoD should substantially strengthen and expand its evaluation efforts in order to further develop a program culture that is driven by outcomes and backed by internal and external evaluations.[39]
    - DoD—like the other SBIR agencies—should be encouraged to develop and provide to Congress a comprehensive annual report on SBIR. This

[33]See Recommendation A-6 in Chapter 2. The U.S. Army Manufacturing Technology (ManTech) Program supports the development and implementation of advanced manufacturing technologies for the production of Army Material.

[34]See Recommendation A-8 in Chapter 2.

[35]See Recommendation A-9 in Chapter 2.

[36]See Recommendation B in Chapter 2.

[37]See Recommendation E-1 in Chapter 2.

[38]See Box 2-2.

[39]See Recommendation C in Chapter 2.

will enhance program accountability. Such a publication will require additional management funding (noted above).
- DoD should develop methodologies and capabilities that allow for a sharing of best practices across its services and defense agencies.
- DoD should enhance existing efforts to develop the collection of data needed to evaluate program outcomes.

- **Provide management funds.** To enhance program utilization, management, and evaluation, as called for above, consideration should be given to the provision of additional program management and evaluation funds. There are three ways that this might be achieved:[40]
  - Additional funds might be allocated internally, within the existing budgets of the services and agencies, as the Navy has done.
  - Funds might be drawn from the existing set-aside for the program to carry out these activities.
  - The set-aside for the program, currently at 2.5 percent of external research budgets, might be modestly increased, with the goal of providing additional resources for management and evaluation to maximize the program's return to the nation.[41]
- **Increase the participation and success rates of woman- and minority-owned firms:**[42]
  - **Improve data collection and analysis.** The Committee strongly encourages the agencies to gather and report the data that would track woman and minority firms as well as principal investigators (PIs), and to ensure that SBIR is an effective road to opportunity.
  - **Encourage participation.** Develop targeted outreach to improve the participation rates of woman- and minority-owned firms, and strategies to improve their success rates.
  - **Encourage emerging talent.** Encourage woman and minority scientists and engineers with the advanced degrees to serve as principal investigators (PIs) and/or senior co-investigators (Co-Is) on SBIR projects.

---

[40]See Recommendation F in Chapter 2.

[41]Each of these options has its advantages and disadvantages. For the most part, the departments, institutes, and agencies responsible for the SBIR program have not proved willing or able to make additional management funds available. Without direction from Congress, they are unlikely to do so. With regard to drawing funds from the program for evaluation and management, current legislation does not permit this and would have to be modified; therefore the Congress has clearly intended program funds to be for awards only. The third option, involving a modest increase to the program, would also require legislative action and would perhaps be more easily achievable in the event of an overall increase in the program. In any case, the Committee envisages an increase of the "set-aside" of perhaps 0.03 percent to 0.05 percent on the order of $35 million to $40 million per year, or roughly double what the Navy currently makes available to manage and augment its program. In the latter case (0.05 percent), this would bring the program "set-aside" to 2.55 percent, providing modest resources to assess and manage a program that is approaching an annual spend of some $2 billion. Whatever modality adopted by the Congress, without additional resources the Committee's call for improved management, data collection, experimentation, and evaluation may prove moot.

[42]See Recommendation G in Chapter 2.

# 1

# Introduction

## 1.1 SBIR—PROGRAM CREATION AND ASSESSMENT

Created in 1982 by the Small Business Innovation Development Act. the Small Business Innovation Research (SBIR) program was designed to stimulate technological innovation among small private-sector businesses while providing the government cost-effective new technical and scientific solutions to challenging mission problems. SBIR was also designed to help to stimulate the U.S. economy by encouraging small businesses to market innovative technologies in the private sector.[1]

As the SBIR program approached its twentieth year of existence, the U.S.

[1]The SBIR legislation drew from a growing body of evidence, starting in the late 1970s and accelerating in the 1980s, which indicated that small businesses were assuming an increasingly important role in both innovation and job creation. David L. Birch, "Who Creates Jobs?" *The Public Interest* 65:3-14, 1981. This evidence gained new credibility with the Phase I empirical analysis by Zoltan Acs and David Audretsch of the U.S. Small Business Innovation Data Base, which confirmed the increased importance of small firms in generating technological innovations and their growing contribution to the U.S. economy. See Zoltan Acs and David Audretsch, *Innovation and Small Firms*, Cambridge MA: MIT Press, 1990. For the importance of small businesses to job creation, see also Steven J. Davis, John Haltiwanger, and Scott Schuh, "Small Business and Job Creation: Dissecting the Myth and Reassessing the Facts," *Business Economics* 29(3):113-122, 1994. More recently, a report by the Organization for Economic Cooperation and Development (OECD) notes that small and medium-sized enterprises are attracting the attention of policy makers, not least because they are seen as major sources of economic vitality, flexibility, and employment. Small business is especially important as a source of new employment, accounting for a disproportionate share of job creation. See Organisation for Economic Co-operation and Development, *Small Business Job Creation and Growth: Facts, Obstacles, and Best Practices*, Paris: Organisation for Economic Co-operation and Development, 1997.

Congress requested that the National Research Council (NRC) of the National Academies conduct a "comprehensive study of how the SBIR program has stimulated technological innovation and used small businesses to meet federal research and development needs," and make recommendations on improvements to the program.[2] Mandated as a part of SBIR's renewal in December 2000, the NRC study has assessed the SBIR program as administered at the five federal agencies that together make up 96 percent of SBIR program expenditures. The agencies are, in decreasing order of program size: the Department of Defense (DoD), the National Institutes of Health (NIH), the National Aeronautics and Space Administration (NASA), the Department of Energy (DoE), and the National Science Foundation (NSF). The SBIR program at DoD is the largest of all the SBIR programs. At $943 million in 2005, DoD accounts for over half the program's funding.

The NRC Committee assessing the SBIR program was not asked to consider if SBIR should exist or not—Congress has affirmatively decided this question on three occasions.[3] Rather, the Committee was charged with providing an evidence based assessment of the program's operations, achievements, and challenges as well as recommendations to improve the program's effectiveness.

## 1.2 SBIR PROGRAM STRUCTURE

Eleven federal agencies are currently required to set aside 2.5 percent of their extramural research and development budget exclusively for SBIR contracts. Each year these agencies identify various R&D topics, representing scientific and technical problems requiring innovative solutions, for pursuit by small businesses under the SBIR program. These topics are bundled together into individual agency "solicitations"—publicly announced requests for SBIR proposals from interested and qualifying small businesses. A small business can identify an appropriate topic it wants to pursue from these solicitations and, in response, propose a project for an SBIR grant, a process now immensely facilitated by the Internet. The required format for submitting a proposal is different for each agency. Proposal selection also varies, though peer review of proposals on a competitive basis by experts in the field is typical. Each agency then selects the proposals that are found best to meet program selection criteria, and awards contracts or grants to the proposing small businesses. Since the SBIR program's inception at DoD, all SBIR awards have been contracts awarded on a competitive basis.

As conceived in the 1982 Act, SBIR's grant-making process is structured in three phases at all agencies:

- Phase I grants essentially fund feasibility studies in which award win-

[2]See Public Law 106-554, Appendix I—H.R. 5667, Section 108.

[3]These are the 1982 Small Business Development Act, and the subsequent multiyear reauthorizations of the SBIR program in 1992 and 2000.

ners undertake a limited amount of research aimed at establishing an idea's scientific and commercial promise. Today, the legislation anticipates Phase I grants as high as $100,000.[4]

- Phase II grants are larger—the legislated amount is $750,000—and fund more extensive R&D to further develop the scientific and commercial promise of research ideas.
- Phase III. During this phase, companies do not receive additional funding from the SBIR program. Instead, grant recipients should be obtaining additional funds from a procurement program (if available) at the agency that made the award, from private investors, or other sources of capital. The objective of this phase is to move the technology from the prototype stage to the marketplace.

### The Phase III Challenge

Obtaining Phase III support is often the most difficult challenge for new firms to overcome. In practice, agencies have developed different approaches to facilitate SBIR grantees' transition to commercial viability; not least among them are additional SBIR grants.[5] The multiple approaches taken to address the Phase III challenge are described in Chapter 5. The Department of Defense has shown considerable initiative in its efforts to enhance commercialization and capture returns for the program. Unlike some major agency participants in the program (e.g., NIH & NSF), DoD seeks to acquire and use many of the technologies and products developed through the SBIR program.

Previous NRC research has shown that firms have different objectives in applying to the program. Some want to demonstrate the potential of promising research but may not seek to commercialize it themselves. Others seek to fulfill agency research requirements more cost-effectively through the SBIR program than through the traditional procurement process. Still others seek a certification of quality (and the investments that can come from such recognition) as they push science-based products towards commercialization.[6]

---

[4]With the agreement of the Small Business Administration, which plays an oversight role for the program, this amount can be substantially higher in certain circumstances and is also often lower, especially with smaller SBIR programs, e.g., EPA or the Department of Agriculture.

[5]The Phase III challenge was explored at a conference convened at the National Academies on June 14, 2005. The proceedings of this conference are reported in National Research Council, *SBIR and the Phase III Challenge of Commercialization*, Charles W. Wessner, ed., Washington, DC: The National Academies Press, 2007.

[6]See Reid Cramer, "Patterns of Firm Participation in the Small Business Innovation Research Program in Southwestern and Mountain States," in National Research Council, *The Small Business Innovation Research Program: An Assessment of the Department of Defense Fast Track Initiative*, Charles W. Wessner, ed., Washington, DC: National Academy Press, 2000.

## 1.3 SBIR REAUTHORIZATIONS

The SBIR program approached reauthorization in 1992 amidst continued concerns about the U.S. economy's capacity to commercialize inventions. Finding that "U.S. technological performance is challenged less in the creation of new technologies than in their commercialization and adoption," the National Academy of Sciences at the time recommended an increase in SBIR funding as a means to improve the economy's ability to adopt and commercialize new technologies.[7]

Following this report, the Small Business Research and Development Enhancement Act of 1992 (P.L. 102-564), which reauthorized the SBIR program until September 30, 2000, doubled the set-aside rate to 2.5 percent.[8] This increase in the percentage of R&D funds allocated to the program was accompanied by a stronger emphasis on the commercialization of SBIR-funded technologies.[9] Legislative language explicitly highlighted commercial potential as a criterion for awarding SBIR grants. For Phase I awards, Congress directed program administrators to assess whether projects have "commercial potential," in addition to scientific and technical merit, when evaluating SBIR applications.

The 1992 legislation mandated that program administrators consider the existence of second-phase funding commitments from the private sector or other non-SBIR sources when judging Phase II applications. Evidence of third-phase follow-on commitments, along with other indicators of commercial potential, was also to be sought. Moreover, the 1992 reauthorization directed that a small business's record of commercialization be taken into account when evaluating its Phase II application.[10]

The Small Business Reauthorization Act of 2000 (P.L. 106-554) extended SBIR until September 30, 2008. It called for a two-phase assessment by the

---

[7]See National Research Council, *The Government Role in Civilian Technology: Building a New Alliance*, Washington, DC: National Academy Press, 1992, p. 29.

[8]For fiscal year 2003, this has resulted in a program budget of approximately $1.6 billion across all federal agencies, with the Department of Defense having the largest SBIR program at $834 million, followed by the National Institutes of Health (NIH) at $525 million. The DoD SBIR program, is made up of 10 participating components: Army, Navy, Air Force, Missile Defense Agency (MDA), Defense Advanced Research Projects Agency (DARPA), Chemical Biological Defense (CBD), Special Operations Command (SOCOM), Defense Threat Reduction Agency (DTRA), National Imagery and Mapping Agency (NIMA), and the Office of Secretary of Defense (OSD). NIH counts 23 separate institutes and agencies making SBIR awards, many with multiple programs.

[9]See Robert Archibald and David Finifter, "Evaluation of the Department of Defense Small Business Innovation Research Program and the Fast Track Initiative: A Balanced Approach," in National Research Council, *The Small Business Innovation Research Program: An Assessment of the Department of Defense Fast Track Initiative*, op. cit, pp. 211-250.

[10]A GAO report had found that agencies had not adopted a uniform method for weighing commercial potential in SBIR applications. See U.S. General Accounting Office, *Federal Research: Evaluations of Small Business Innovation Research Can Be Strengthened*, AO/RCED-99-114, Washington, DC: U.S. General Accounting Office, 1999.

National Research Council of the broader impacts of the program.[11] The goals of the SBIR program, as set out in the 1982 legislation, are: "(1) to stimulate technological innovation; (2) to use small business to meet federal research and development needs; (3) to foster and encourage participation by minority and disadvantaged persons in technological innovation; and (4) to increase private sector commercialization innovations derived from federal research and development.

## 1.4 STRUCTURE OF THE NRC STUDY

This NRC assessment of SBIR has been conducted in several ways. In an exceptional step, at the request of the agencies, a formal research methodology was developed by the NRC. This methodology was then reviewed and approved by an independent National Academies panel of experts.[12] As the research began, information about the program was also gathered through interviews with SBIR program administrators and during two major conferences where SBIR officials were invited to describe program operations, challenges, and accomplishments.[13] These conferences highlighted the important differences in the goals, and practices of the SBIR program at each agency. The conferences also explored the challenges inherent in assessing such a diverse range of program objectives and practices and the limits of using common metrics across agencies with significantly different missions and objectives.

Implementing the approved research methodology, the NRC Committee deployed multiple survey instruments and its researchers conducted a large number of case studies that captured a wide range of SBIR firms. The Committee then evaluated the results and developed both agency-specific and overall findings and recommendations for improving the effectiveness of the SBIR program at each agency. This report includes a complete assessment of the operations and achievements of the SBIR program at DoD and makes recommendations as to how it might be further improved.

---

[11]The current assessment is congruent with the Government Performance and Results Act (GPRA) of 1993: <*http://govinfo.library.unt.edu/npr/library/misc/s20.html*>. As characterized by the GAO, GPRA seeks to shift the focus of government decision making and accountability away from a preoccupation with the activities that are undertaken—such as grants dispensed or inspections made—to a focus on the results of those activities. See <*http://www.gao.gov/new.items/gpra/gpra.htm*>.

[12]The SBIR methodology report is available on the Web. National Research Council, *An Assessment of the Small Business Innovation Research Program—Project Methodology*, Washington, DC: The National Academies Press, 2004, accessed at <*http://books.nap.edu/catalog.php?record_id=11097#toc*>.

[13]The opening conference on October 24, 2002 examined the program's diversity and assessment challenges. For a published report of this conference, see National Research Council, *SBIR: Program Diversity and Assessment Challenges*, Charles W. Wessner, ed., Washington, DC: The National Academies Press, 2004. The second conference, held on March 28, 2003 was titled, "Identifying Best Practice." The conference provided a forum for the SBIR Program Managers from each of the five agencies in the study's purview to describe their administrative innovations and best practices.

## 1.5 SBIR ASSESSMENT CHALLENGES

### Program Diversity and Flexibility

At its outset, the NRC's SBIR study identified a series of assessment challenges that must be addressed. As the October 2002 conference made clear, the administrative flexibility found in the SBIR program makes it difficult to make cross-agency assessments. Although each agency's SBIR program shares the common three-phase structure, the SBIR concept is interpreted uniquely at each agency. At DoD, the program is spread across the three services and seven agencies involving widely different missions, ranging from missile defense to Navy submarines to Army support for special forces to the special needs of DARPA.

This flexibility is a positive attribute in that it permits each agency to adapt its SBIR program to the agency's particular mission, scale, and working culture. For example, NSF operates its SBIR program differently than DoD because "research" is often coupled with procurement of goods and services at DoD but normally not at NSF. Programmatic diversity means that each agency's SBIR activities must be understood in terms of their separate missions and operating procedures. While commendable in itself, this diversity of objectives, procedures, mechanisms, and management makes an assessment of the program as a whole more challenging.

### Nonlinearity of Innovation

A second challenge concerns the linear process of commercialization implied by the design of SBIR's three phase structure.[14] In the linear model, illustrated in Figure 1-1, innovation begins with basic research supplying a steady stream of fresh and new ideas. From among these ideas, those that show technical feasibility become innovations. Such innovations, when further developed by firms, can become marketable products driving economic growth.

As NSF's Joseph Bordogna observed at the launch conference, innovation almost never takes place through a protracted linear progression from research to development to market. Research and development drives technological innovation, which, in turn, opens up new frontiers in R&D. True innovation, Bordogna noted, can spur the search for new knowledge and create the context in which the next generation of research identifies new frontiers. This nonlinearity, illustrated in Figure 1-2, underscores the challenge of assessing the impact of the SBIR

[14]This nonlinear perception was underscored by Duncan Moore: "Innovation does not follow a linear model. It stops and starts." See the National Research Council, *SBIR: Program Diversity and Assessment Challenges*, op. cit.

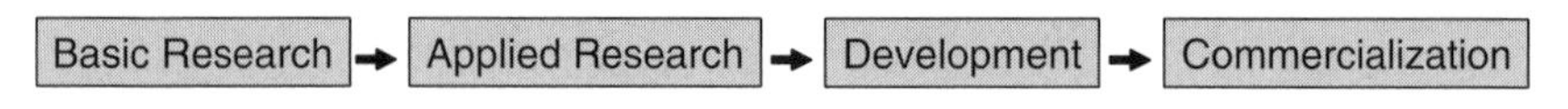


**FIGURE 1-1** The Linear Model of Innovation.

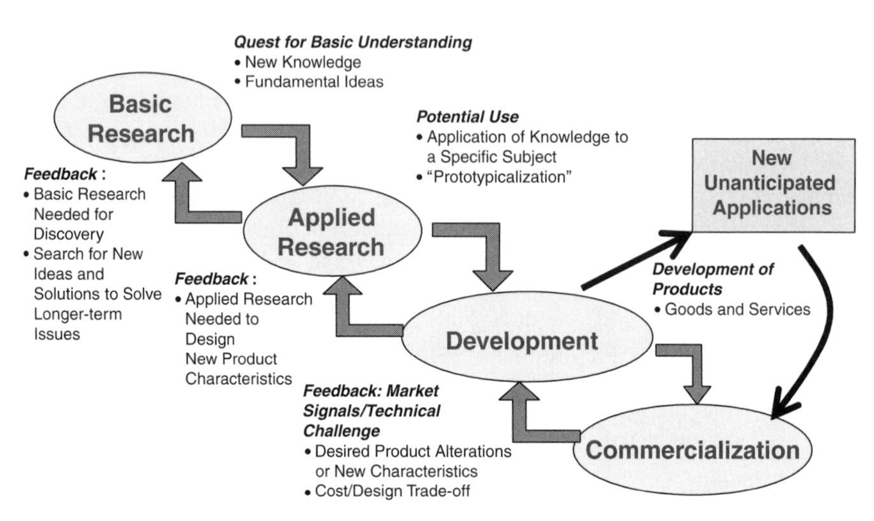


**FIGURE 1-2** A Feedback Model of Innovation.

program's individual awards. Inputs do not match up with outputs according to a simple function.[15]

## Measurement Challenges

A third assessment challenge relates to the measurement of outputs and outcomes. Program realities can and often do complicate the task of data gathering. In some cases, for example, SBIR recipients receive a Phase I award from one agency and a Phase II award from another. In other cases, multiple SBIR awards may have been used to help a particular technology become sufficiently mature to reach the market. Also complicating matters is the possibility that for any particular grantee, an SBIR award may be only one among other federal and

---

[15]For a higher level view that pure research and applied research can be considered as independent variables rather than as the extremes of a linear dichotomy of pure vs. applied research, see Donald E. Stokes, *Pasteur's Quadrant, Basic Science and Technological Innovation*, Washington, DC: Brookings Institution Press, 1997.

nonfederal sources of funding. Causality can thus be difficult, if not impossible, to establish.

The task of measuring outcomes is also made harder because companies that have garnered SBIR awards can also merge, fail, or change their name before a product reaches the market. In addition, principal investigators or other key individuals can change firms, carrying their knowledge of an SBIR project with them. A technology developed using SBIR funds may eventually achieve commercial success individually, at an entirely different company than the one that received the initial SBIR award.

### Gauging Commercial Success

Complications plague even the apparently straightforward task of assessing commercial success. For example, research enabled by a particular SBIR award may take on commercial relevance in new unanticipated contexts. At the launch conference, Duncan Moore, former Associate Director of Technology at the White House Office of Science and Technology Policy (OSTP), cited the case of SBIR-funded research in gradient index optics that was initially considered a commercial failure when an anticipated market for its application did not emerge. Years later, however, products derived from the research turned out to be a major commercial success.[16] Today's apparent dead end can sometimes be a lead to a major achievement tomorrow, while others are, indeed, dead ends. Yet, even technological dead ends have their value, especially if they can be determined for the low costs associated with an SBIR award.

Gauging commercialization is also difficult when the product in question is destined for public procurement. The challenge is to develop a satisfactory measure of how useful an SBIR-funded innovation has been to an agency mission. A related challenge is determining how central (or even useful) SBIR awards have proved to be in developing a particular technology or product. Often, multiple SBIR awards and other funding sources contribute to the development of a product or process for DoD. In some cases, the Phase I award can meet the agency's need—completing the research with no further action required. In other cases, Phase II awards, supplemental funding, and substantial management and financial resources are required for "success."

Measurement challenges are substantial. For example, one way of measuring commercialization success is to count product sales. Another is to focus on the products developed using SBIR funds that are procured by DoD. In practice, however, large procurements from major suppliers are typically easier to track than products from small suppliers such as SBIR firms. In other cases, successful Phase II awards are just that—they meet the agency need and no further commer-

[16]Duncan Moore, "Turning Failure into Success" in National Research Council, *SBIR: Program Diversity and Assessment Challenges*, op. cit., p. 94.

cialization takes place. In other cases, substantial commercialization occurs, and then ceases as a promising firm or technology is acquired by a defense supplier.

Moreover, successful development of a technology or product does not always translate into successful "uptake" by the procuring agency. Often, the absence of procurement may have little to do with the product's quality or the potential contribution of SBIR. Small companies, especially new entrants to the program, entail greater risk for program officers. Perceived uncertainties about reliability, timeliness of supply, and risks of program delays all militate against acquisition of successful technologies from new, unproven firms.

## Understanding and Anticipating Failure

Understanding failure is equally challenging. By its very nature, an early-stage program such as SBIR should anticipate a significant failure rate. The causes of failure are many. The most straightforward, of course, is *technical failure*, where the research objectives of the award are not achieved. In some cases, the project can be a technically successful but a *commercial failure*. This can occur when a procuring agency changes its mission objectives and hence its procurement priorities. NASA's new Mars Mission is one example of a *mission shift* that may result in the cancellation of programs involving SBIR awards to make room for new agency priorities. Cancelled weapons system programs at the Department of Defense can have similar effects.

Technologies procured through SBIR may also *fail in the transition to acquisition*. Some technology developments by small businesses do not survive the long lead times created by complex testing and certification procedures required by the Department of Defense. Indeed, small firms encounter considerable difficulty in surmounting the long lead times, high costs, and complex regulations that characterize defense acquisition. In addition to complex federal acquisition procedures, there are strong disincentives, noted above, for high-profile projects to adopt untried technologies. Technology transfer in commercial markets can be equally difficult. A *failure to transfer to commercial markets* can occur even when a technology is technically successful if the market is smaller than anticipated, competing technologies emerge or are more competitive than expected, or the product is not adequately marketed. Understanding and accepting the varied sources of project failure in the high-risk, high-reward environment of cutting-edge R&D is a challenge for analysts and policy makers alike.

## Evaluating SBIR: "Compared to What?"

This raises the issue concerning the standard by which SBIR programs should be evaluated. An assessment of SBIR must take into account the expected distribution of successes and failures in early-stage finance. As a point of comparison, Gail Cassell, Vice President for Scientific Affairs at Eli Lilly, has noted that only one in ten innovative products in the biotechnology industry will turn out to be a commercial success.[17] Similarly, venture capital funds often achieve considerable commercial success on only two or three out of twenty or more investments.[18]

In short, commercial success tends to be concentrated. Yet, commercial success is not the only metric of the program. At the Defense Department, SBIR can and does provide a variety of valuable services and products that do not achieve widespread commercial success, even if they do have sales or licensing revenue.

In setting metrics for SBIR projects, therefore, it is important to have a realistic expectation of the success rate for competitive awards to small firms investing in promising but unproven technologies. Similarly, it is important to have some understanding of what can be reasonably expected—that is, what constitutes "success" for an SBIR award, and some understanding of the constraints and opportunities successful SBIR awardees face in bringing new products to market. This is especially relevant in the case of a constrained, regulation-driven market such as the defense procurement market. From the management perspective, the rate of success also raises the question of appropriate expectations and desired levels of risk taking. A portfolio that always succeeds would not be pushing the technology envelope. A very high rate of "success" would, thus, paradoxically suggest an inappropriate use of the program. Even when technical success is achieved, as noted above, it does not automatically transfer into commercial success for a variety of reasons related to the defense mission and to procurement procedures. Understanding the nature of success and the appropriate benchmarks for a program with this focus is therefore important to understanding the SBIR program and the approach of this study.

---

[17]Gail Cassell, "Setting Realistic Expectations for Success," in National Research Council, *SBIR: Program Diversity and Assessment Challenges*, op. cit., p. 86.

[18]SBIR awards often occur earlier in the technology development cycle than where venture funds normally invest. Nonetheless, returns on venture funding tend to show the same high skew that characterizes commercial returns on the SBIR awards. See John H. Cochrane, "The Risk and Return of Venture Capital," *Journal of Financial Economics* 75(1):3-52, 2005. Drawing on the VentureOne database Cochrane plots a histogram of net venture capital returns on investments that "shows an extraordinary skewness of returns. Most returns are modest, but there is a long right tail of extraordinary good returns. Fifteen percent of the firms that go public or are acquired give a return greater than 1,000 percent! It is also interesting how many modest returns there are. About 15 percent of returns are less than 0, and 35 percent are less than 100 percent. An IPO or acquisition is not a guarantee of a huge return. In fact, the modal or "most probable" outcome is about a 25 percent return." See also Paul A. Gompers and Josh Lerner, "Risk and Reward in Private Equity Investments: The Challenge of Performance Assessment," *Journal of Private Equity* 1(Winter 1977):5-12. Steven D. Carden and Olive Darragh, "A Halo for Angel Investors," *The McKinsey Quarterly* 1, 2004 also show a similar skew in the distribution of returns for venture capital portfolios.

## 1.6 SBIR ASSESSMENT RESULTS

Drawing on interviews, multiple survey instruments and case studies, and overcoming many of the research challenges identified above, the NRC Committee has developed a number of findings and practical recommendations for improving the effectiveness of the SBIR program at the Department of Defense.

The Committee found that the SBIR program at DoD is, in general, meeting the legislative and mission-related objectives of the program. The program is contributing directly to enhanced capabilities for the Department of Defense and the needs of those charged with defending the country.

Further, the Committee found that the DoD program also provides substantial benefits for small business participants in terms of market access, funding, and recognition. The program supports a diverse array of small businesses contributing to the vitality of the defense industrial base while providing greater competition and new options and opportunities for DoD managers. In addition, the Committee noted that the DoD SBIR program is generating significant intellectual capital, contributing to new scientific and technological knowledge, and generating numerous publications and patents.

The Committee's recommended improvements to the program have been designed to enable the DoD SBIR managers to address the program's congressional goals more efficiently and effectively. These include further work to improve the Phase III transition by (among other approaches) changing incentives faced by program managers so that they are motivated to make better use of the SBIR program. The Committee also recommends that additional funding should be provided for program management and assessment in order to encourage and support the development of an innovative and results-oriented SBIR program. The Committee's complete findings and recommendations are listed in Chapter 2.

Chapter 3 provides a comprehensive overview of the distribution of SBIR awards by DoD, providing a basis (as drawn out in Chapter 4) for understanding program outcomes. Chapter 5 describes the Phase III challenge of commercialization at DoD. Chapter 6 describes the diversity of management structures as well as current practices and recent reforms found among the different services and agencies that fund SBIR programs at DoD. Together, this report provides the most detailed and comprehensive picture to date of the SBIR program at the Department of Defense.

# 2

# Findings and Recommendations

## NRC STUDY FINDINGS

The SBIR program at the Department of Defense is meeting the legislative and mission-related objectives of the program. The program is contributing directly to enhanced capabilities for the Department of Defense and the needs of those charged with defending the country.

A. **A substantial percentage of SBIR projects at DoD commercialize.**[1]

1. **The NRC Phase II Survey, which was sent to all firms with Phase II awards from 1992 to 2002, provides evidence of substantial, if highly skewed, commercialization.**[2]

   - Nearly half (46 percent) of respondents indicated that the surveyed SBIR project had reached the marketplace (i.e., they reported more than $0 in sales and licensing revenues from the project by May 2005, which is the closing date of the survey).[3]

[1]All data in this section are drawn from the NRC Phase II Survey, unless otherwise stated. Commercialization refers here to the extent to which projects generate outcomes that have market value. Commercialization in the context of DoD also refers to the take-up of projects within DoD, often (but not always) in the context of Phase III funding from non-SBIR resources. This aspect of commercialization is taken up in Section 4.3.

[2]See Appendix B for a detailed description of the survey, response rate, and related issues. For DoD, the response rate was 42 percent of the awards contacted. See also National Research Council, *Assessment of the Small Business Innovation Research Program—Project Methodology*, Washington, DC: The National Academies Press, 2004, accessed at *<http://books.nap.edu/catalog.php?record_id=11097#toc>*.

[3]See Figure 4-1 (NRC Phase II Survey, Question 1 and Question 4).

- Of the 420 projects reporting some sales, just under 1 percent reported sales greater than $50 million, another 9.2 percent of projects reporting some commercialization, indicated sales between $5 million and $50 million.[4]

- In addition, 17.6 percent of respondents reported sales by licensees of their technology, with three reporting licensee sales of greater than $50 million.[5,6]

- For projects that have received sales, survey responses indicate that 87.6 percent of *first sales* occurred within 4 years of the Phase II award date.[7] Interviews and cases, however, support the view that the *bulk of sales* will be realized in the longer run—that is, beyond the date of first sale.

- These figures, while positive, necessarily reflect the concentration and skewed outcomes often associated with early-stage funding and the special challenge of the procurement process. The figures also understate, perhaps substantially, the amount of commercialization ultimately to be generated from the funded projects. It is important to recognize that these data constitute only a snapshot of sales and licensing revenues, as of May 2005. Projects completed in more recent years will continue to generate revenues well into the future. Consequently, the data aggregated for the May 2005 snapshot necessarily underreports the eventual return from the SBIR Phase II awards that were made during the latter part of the study period (1992–2002).[8]

---

[4]See Figure 4-2 (NRC Phase II Survey, Question 4).

[5]This type of "skew"—in which a majority of projects fail or are minimally successfully while a small proportion generates large revenues—is typical of early-stage finance and has been noted in previous Academy research. See National Research Council, *The Small Business Innovation Research Program: An Assessment of the Department of Defense Fast Track Initiative*, Charles W. Wessner, ed., Washington, DC: National Academy Press, 2000. See also Joshua Lerner, "Public Venture Capital: Rationales and Evaluation," in National Research Council, *The Small Business Innovation Research Program: Challenges and Opportunities*, Charles W. Wessner, ed., Washington, DC: National Academy Press, 1999.

[6]NRC Phase II Survey, Question 4.

[7]Ibid.

[8]The total eventual return from these awards is estimated to be approximately 50 percent higher than the data captured at the time of the survey. (For an explanation of the methodology underlying this analysis, see Chapter 4.) This suggests that the actual sales and licensing revenues that will in the end be generated by projects funded during the study period, on average, are approximately $2.2 million, and about $5.6 million for each project that did report some sales or licensing revenues. For DoD as a whole, the SBIR Program Manger Michael Caccuitto, reports that the amount of commercialization generated from SBIR projects now leads the total amount spent on SBIR, with about a 4-year lag from the year of Phase II award. See National Research Council, *SBIR and the Phase III Challenge of Commercialization*, Charles W. Wessner, ed., Washington, DC: The National Academies Press, 2007. This corresponds with the findings of the 1992 GAO report's assessment of commercialization, which found that not enough time had elapsed since the program's inception for projects to mature. See U.S.

2. **Success in attracting further research funding for ongoing development offers evidence indicating that a project is on the path to commercialization, even if no sales have yet been made.**

    ◦ Over one-quarter of projects that received additional funding reported the acquisition of additional funds from other federal sources and 13.2 percent reported funding from other companies.[9] This suggests significant interest in these projects—not least from DoD, and possibly also among the prime contractors (a likely source of funding from "other companies").

    ◦ Venture capital is not widely available to companies primarily focused on the Defense market. Hurdles associated with regulations in federal acquisition, and the limited size of many defense markets tends to limit venture funds' interest in the DoD market. Only 30 projects—3.8 percent of respondents with some additional funding—reported receiving venture capital,[10] although the average VC investment that was received is much higher than the average investment received in each other category, at more than $5 million per project.[11]

3. **Additional SBIR awards are a further signal of commercial potential.**

    ◦ Given that SBIR is a highly competitive program, the acquisition of related SBIR awards also suggests that a project is moving along the development path toward commercialization, not least because commercialization potential has become a significant component in the decision to make an SBIR award. 43.5 percent of respondents indicated that they had received at least one additional related SBIR award.[12]

B. **SBIR is in broad alignment with the needs of the DoD agencies and components.**

1. **DoD's SBIR program has contributed to significant enhancements of its mission capabilities.**

    ◦ A central mission of the DoD SBIR program is to use the inventiveness of small companies to solve DoD's technical problems, and to develop new technologies that can be applied to the weapons and logistics systems that are eventually used by the Armed Forces.[13]

---

General Accounting Office, *Small Business Innovation Research Program Shows Success But Can Be Strengthened*, RCED–92–32, Washington, DC: U.S. General Accounting Office, 1992.

[9]Derived from NRC Phase II Survey, Question 23. See also Table App-A-37.

[10]Ibid.

[11]NRC Phase II Survey, Question 23. See also Table App-A-37. Similar to sales, the amount of venture funding is skewed with only eight projects reporting $5 million or more in venture funding.

[12]See Table 4-12.

[13]Interviews with SBIR program managers.

**BOX 2-1**
**ArmorWorks, Inc.—Body Armor in Iraq**

Technologies developed from SBIR-funded research efforts were used in the design of Small Arms Protective Inserts (SAPI) Body Armor Plates used in the Interceptor Vest currently being worn by U.S. service men and women in the Middle East. ArmorWorks has been awarded more than $50 million in contracts from the Army and Marines to produce SAPI plates for body armor, making them a leading producer for the U.S. military. To date, some 350,000 SAPI plates have been produced for the Department of Defense.

ArmorWorks also manufactures vehicle armor. The company's SBIR research contributed to the design of HMMWV and add-on armor kits for trucks currently in use in the Middle East. The vehicles armor produced by ArmorWorks has a number of valuable features for the battlefield, including easy installation (requires no vehicle modification or special tools) and field configurability (contains simplified installation to allow for reconfiguration for specific missions). ArmorWorks recently received another $30 million contract from the U.S. Defense Logistics Agency to produce and deliver Kevlar inserts to protect U.S. troops in Iraq and Afghanistan against small arms fire.

---

SOURCE: DoD SBIR Success Stories, *<http://www.dodsbir.net/SuccessStories/armorworks.htm>*.

- Improved mission capabilities in the context of DoD relate to maintaining technological dominance in battle space conditions, increased responsiveness to new, unexpected situations, such as responding to improvised explosive devices (IEDs), and reductions in the cost of operations and support systems.
- DoD SBIR program managers also speak favorably about the creativity of the small- and medium-sized firms that comprise the SBIR community.[14]

2. **DoD appears to be making a concerted effort to ensure that SBIR projects are aligned with the needs of the weapons system managers.**[15]
   - The topic development process has since 1999 incorporated important elements specifically designed to align acquisitions needs and SBIR topics.[16]

---

[14]The agency's Technical Point of Contacts (TPOCs) were surveyed by the NRC. See NRC Project Manager Survey in National Research Council, *An Assessment of the SBIR Program at the National Aeronautics and Space Administration*, Charles W. Wessner, ed., Washington, DC: The National Academies Press, 2009.

[15]A number of initiatives in this direction are described in Chapter 5: The Phase III Challenge.

[16]See Section 6.2.

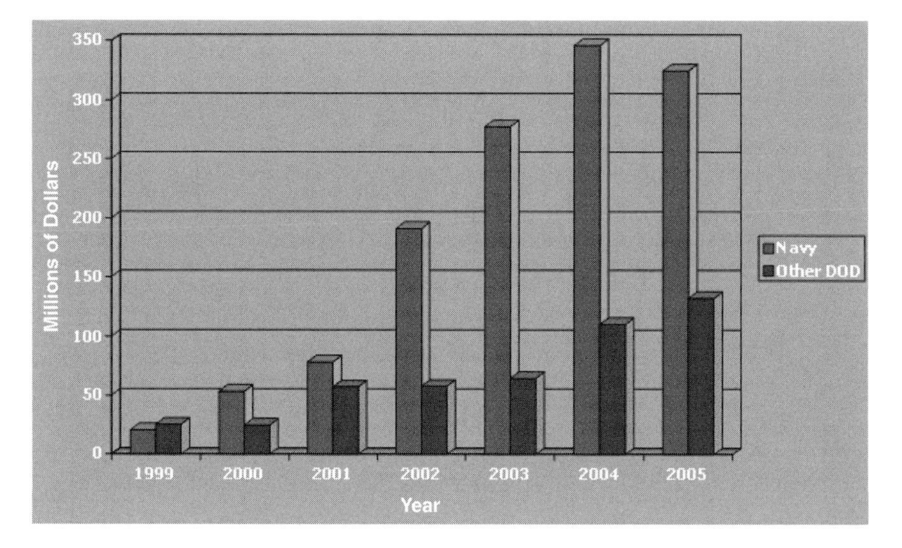


**FIGURE 2-1** Reported Phase III contract awards value 1999–2005.
SOURCE: John Williams, Navy, April 7, 2005.

- According to DoD managers, 60 percent of topics published in the SBIR solicitation are now sponsored or otherwise supported by acquisition agencies.[17]
- Based on DoD data, identified Phase III contracts[18] now total more than $450 million[19] annually, and have been growing rapidly in recent years.

3. **SBIR offers an unusual degree of execution year flexibility and a short**

[17]The Navy finds that 80 percent of its topics are acquisition linked. Comments by Dr. Holland and Mr. Caccuitto at the NRC Phase III Symposium, June 2005. See National Research Council, *SBIR and the Phase III Challenge of Commercialization*, op. cit.

[18]Many Phase III contracts are not captured effectively by the DD350 reporting system, which undercounts the number and size of Phase III contracts awarded, to a varied degree at different DoD components. See Chapter 5. In addition, commercialization occurring intraindustry, at a subcontract level, or in the commercial marketplace (and then perhaps finding its way to DoD) is not measured by this data source. The DD350 has been replaced by the Federal Procurement Data System—Next Generation (FPDS-NG) as the system tracking, and thus characterizing, all prime contract actions. This system is now standard for the federal government, with a small number of exceptions.

[19]$450 million is based on the DD 350: however, the 2005 DoD Annual SBIR Report to SBA lists by contract $565 million in known Phase III contracts and that listing is considered to be incomplete. It includes only contracts known to the SBIR program managers. It should be kept in mind that these contracts are for direct sales to DoD or direct further DoD R&D funding. Sales and further R&D from DoD primes are not included. The NRC Phase II Survey and the DoD Commercialization data indicate the DoD sales and funding are actually less than half of the total commercialization from DoD SBIR.

**planning horizon, permitting the program to rapidly address urgent mission needs.**

- SBIR-supported innovations have contributed to enhanced U.S. combat capabilities, and provided technological solutions to meet sudden, unexpected challenges to our military. Indeed, the high degree of flexibility characteristic of small firms means that SBIR has provided DoD with an increased number of suppliers capable of quickly responding on short notice to unexpected battlefield situations.[20]
- For example, the Navy has taken advantage of this flexibility to issue a "quick response IED topic" in 2004, and made 38 Phase I awards within 5 months of topic development. These developed into 18 Phase II awards, and results from these will be available in 2006–2007. The first prototypes were expected in Iraq in the Fall of 2006.
- By contrast, most RTD&E accounts require considerable forward planning.

4. **SBIR increases the number of potential suppliers for new technologies, and also creates new opportunities for these firms to partner together in new undertakings.**
   - Used effectively, SBIR can act as a low-risk, low-cost technology probe and a search tool for finding new technology suppliers. It has helped DoD personnel learn about new technologies, new applications, and a new set of high-tech firms with whom they would not otherwise have contact.[21]
   - The laws governing the SBIR program permit the use of sole source procedures when federal agencies acquire technologies developed with SBIR funding. This allows substantially faster acquisition than through standard channels and acts as a powerful incentive for SBIR firms and their partners.
5. **The quality of SBIR-funded research is broadly comparable to that of other non-SBIR research according to the NRC Project Manager Survey.**

[20]See, for example, the case of ArmorWorks, Inc., in Box 2-1.

[21]Firms express this outcome as follows: Had it not been for SBIR, their business with DoD services or agencies would not have developed. Services likely would have stayed with their pre-existing sources of supply. Program managers are normally too busy administering multiple contracts to search out or respond attentively to new sources of technology. Their propensity is to hire a contractor to solve problems rather than seek out the most technologically innovative performer. The SBIR program requires that program managers become involved with small firms, to look at technical options, and to allow for increased competition in the selection of R&D performers.

- In the NRC Project Manager Survey, 53 percent of Technical Points of Contact (TPOC) respondents indicated that the specific SBIR project identified in the survey produced results that were useful to them and that they had followed up on this work with other research. SBIR projects are normally part of a wider portfolio of research responsibilities handled by TPOCs.[22]
- The NRC Project Manager Survey also indicated that the quality of SBIR funded research is comparable to non-SBIR research they manage. Normalized survey scores indicate that the quality of SBIR research is equivalent to that of other research at DoD.[23]

6. **The Department has devoted considerable recent effort to strengthen the critical connection between SBIR and the acquisition programs through Phase III. While this focus is to be commended, the Phase III process can be considerably widened and improved.**
   - A striking aspect of SBIR Phase III at DoD is the extraordinarily uneven character of outcomes and activities between services, and between components within services.
   - Notably, Phase III transitions at PEO SUBS account for approximately 86 percent of all Navy Phase III contracts, and Navy in turn accounts for about 70 percent of all DoD Phase III contracts, as captured by the DD350 forms completed by contracting officers.[24]
   - This skew partly results from the additional effort made by Navy to ensure that DD350 forms are completed and accurately reflect SBIR contributions, which in turn reflects different views of the importance of supporting SBIR, as evidenced by the amount of resources, staff, and funding that services and components allocate for program support.
   - At some components, such as Navy, senior management recognizes the potential value of SBIR and has supported extensive efforts to build effective bridges between SBIR and the acquisition programs. At other components, efforts have been less well supported, and on the basis

[22]See the related discussion in Section 4.3.1.2.

[23]The scores were normalized scores by removing the outliers in the top and bottom 5 percent of scores. Statistical procedures often assume that the variables are normally distributed. A significant violation of the assumption of normality can seriously increase the chances of a Type I (overestimation) or Type II (underestimation) error. Nonnormality can occur in the presence of outliers (scores that are extreme relative to the rest of the sample). Removing the outliers can improve the normality of the distribution. See C. M. Judd and G. H. McClelland, *Data Analysis: A Model-Comparison Approach*. San Diego, CA: Harcourt Brace Jovanovich, 1989.

[24]See the discussion in Section 5.2, including Figure 5-1.

of the data presented in Chapter 5, they appear to be considerably less effective.

7. **There is no effective and comprehensive tracking system within DoD to follow SBIR-funded technologies to their final outcome.**

    - The Company Commercialization Report is self-reported data, and must be updated only when a company applies for a new DoD SBIR.

    - The agency's DD350 reporting system may substantially undercount Phase III awards, as contracting officers must be specifically trained to capture this data correctly.[25]

    - The lack of a reliable and effective tracking system for SBIR awards that would identify follow-on funding sets back efforts to assess the impact of the program and to document its successes.

C. **SBIR awards made by DoD support small businesses in a number of important ways.**

1. **SBIR awards have had a substantial impact on participating companies.**

    - **Company Creation.** Just over 25 percent of companies responding to the NRC Firm Survey indicated that they were founded entirely or partly because of a prospective SBIR award.[26]

    - **The Decision to Initiate Research.** Only 13 percent of DoD project respondents thought that their project would "definitely" or "probably" have gone ahead without SBIR funding. Over two-thirds (about 70 percent) thought they definitely or probably would not have initiated the research; most of those who anticipated that their project would have gone ahead without the award acknowledged the likelihood of substantial delays without the award.[27]

    - **Company Growth.** Almost half (48 percent) of the respondents in-

[25]There are multiple limitations to current systems for tracking SBIR awards. The DD350 reporting system can be used to extract some SBIR data; however, the system was not designed to gather SBIR award data, and is used differently by the services and agencies. For example the Navy reports Phase II Enhancements as Phase III, while the other services report such awards as Phase II. The Phase I and Phase II data in the DD350 does not match the SBIR budget or DoD Annual Report to SBA. The Annual Report to SBA, prepared by the DoD SBIR program managers, accounts for the budgeted SBIR funding. The SBIR program managers also have less knowledge of the Phase III awards, since these are made with funds that are not under control of the SBIR program.

[26]See Table 4-15 (NRC Firm Survey, Question 1). Data reported in Table 4-15 are for firms with at least one DoD award. NRC Firm Survey results reported in Appendix B are for all agencies (DoD, NIH, NSF, DoE, and NASA).

[27]See Figure 4-10 (NRC Phase II Survey, Question 13).

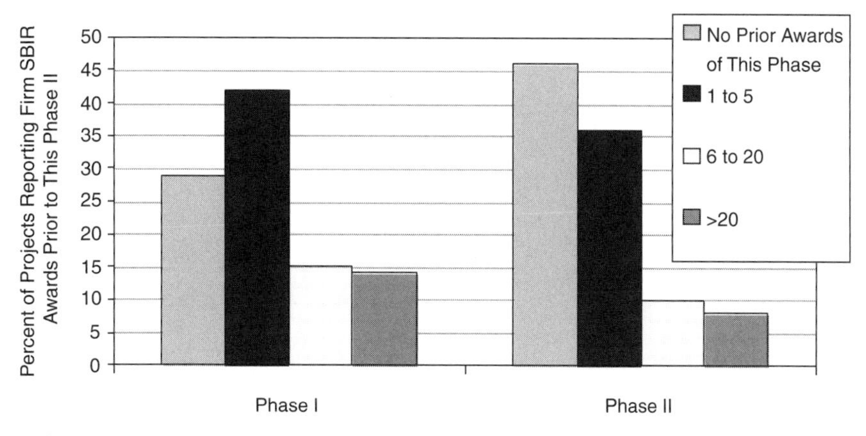


**FIGURE 2-2** New winners at DoD.
SOURCE: NRC Phase II Survey.

dicated that more than half the growth experienced by their firm was directly attributable to SBIR.[28]

- **Partnering.** SBIR funding is often used by small firms to gain access to outside resources, especially academic consultants and, often, to seek company partners.

2. **SBIR Awards Attract Participation by New Firms.**
   - FY2005 data from DoD show that 29 percent of Phase I awards went to firms that had not previously won a DoD Phase II SBIR award.
   - DoD data indicate that a further 50 percent of awards went to companies with five or fewer previous Phase II awards. Only 13 percent of awards went to companies with at least six Phase II awards.[29]
   - Data from the Naval Air Systems Command (NAVAIR) show that about half of all its Phase I contracts go to companies that have never won an SBIR award from NAVAIR before. In addition, about 40–45 percent of Phase II contracts go to newly participating firms.[30]

D. **Multiple SBIR awards serve multiple objectives for firms and agencies.** A small number of companies receive multiple awards. A few companies have

[28]See Table 4-16.

[29]Michael Caccuitto and Carol van Wyk, "Enhancing the Impact of Small Business Innovation Research (SBIR) Program: The Commercialization Pilot Program," Presentation, September 27, 2006.

[30]See Section 3.2.2.

received more than 100 awards at DoD (over 20 years), and are higher in commercialization (on average).[31]

1. **The myth of the linear innovation model:**

    - The program has often encountered criticism with regard to firms—sometimes labeled SBIR "mills"—that are described as winning large numbers of awards but that achieve lower levels of commercialization than firms with significantly fewer awards.[32] Implicit in this view of multiple-award winners is the simple linear model of the program. In reality, multiple awards serve a variety of functions, as noted below.

2. **Role of multiple awards.** More than one award is often required to develop a technology or firm capability, quite apart from the commercialization of a complete product. These needs reflect the complex and differing characteristics of firms.

    - **Diversity of firm objectives.** Reflecting the diversity of the program's objectives and of the participants, some firms approach the SBIR award process at different stages of development and with different objectives. Some firms are developing technology concepts; some firms see their vocation as contract research organizations; others actively seek to develop commercial products, either for public agencies or for the marketplace.[33]

    - **Diversity in firm strategy.** For example, investigator-led firms, limited in size and focused on a single concept may seek multiple awards as they advance research on a promising technology.[34] For firms that

[31]It is important to keep in mind the difficulties in tracking companies over time. Companies regularly change names, locations, even employer identification numbers, which makes tracking them across time within the DoD awards database difficult.

[32]See the 1992 GAO report, U.S. General Accounting Office, *Small Business Innovation Research Program Shows Success But Can Be Strengthened*, RCED–92–32, op. cit. That report focused on firms that had received twenty or more awards and on a single program metric, that is, commercialization. Congressional legislation later fixed fifteen awards in a five-year period as the level where efforts to commercialize should be taken into account.

[33]See Reid Cramer, "Patterns of Firm Participation in the Small Business Innovation Research Program in the Southwestern and Mountain States," in National Research Council, *The Small Business Innovation Research Program: An Assessment of the Department of Defense Fast Track Initiative*, op. cit., p. 151. The author describes the incremental nature of technical advance, which sometimes necessitates several awards. See also John T. Scott, "An Assessment of the Small Business Innovation Research Program in New England: Fast Track Compared with Non-Fast Track," in National Research Council, *The Small Business Innovation Research Program: An Assessment of the Department of Defense Fast Track Initiative*, op. cit., p. 109 for a discussion of Foster-Miller, Inc.

[34]Ibid. The mirror image of this approach is the program manager who makes several awards for similar technologies among different companies. In fact, it is not uncommon to have multiple awardees on the same Phase I topic. For an example, see the Navy's SBIR Web site selections page for their FY-06.1 awardees at <*http://www.navysbir.com/06_1selections.html*>. This page not only shows

carry out research as a core activity, success is often measured in multiple contract awards.

3. **Addressing agency missions.** Some firms, mainly at DoD, have won large numbers of awards over the life of the program. Yet, even with many awards, there is nothing intrinsically wrong with a process that makes high-quality research available to the department at relatively low cost. Some of this research is intrinsically noncommercial, but may have considerable value.[35]

   - **Identifying dead ends.** Inexpensive exploration of new technological approaches can be valuable, particularly if they limit expenditure on technological dead ends. For research oriented firms, the key issue is the quality of the research and its alignment with service and agency needs.[36] Each of the seven most frequent winners, who have received over 100 Phase II awards since the program inception, has a large number of researchers who submit proposals. The high number of quality proposals can produce a high number of awards. Some successful applicants use spin-off firms to commercialize the results of their SBIR awards.

   - **Providing solutions.** In some cases, firms respond to an agency solicitation and "solve" the problem, provide the needed data, or propose a solution that can then be adopted by the agency with no further "commercialization" revenues for the firm.[37]

---

several awardees for each topic, but if you click on the "Details" link, you can see the differences in companies' approaches to the topics.

[35]To secure additional awards, a small company has to submit its proposal for follow-on research through the regular review process. These awards are relatively small in amount—the normal Phase I and Phase II awards would total $850,000. As a point of comparison, the top three U.S. prime contractors in 2007 garnered over $20.5 billion in defense revenues. See *Washington Technology*, "Top 100 Federal Prime Contractors: 2004," May 14, 2007.

[36]For example, Foster-Miller, the most frequent SBIR award winner, provided armor for Humvees and aircraft and developed robots for use in Iraq to identify roadside improvised explosive devices. (This company has since been acquired, making it ineligible to participate in SBIR. See *Financial Times*, "Qinetiq set to make its first US acquisition," September 8, 2004.) Their LAST® Armor has had sales in excess of $170 million.

[37]There are cases where a small business successfully completes the requirements and objectives of a Phase II contract, meeting the needs of the customer (e.g., by delivering specialized software or hardware), without gaining additional commercialization revenues. For example, Aptima, Inc., designed an instructional system to improve boat-handling safety by teaching the use of strategies that mitigate shock during challenging wave conditions. A secondary goal was to demonstrate how an innovative learning environment could establish robust skill levels while compressing learning time. Phase I of the project developed a training module, and in Phase II, instructional material, including computer animation, videos, images, and interviews were developed. The concept and the supporting materials were adopted as part of the introductory courses for Special Operations helmsmen with the goal of reducing injuries and increasing mission effectiveness.

- **Developing technologies.** Those firms that seek to develop commercial products may, in an initial phase, seek multiple awards to rapidly develop a technology. For the high-growth firms, this period is limited in time, before private investment becomes the principal source of funding.[38]

- **Flexibility and speed.** Some multiple-award winners have provided the highly efficient and flexible capabilities needed to solve pressing problems rapidly. For example, Foster-Miller, Inc., responded to needs of U.S. forces in Iraq by developing and the manufacturing add-on armor for Humvees that provide added protection from insurgent attacks.[39]

4. **There is evidence that companies winning multiple awards commercialize their projects at least as effectively as firms with fewer awards. The capacities built up through multiple awards can also enable them to meet agency needs in a timely fashion.**

   - **Commercialization success.** Aggregate data from the DoD commercialization database indicates that the companies winning the most awards generate more commercialization per award than those winning few awards.[40] The 27 firms with more than 50 total Phase II awards account for 16.4 percent of all awards as reported through the CCR database, and for 30 percent of all reported commercialization.[41] Among these, firms with 50–75 Phase II awards were the most successful.

   - **Meeting agency needs.** Case studies show that some companies that have substantial numbers of awards have successfully commercialized products and have also met the needs of sponsoring agencies in other ways.[42]

---

[38]For a discussion of Martek as an example, see Maryann P. Feldman, "Role of the Department of Defense in Building Biotech Expertise," in National Research Council, *The Small Business Innovation Research Program: An Assessment of the Department of Defense Fast Track Initiative*, op. cit., pp. 266-268. See also Reid Cramer, "Patterns of Firm Participation in the Small Business Innovation Research Program in the Southwestern and Mountain States," in National Research Council, *The Small Business Innovation Research Program: An Assessment of the Department of Defense Fast Track Initiative*, op. cit., pp. 146-147, who discusses several firms that realized commercial success after several awards.

[39]Foster-Miller's LAST® Armor, which uses Velcro-backed tiles to protect transport vehicles, helicopters and fixed wing aircraft from enemy fire, was developed on two Phase I SBIRs and a DARPA Broad Agency Announcement. The technology has helped improve the safety of combat soldiers and fliers in Bosnia and Operation Desert Storm. Access at *<http://www.dodsbir.net/SuccessStories/fostermiller.htm>*.

[40]See Table 3-5.

[41]CCR table provided by the database contractor, BRTRC, December 18, 2006.

[42]See, for example, the case studies of Creare and Foster-Miller. The latter responded to needs in Iraq by providing add-on ceramic armor for HMMWVs. As noted above, contract research can

- **Graduation.** Some multiple-award winners eventually "graduate" from the program, either by exceeding the 500-employee limit to qualify as a small firm or by being acquired by another firm. Successful firms such as Digital Systems Resources and Martek have provided valuable products, shown commercial success, and also received numerous awards.[43]

- **Shifting revenues.** Some firms with multiple awards show a declining percentage of revenues over time. Radiation Monitoring Devices, for example, testified that it currently generates only about 16 percent of company revenues from SBIR.[44] In general, the larger the firm, the lower the percentage of revenues reported from SBIR awards.

- **Company creation.** Some multiple winners—like Optical Sciences, Creare, and Luna Innovations—frequently spin off companies. Creating new firms is a valuable contribution of the SBIR program especially with regard to the defense industrial base. Newly created firms create new opportunities for defense contractors, greater competition, and permit more rapid development of new defense solutions.

E. **While the DoD SBIR program supports woman- and minority-owned businesses, the steady decline in the share of Phase I awards to minority-owned businesses (falling to below 10 percent in 2004 and 2005) is a matter for concern and further review.**[45]

1. **A caveat on measurement.** A stated objective of the SBIR program is to expand opportunities for women and minorities in federal S&T. One way to measure program performance in this area is to review the share of awards being made to woman- and minority-owned firms. In doing so, we must keep in mind the overall percentage of the population is a less relevant benchmark than are the number of science, technology, and engineering graduates, the demographics of high-tech firm ownership, as

---

be valuable even in the absence of commercialization. Agency staff report that SBIR fills multiple needs, many of which do not show up in sales data. For example, efficient probes of a technology's potential, conducted in a relatively short time frame within a limited budget may save substantial time and resources. A key question to ask with regard to this and other aspects of the program is "compared to what?"—that is, what realistically are the probably alternative modes of exploration open to program managers?

[43]VIASAT, Inc., garnered 24 Phase II awards in growing from a three-person start-up in 1986 to over 500 employees in 2001 (thus ineligible for SBIR). It has grown to over $350 million in annual revenue, mostly in sales to DoD and DoD prime contractors. Products include some of the most critical communications systems in DoD.

[44]Michael Squillante, Vice President, Radiation Monitoring Devices, private communication, June 2004.

[45]The absolute number of DoD Phase I awards to minorities has grown by 12 percent since 1995, while the overall number of DoD Phase I awards grew by 57 percent.

well as other variables such as the greater difficulty faced by these groups in accessing capital from other sources.

2. **Overall results meet congressional objectives.** Overall, the DoD SBIR program awards about 20 percent of Phase I awards to either woman- or minority-owned firms.[46] (See Figure 2-3.)

3. **Awards to woman-owned firms continue to increase.** Figure 2-3 shows two divergent trends. Phase I awards to woman-owned firms continue to increase, increasing even as a percentage of the rising number of overall awards. However, the share awarded to minority-owned firms declined quite substantially since the mid-1990s, and fell below 10 percent for the first time in 2004 and 2005.

4. **Decline in award shares for minority-owned firms.** Data on Phase II awards suggest that the decline in Phase I award shares for minority-owned firms has been reflected in Phase II. It is an open question whether the increase in awards to minority-owned firms in 2005 is the start of a reversal of this trend (see Figure 2-4).

5. **No apparent bias for or against woman- and minority-owned firms in Phase II awards.** These data also indicate that both woman- and minority-owned firms are converting Phase I awards into Phase II at a rate very close to that of all award winners. This suggests that these firms are being invited to compete for Phase II awards at about the same rate as other firms, and that as a result there is no discernable bias for or against woman- and minority-owned firms in the selection of Phase II awards.

F. **The SBIR program at DoD generates considerable new technical knowledge, and is helping to expand the nation's science and technology base.**

1. **A key element in award selection is the technical merit and innovative character of the proposal. This criterion is applied to ensure that projects that do receive awards have the potential to generate new knowledge.**

    - In general, the SBIR program at DoD is highly competitive; since 1992 about 15 percent of Phase I proposals have been funded. Approximately half of Phase I winners receive Phase II contracts.

2. **The DoD SBIR program contributes new scientific and technological knowledge in several forms.**

    - **These include intellectual property rights** secured by the inventing firm. Intellectual property rights create publication and licensing op-

[46]Award figures were provided by DoD.

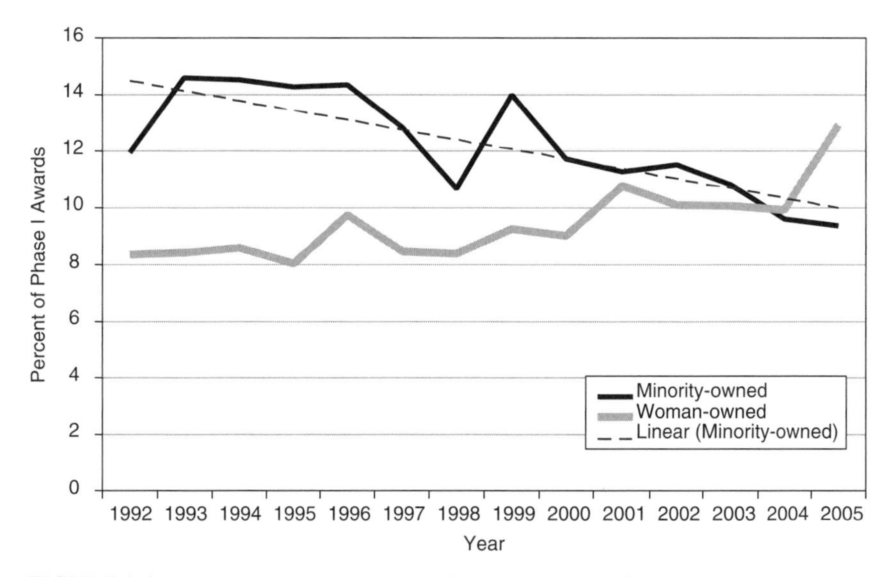


**FIGURE 2-3** Phase I awards to woman- and minority-owned firms.
SOURCE: DoD awards database.

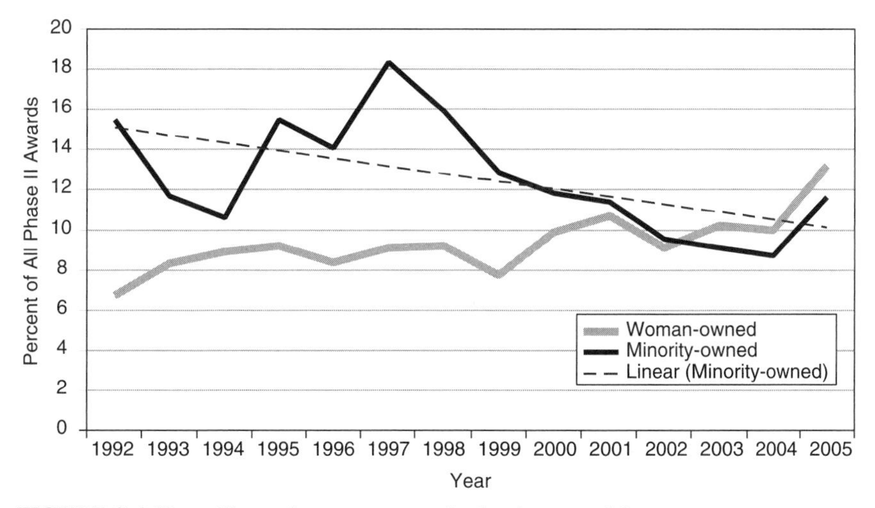


**FIGURE 2-4** Phase II awards to woman- and minority-owned firms.
SOURCE: DoD awards database.

portunities. Publications in the scientific and technical literatures and presentations at professional and technical meetings help the knowledge created by intellectual property rights to spill over into the public domain.

- **SBIR companies have generated numerous patents and publications**, the traditional measures of knowledge activity. 34.4 percent of DoD projects surveyed by NRC generated at least one patent application, and just over 25 percent had received a patent.[47] (There is typically a 2–4 year lag between patent application and patent award or declination for recent patents.)
- **Patents.** For many firms, patents are an essential means of establishing intellectual property rights. This is especially important for firms planning to sell or license their technology for use in larger weapons and logistics systems where the immediate purchaser of their innovation will be a far larger defense contractor.
- **Trade secrets.** Other successful DoD awardees, however, see patents as less important. Some firms prefer to use trade secrets, saving the costs of filing and defending the patent. Some firms also see little economic profit from a patent if target market is primarily the federal government, which under SBIR has the right to its use royalty free. Thus a focus on patent data alone will understate the intellectual property generated through the DoD SBIR program.
- **Published articles.** Forty-two percent of respondents had published at least one peer-reviewed article based on the SBIR project surveyed by the NRC, and 3.9 percent had generated more than five articles.[48,49]

3. **SBIR supports the transfer of technology from the university into the marketplace.**
   - Responses to the NRC Phase II and Firm Surveys suggest that SBIR awards are supporting the transfer of knowledge, firm creation, and partnerships between universities and the private sector:
   - In more than 66 percent of responding companies (at all agencies), at least one founder was previously an academic[50];

[47]See Table 4-18.

[48]Without detailed identifying data on these patents and publications, it was not possible to apply bibliometric and patent analysis techniques to assess their relative importance.

[49]NRC Phase II Survey, Question 18.

[50]NRC Firm Survey, Question 2.

- About 36 percent of founders (at all agencies) were most recently employed as academics before the creation of their company.[51]
- At DoD, 13.6 percent of reporting projects had university faculty as subcontractors or consultants on the project, at 12.5 percent a university was itself a subcontractor, at 9.2 percent university facilities were used, and at 11.4 percent graduate students worked on the project.[52]

4. **There is anecdotal evidence concerning beneficial "indirect path" effects of SBIR.**
   - These indirect effects refer to the existence of projects that provide investigators and research staff with knowledge that may later become relevant in a different context—often in another project or even another company. While these effects are not easily measurable, comments made during interviews and case studies suggest they exist.[53]

G. **DoD has taken steps to improve program performance, with some significant successes.[54] However, implementation of these best practices across the services and agencies is uneven.**

1. **Improved information flows.**
   - DoD has a very extensive SBIR/STTR Web site, with exceptional support for potential SBIR applicants.
   - The Pre-Release Program provides detailed guidance on specific topics.
   - The Help Desk Program removes a significant burden from military staff and provides better service using professional civilian contractors.
2. **Efforts to address funding gaps and timeline issues.**
   - DoD and its components have made a number of efforts to address the gaps between topic conceptualization and Phase I funding, Phase I and Phase II, and the "TRL Gap" that often emerges after Phase II, before acquisitions programs can accept a technology as ready for acquisition.
3. **Acquisition alignment and Phase III.** DoD is increasingly aware that

[51]See NRC Firm Survey, Question 3.

[52]See Table 4-19. (NRC Phase II Survey, Question 32).

[53]For a discussion of the indirect path for awards made under the Advanced Technology Program, see Rosalie Ruegg, "Taking a Step Back: An Early Results Overview of Fifty ATP Awards," in National Research Council, *The Advanced Technology Program: Assessing Outcomes*, Charles W. Wessner, ed., Washington, DC: National Academy Press, 2001.

[54]The agency has a long history of initiatives, dating back to the 1995 PAT report. These initiatives are described in more detail in Chapters 5 and 6.

**BOX 2-2**
**Best Practices for SBIR**

A major strength of the SBIR program is its flexible adaptation to the diverse objectives, operations, and management practices at the different agencies. In some cases, however, there are examples of best practice that should be examined for possible adoption by other agencies. Examples of these best practices include:

**DoD. The Pre-release period.** DoD announces the contents of its upcoming solicitations some time before the official start date of the solicitation. By attaching detailed contact information, prospective applicants can talk directly to the technical officers in charge of specific topics. This helps companies determine whether they should apply and gives the prospective applicant a better understanding of the agency needs and objectives. This informal approach provides an efficient mechanism for information exchange. Federal Acquisition Regulations prevent such discussion after formal release.

**DoD. Help Desk and Web support.** DoD maintains an extensive and effective web presence for the SBIR program, which can be used by companies to resolve questions about their proposals. In addition, DoD staffs a Help Desk aimed at addressing nontechnical questions. This is appreciated by companies, and is strongly supported by program staff because it reduces the burden of calls on technical staff.

**DoD. Commercialization tracking.** DoD's approach requires companies with previous Phase II awards to enter data into a commercialization tracking database each time these companies apply for SBIR awards at DoD. The database captures outcomes (both financial, such as sales and additional funding by source, and other benefits resulting from SBIR; e.g., public health, cost savings, improved weapon system capa-

success in Phase III requires strong support from program offices, and long-term alignment between SBIR activities and program needs. Efforts to improve topic alignment have been underway for several years.

4. **Administrative funding.** The Navy has taken the lead in providing extended administrative funding and support to its SBIR program. It may not be a coincidence that Navy's Phase III results—as reported in the DD350 forms—are better than those of all other services and components combined.[55]

5. **Data collection and analytic capabilities.** Data collection, reporting requirements, and analytic capability have all been improved. The CCR database represents the most comprehensive source of data on outcomes from SBIR projects at any of the agencies. The NRC study and other

[55]See Box 2-3 discussing the Navy's approach.

bility, etc.) from these companies for all their previous SBIR awards, including those at other agencies. It also captures information on firm size and growth since entering the SBIR program, as well as the percent of annual revenue derived from SBIR awards. These historical results of prior awards are then used in proposal evaluation.

Non-DoD agencies should consider adapting both this approach and the DoD technology and contributing to the DoD database. This would provide a unified tracking system. Adaptations could be made to track additional data for specific agencies, but this would provide a cost-effective approach to enhance data collection on award outcomes.

**Multiple agencies. Gap-reduction strategies.** The agencies have, to different degrees, recognized the importance of reducing funding gaps. While details vary, best practice would involve development of a formal gap-reduction strategy with multiple components covering application, selection, contract negotiation, the Phase I–Phase II gap, and support after Phase II.

**DoD Phase II Enhancement (and NSF Phase IIB).** The matching-fund approach adopted by NSF for Phase IIB and DoD for Phase II Enhancement might be explored at other agencies. The NSF matching requirement represents an important tool for helping companies to enter Phase III at nonprocurement agencies. The DoD funding match by acquisition programs provides a transition link into Phase III contracts with the agency.

**DoD-Navy. Technology Assistance Program.** The Navy has developed the most comprehensive suite of support mechanisms for companies entering Phase III, and has also developed new tools for tracking Phase III outcomes. These are important initiatives, and other components and agencies should consider them carefully.

recent reviews represent a positive effort to connect data and analysis to practice.[56]

6. **Phase III results remain uneven across the services and among components within the services.**
   - This suggests that other components and other services could improve the performance of their programs.
   - Interviews suggest that some elements of the department have not fully integrated SBIR within their own program missions and have not

[56]See also a recent RAND Report requested by DoD: Bruce Held, Thomas Edison, Shari Lawrence Pfleeger, Philip S. Anton, and John Clancy, *Evaluation Recommendations for Improvement of the Department of Defense Small Business Innovation Research (SBIR) Program*, Santa Monica, CA: RAND Corporation, 2006.

provided SBIR with the resources and management attention needed to maximize its effectiveness.[57]

- The 2005 NRC Symposium on SBIR Phase III contributed to the awareness of the SBIR program's potential, the challenge that promising products face in the Phase III transition, and the need for additional efforts to "team" across agencies, with SBIR program managers, Program Executive Officers, and prime contractors.[58]

H. **Prime contractors are taking a positive approach towards the SBIR program.**

1. **Increased interest in SBIR.** As the program has grown in size and performance, it has garnered greater attention from the DoD upper management and, importantly, the prime contractors.

   - There is considerable evidence that prime contractor interest in—and engagement with—the SBIR program has been growing rapidly in recent years (see Chapter 6). Concretely, this is reflected in growing contractual linkages.[59]

2. **Interest has been followed by action.** This increased focus on the prospective contributions of the SBIR program by the prime contractors appears to represent a significant positive endorsement of the contributions of the program.

   - Steps taken by the prime contractors to integrate SBIR within their

[57]Some DoD program managers see the SBIR program as a "tax," that is an unwarranted allocation of funds to small business, one that hinders effective R&D program management and one that is time consuming to manage and without funds to cover the cost of management. For a more positive view of the program, see the discussion of the Navy program in the Findings section. The recommendations address program perception and incentives.

[58]Following the National Academies meeting on the SBIR commercialization challenge, the Senate Small Business & Entrepreneurship Committee proposed legislation that established a Commercialization Pilot Program. See Section 252 of the 2006 National Defense Authorization Act. The bill was passed in bipartisan spirit by the Senate Committee on Small Business & Entrepreneurship (SBE) under the leadership of the Committee Chair, Olympia Snowe (R-ME) and Ranking Member, John Kerry (D-MA). Further reflecting the growing appreciation of the program's role and the increased focus on Phase III, Dr. Finley has described the SBIR program as a means of accelerating innovation and putting better equipment into the hands of the war-fighter. See remarks by Deputy Under Secretary of Defense for Acquisition and Technology Finley at the Small Business Technology Coalition Conference on September 27, 2006.

[59]Raytheon, for example, estimates that the value of technology leveraged through SBIR jumped from $3.8 million in 2004 to $11.6 million in 2005, and looks set to grow as rapidly in 2006. Raytheon is involved with 36 Phase I projects, 17 Phase II projects, 4 Phase III projects (with three more in the works), and has been a subcontractor on other projects. See Lani Loell, SBIR Program Manager Raytheon Integrated Defense Systems, Presentation to SBTC SBIR in Transition Conference, September 27, 2006, Washington, DC.

strategic roadmaps reveal that they see the program contributing to technological innovations that further the Defense mission.[60]

- At the Academies' Phase III conference, representatives of prime contractors stated that there was already a substantial amount of prime involvement with the SBIR program. Moreover, several of the primes affirmed that they had made significant efforts to increase their levels of involvement.
- For example, Boeing had recently decided to increase its emphasis on SBIR.
- Similarly, at Raytheon, some divisions (e.g., Integrated Defense Systems) had formal working arrangements with SBIR for several years.

## NRC STUDY RECOMMENDATIONS

**As noted in the Findings section above, the Department of Defense has an effective SBIR program. The recommended improvements listed below should enable the DoD SBIR managers to address the four mandated congressional goals in a more efficient and effective manner.**

A. **Improve the Phase III transition.** DoD should continue to expand its work on improving the Phase III transition (the transition from SBIR-funded Phase I and Phase II research to commercialization—especially testing and evaluation funded by other DoD sources). It is important to recognize that the transition of new technologies is a complex process requiring teaming across areas of responsibility, additional resources, and often coping with some element of additional risk.[61] Areas for possible action include:

   1. **Aligning incentives.**
      - For the SBIR program to achieve its full potential, better incentives are required. Expansion of positive incentives for program officers to utilize the SBIR program for their own research needs, beyond the current requirements for their involvement in topic development.
      - Management needs to improve incentives so that acquisitions officers perceive reduced risks and enhanced benefits from participating in the program.
   2. **Increasing resources.** SBIR managers need greater resources to "match" program funds to encourage uptake. In addition to increased Phase III

[60] See Section 5.4.4 and the discussion in Chapter 6.

[61] See the discussion of these problems and potential remedies in the section on Phase III Transition in this volume. See also National Research Council, *SBIR and the Phase III Challenge of Commercialization*, op. cit.

SBIR funding, linkages with other programs (e.g., ManTech) might be enhanced to facilitate the Phase III transition.

3. **Developing an evaluation culture.** Agency and service managers should have effective data collection and analysis as performance metrics.
4. **Involving acquisition officers.** Active participation by acquisition officers is key to successful Phase III transitions.
    - Acquisition officers control the funding, and their involvement is important for successful commercialization of SBIR technologies. A cultural shift in program participation and use seems to have occurred at Navy once Program Executive Offices (PEO) became active champions of SBIR involvement in acquisitions.
    - Senior management support and encouragement, better information flows, improved PEO education about SBIR, and additional incentives for PEOs to use SBIR are all elements of an effective overall program.
5. **Integrating with roadmaps.** The long technology development and acquisition cycle for major weapons and logistics systems means that effective Phase III transition requires early integration of SBIR topics and firms into the planning process.[62]
6. **Linking SBIR with other programs.** Linkages with other programs (e.g., ManTech) might be enhanced to facilitate the Phase III transition.
7. **Improving outreach and matchmaking.** There are significant barriers to the flow of information among SBIR firms, prime contractors, and acquisitions offices. Effective transition requires that these barriers be overcome, most likely through implementation of a range of activities, including improved electronic communications methods and matchmaking services like the Navy Opportunity Forum. In particular, efforts should be made, as appropriate, to align the SBIR program with the needs of the prime contractors responsible for the development of major systems.
8. **Connecting with the primes.** The growing interest among prime contrac-

[62]In the case of the semiconductor industry, the industry perceived early on that problems of coordination could arise with a complex technology, multiple participants, and many ways of proceeding. This realization led to cooperative efforts led by SEMATECH to develop a technology roadmap setting out the relationships among science, technology, and applications as a point of reference for the researchers, technologists, project managers, suppliers, and users involved in and affected by the consortium's work. As a general approach, roadmaps can advance similar coordination functions in other industry partnerships and in this way contribute to more efficient and more cooperative research. See National Research Council, *Government-Industry Partnerships for the Development of New Technologies: Summary Report*, Charles W. Wessner, ed., Washington, DC: The National Academies Press, 2003.

tors of the SBIR program's outputs and opportunities for partnering with SBIR companies should be encouraged. Consideration should be given to performance incentives to further encourage development of SBIR supported technologies.

9. **Assessing and expanding commercialization programs.** Commercialization programs that provide training, counseling, and networking opportunities should be assessed and, as appropriate, expanded.

B. **DoD should take immediate steps to enhance the perception of the SBIR program's potential and accomplishments, promoting SBIR as an opportunity.**

   - A key element in the program's operation is the attitude taken towards the program by the different levels of management in the Defense research and development community.
   - Where SBIR is seen as an unwarranted intrusion on program management, a "tax" on R&D resources, it is less likely to be effectively aligned with service needs and less likely to have the resources to develop and ultimately insert the results of successful Phase II technologies in weapons and logistics systems and other programs.
   - When the program is seen as an effective tool to engage the ingenuity of small companies in support of the Defense mission, with shorter lead times and more flexibility, it is much more likely to have its results adopted and incorporated.
   - In short, there is an element of circularity in developing measures to enhance program effectiveness and management's guidance and rewards for those managers who use the program effectively. Providing the resources and incentives for managers to see opportunity rather than obligation may well enhance program effectiveness.

C. **DoD should substantially strengthen and expand its evaluation efforts in order to further develop a program culture that is driven by outcomes, data, and internal and external evaluations.**

   - Efforts to identify outcomes should be improved, and evaluations[63] should

[63]For example, each SBIR award has a DoD technical monitor who serves as the contracting officer's technical representative (COTR). This individual monitors the contractor's performance during Phase I and recommends or issues the invitation for Phase II, monitors performance of Phase II, and receives the contractor's Phase I and II reports. Despite the importance of the TPOC's role, there is no systematic attempt to use the technical monitor to evaluate the quality of the SBIR efforts and to facilitate Phase III. Since these individuals are not funded separately for SBIR, dissemination of SBIR outcomes and assistance in transition is a function of individual initiative and competing responsibilities. DoD should consider training, funding, and making better use of this valuable asset.

be connected much more directly to program management. More attention should be devoted to the role and contributions of the Contracting Officer's Technical Representative (COTR). It is important that DoD create the capability to use outcomes data to help assess best practice.

- New mechanisms need to be developed that allow for the efficient design, implementation, and subsequent assessment of pilot programs.
- Efforts should be increased to make sure that appropriate metrics and benchmarks are adopted and implemented by all units, components, and Services.

D. **DoD should encourage and support the development of a results-oriented SBIR program with a focused evaluation culture.**

1. **Effective oversight requires additional staff and funding. Effective management of a data-driven SBIR program requires the regular collection of higher-quality data and systematic assessment. Currently, sufficient resources are not available for these functions. Additional funding should be provided for program management and assessment.**
   - This funding should also be used to provide management oversight, including site visits, improved data collection and analysis, regular reporting, program review, and systematic third-party assessments.
2. **To help foster an active evaluation culture, DoD should consider:**
   - **Preparing an expanded annual SBIR program report.** DoD should prepare an annual SBIR program report, which gathers all relevant data about awards, outcomes, program activities, and management initiatives. In particular, the Department should publish detailed data annually about Phase III take-up at each service, and at each component within each service, as well as providing information about program initiatives.
   - **Commissioning regular assessments.** SBIR programs at both DoD and the individual DoD components should seek to enhance a data-driven management approach, with regular assessment supporting policy development and program management.
   - **Instituting systematic and objective, outside review.** The internal assessment program should be supported and supplemented by systematic, objective outside review and evaluation, as envisaged in the reauthorization legislation.
   - **Convening an advisory board.** DoD should consider development of a formal advisory board, which would receive the annual program

report and provide its own supplementary review of the report and management practices on an annual basis to senior DoD officials in charge of the SBIR program, or possibly to a subpanel of the Defense Sciences Board.

3. **DoD should consider greater internal review and adoption of best practices.**

   Such an assessment would identify best practices within DoD and develop mechanisms for encouraging other components to implement these practices within their SBIR programs.

   - One important example of best practice might be one focused on the Phase III transition at the Navy (see Box 2-3).
   - Additional research should be undertaken to address, *inter alia*, three questions related to Navy practice.
   - What unique factors make the Navy SUBS program successful?
   - What role is played by initiatives at the service level in supporting the SUBS program?
   - What elements of that success can best be transferred elsewhere in DoD as best practices to be followed? What changes will be needed to make those transfers successful?

E. **DoD should encourage and support pilot programs that evaluate new tools for improving the program's overall performance.**

1. **Innovation through pilot programs.** Making changes initially through pilot programs allows DoD to alter selected areas on a provisional basis; a uniform approach is unlikely to work well for all components of a program that funds highly diverse projects with very different capital requirements and very different product development cycles.

2. **Some possible pilot projects include:**

   - **Small Phase III awards.** These could be a key to bridging the financing 'Valley of Death' that many firms face in converting research to innovation to products.[64] NASA for example sometimes provides a small Phase III award—perhaps enough money to fly a demonstration payload—for a technology not ready for a full Phase III. These might

[64]See Lewis Branscomb and Philip Auerswald, *Between Invention and Innovation: An Analysis of Funding for Early-Stage Technology Development*, NIST GCR 02-841, Gaithersburg, MD: National Institute of Standards and Technology, Prepared for the Economic Assessment Office, Advanced Technology Program, November 2002. See also National Research Council, *SBIR and the Phase III Challenge of Commercialization*, op. cit.

**BOX 2-3**
**Lessons from the Navy Model**

Many of the issues we identify with regard to the SBIR program at the Department of Defense have been addressed, with considerable success, by the Navy SBIR program. Keeping in mind the appropriate caveats concerning different agency needs, operating conditions, and cultural traditions, a number of aspects of the Navy program address these concerns. Key features of the Navy program include:

**Positive acceptance of the program.** Navy PEO's and program managers increasingly appear to see the SBIR program as a useful tool in meeting mission objectives, as acquisition staff are drawn increasingly into topic development and SBIR project management.

**Top management focus.** One reason for the positive perception of the program's utility is that the Navy provides significant management attention, particularly at the program executive officer (PEO) level, to the integration of SBIR into technology development to meet program needs. Strong leadership from the Navy hierarchy emphasizes the potential of SBIR for Navy missions.

**Administrative funding and activities.** The Navy provides substantial additional funding, now on the order of $20 million per annum, to operate the program. These funds meet a variety of needs ranging from additional professional staff support, funding for the Technology Assistance Program, and resources for the Navy Opportunity Forum that helps match SBIR companies with potential customers.

be combined with milestones or gateways to additional rounds of Phase II funding.

- **Unbundling larger contracts.** Organizing larger contracts into smaller components would tend to open more Phase III opportunities for SBIR firms.[65]
- **Redefining testing and evaluation within SBIR.** DoD could pilot adoption of a wider view of RDT&E, so that SBIR projects could qualify for limited testing and evaluation funding. That in turn would help fund improvements in readiness levels.
- **"Spring loading" Phase III,** by putting in place Phase II milestones that could help to trigger initial Phase III funding. This could possibly

[65]The recent focus for DoD research, development, and engineering centers has been to take many small support contracts, which were then done by small business and bundle them into a large solicitation, which can only be won by a firm which has significant resources. Bundling limits options for follow-on engineering support by SBIR firms. An alternative to unbundling is to provide incentives within large omnibus support contracts that make subcontract awards to SBIR firms.

**Emphasis on Phase III funding and process.** Navy PEO's have embraced the challenge of maturing innovative technologies (including SBIR products), which requires serial funding for the many testing, evaluation and demonstration steps that precede acquisition.

**Demonstration effects and program integration.** The strong Phase III take-up recently demonstrated at Navy suggests that acceptance of the program as a valid and useful component in the Navy's overall technology development strategy creates a virtuous cycle. Successfully transitioned technologies such as the SAVI logistics tracking system, the DSR sonar, the ACR's "Silver Fox" reconnaissance UAV, and cost-saving diagnostic technologies provide powerful demonstration effects, underscoring the potential contributions of the program to meet a broad range of Navy needs.

**Documented achievement.** One of the distinctive features of the programs at Navy is that it successfully documents its accomplishments.[a] Data from DD350 reports shows that Navy's Phase III contracts grew from $50 million in 2000 to $350 million in 2005. While these growing achievements may in part reflect unique or superior record keeping as compared with other services, this in itself reflects successful adoption of a data-driven assessment culture. The data gathered provides superior information and improved understanding of the operation and potential of the program.

[a]One factor contributing to the Navy's assessment culture is the tradition of research excellence by the Office of Naval Research, one of the nation's preeminent federal research funding agencies.

occur in the context of larger, staged, Phase II awards in which additional stages fund more Demonstration and T&E, where non-SBIR funds or resources are leveraged.

3. **Evaluating pilots.** DoD should develop a formal mechanism for designing, implementing, and evaluating pilot programs. Pilot programs allow agencies to investigate program improvements at lower risk and potentially lower cost. Effective pilot programs require rigorous design and evaluation, clear metrics for success, and the necessary resources and internal support.

4. **A flexible approach is required.** In some cases, pilot programs may require waivers from SBA for activities not otherwise permitted under the SBA guidelines.[66] SBA should be encouraged to take a highly flexible view of all agency proposals for pilot programs.

[66]In the past, SBA has shown commendable flexibility in allowing agencies to deviate from standard award sizes to accommodate the needs of the technology in question and to devise program innovations, such as the NSF Phase IIB, to provide incentives for commercialization.

F. **Provide additional management resources.** To carry out the measures recommended above to improve program utilization, management, and evaluation, the program should be provided with additional funds for management and evaluation.

1. Effective oversight relies on appropriate funding.[67] A data-driven program requires high quality data and systematic assessment. As noted above, sufficient resources are not currently available for these functions.

2. Increased funding is needed to provide effective oversight, including site visits, program review, systematic third-party assessments, and other necessary management activities.

3. In considering how to provide additional funds for management and evaluation, there are three ways that this might be done:

    - Additional funds might be allocated internally, within the existing budgets of the services and agencies, as the Navy has done.

    - Funds might be drawn from the existing set-aside for the program to carry out these activities.

    - The set-aside for the program, currently at 2.5 percent of external research budgets, might be marginally increased, with the goal of providing management resources necessary to maximize the program's return to the nation.[68]

G. **DoD should take steps to increase the participation and success rates of woman- and minority-owned firms in the SBIR program.**

---

[67]According to recent OECD analysis, the International Benchmark for program evaluation of large SME and Entrepreneurship Programs is between 3 percent for small programs and 1 percent for large-scale programs. See "Evaluation of SME Policies and Programs: Draft OECD Handbook," OECD Handbook, CFE/SME(2006)17.

[68]Each of these options has its advantages and disadvantages. For the most part, the departments, institutes, and agencies responsible for the SBIR program have not proved willing or able to make additional management funds available. Without direction from Congress, they are unlikely to do so. With regard to drawing funds from the program for evaluation and management, current legislation does not permit this and would have to be modified; the Congress has clearly intended program funds to be for awards only. The third option, involving a modest increase to the program, would also require legislative action and would perhaps be more easily achievable in the event of an overall increase in the program. In any case, the Committee envisages an increase of the "set-aside" of perhaps 0.03 percent to 0.05 percent on the order of $35 million to $40 million per year, or roughly double what the Navy currently makes available to manage and augment its program. In the latter case (0.05 percent), this would bring the program "set-aside" to 2.55 percent, providing modest resources to assess and manage a program that is approaching an annual spend of some $2 billion. Whatever modality adopted by the Congress, without additional resources the Committee's call for improved management, data collection, experimentation, and evaluation may prove moot. See also the National Research Council, *SBIR and the Phase III Challenge of Commercialization*, op. cit.

1. **Encourage participation.** Develop targeted outreach to improve the participation rates of woman- and minority-owned firms, and strategies to improve their success rates. These outreach efforts and other strategies should be based on causal factors determined by analysis of past proposals and feedback from the affected groups.[69]

2. **Encourage emerging talent.** The number of women and, to a lesser extent, minorities graduating with advanced scientific and engineering degrees has been increasing significantly over the past decade, especially in the biomedical sciences. This means that many of the woman and minority scientists and engineers with the advanced degrees usually necessary to compete effectively in the SBIR program are relatively young and may not yet have arrived at the point in their careers where they own their own companies. However, they may well be ready to serve as principal investigators (PIs) and/or senior co-investigators (Co-Is) on SBIR projects. Over time, this talent pool could become a promising source of SBIR participants.

3. **Improve data collection and analysis.** The Committee also strongly encourages the agencies to gather and report the data that would track woman and minority firms as well as principal investigators (PIs), and to ensure that SBIR is an effective road to opportunity.

[69]This recommendation should not be interpreted as lowering the bar for the acceptance of proposals from woman- and minority-owned companies. Rather it should be seen as assisting them to become able to meet published criteria for grants at rates similar to other companies on the basis of merit, and to ensure that there are no negative evaluation factors in the review process that are biased against these groups.

# 3

# SBIR Awards at DoD

## 3.1 INTRODUCTION

This chapter reviews SBIR awards made by DoD, based on data provided by the department. All awards at DoD are made in the form of contracts, and as such require a deliverable. At a minimum, this means a final report; in some cases, a prototype or working model is also delivered.

The chapter provides a comprehensive overview of the distribution of awards, giving a context into which questions regarding outcomes and program management can best be placed.

This is especially important because of the very decentralized character of the DoD SBIR program. Each agency or service in effect operates its own program within the guidelines sets by the Small Business Administration (SBA) and Defense Research & Engineering (DDR&E). Program objectives, mechanisms, and assessment are all developed and implemented primarily at the service and agency level.

Overall, the number of awards made at DoD has grown sharply in recent years, reflecting increases in the department's R&D budget.

## 3.2 NUMBER OF PHASE I AWARDS

While SBIR funding for DoD has substantially increased in recent years (see Figure 3-1), the number of Phase I awards awarded has not. The number of Phase I awards remained relatively constant from 1993 to 2001 before increasing substantially in 2002. Since then the number of Phase I awards has again stayed relatively flat. (See Table 3-1.) The substantial (65 percent) increase in the number of Phase I awards made by DoD in 2002 resulted from a number of factors.

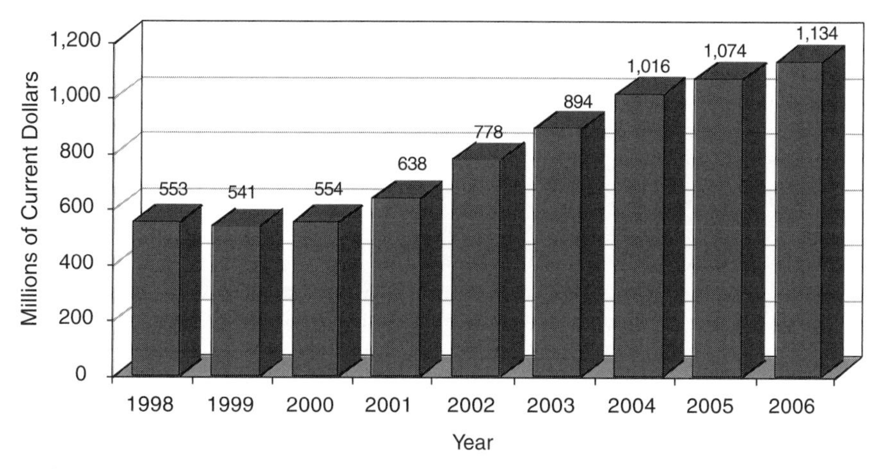


**FIGURE 3-1** SBIR funding at DoD, FY1998–2006.
SOURCE: Michael Caccuitto, DoD SBIR/STTR Program Administrator, and Carol Van Wyk, DoD CPP Coordinator, Presentation to SBTC SBIR in Rapid Transition Conference, September 27, 2006, Washington, DC.

After several years of relatively constant funding, the DoD R&D set-aside increased by 15 percent in FY2001. Cautiously, DoD awarded less than 8 percent more Phase I contracts. In 2002, DoD received a further increase of 22 percent in R&D funding and it became clear that the FY2003 and FY2004 DoD R&D budgets were likely to grow even further.

**TABLE 3-1** SBIR Awards at DoD, 1992–2005

| Year | Number of Phase I Awards | Number of Phase II Awards | Total |
|---|---|---|---|
| 1992 | 1,065 | 433 | 1,498 |
| 1993 | 1,303 | 591 | 1,894 |
| 1994 | 1,370 | 406 | 1,776 |
| 1995 | 1,262 | 575 | 1,837 |
| 1996 | 1,372 | 611 | 1,983 |
| 1997 | 1,526 | 638 | 2,164 |
| 1998 | 1,286 | 672 | 1,958 |
| 1999 | 1,393 | 568 | 1,961 |
| 2000 | 1,220 | 626 | 1,846 |
| 2001 | 1,310 | 702 | 2,012 |
| 2002 | 2,162 | 661 | 2,823 |
| 2003 | 2,113 | 1,078 | 3,191 |
| 2004 | 2,075 | 1,173 | 3,248 |
| 2005 | 2,344 | 998 | 3,342 |
| Total | 21,801 | 9,732 | 31,533 |

SOURCE: DoD awards database.

DoD responded by increasing the number of SBIR topics by about 10 percent in FY2002, but received about 75 percent more proposals as the private, venture-funding technology bubble burst and small technology companies sought new sources of funding. The confluence of increased funding available, more topics, and more demand led to a significantly higher number of Phase I awards.

Agency-specific factors also played a part. In 2001, the Missile Defense Agency (MDA) was seeking to exit the SBIR program. This led to a reduced number of MDA contracts in FY2001 and the "loaning-out" of MDA set-aside funding for use by other agencies. When this exit strategy was rejected by DoD, MDA found that the low number of Phase I contracts it awarded in FY2001 resulted in a reduced number of Phase II contracts in FY2002. But because MDA was now fully committed to spending its entire SBIR set-aside, it had to give out an extra-large number of Phase I contracts in FY2002.

The substantial increase in Phase I contracts in 2002 helps to explain the 59 percent increase in the number of Phase II awards between 2001 and 2003. Since this step jump, numbers have increased only slightly (see Table 3-1).

### 3.2.1 Phase I—Median Award Size

Figure 3-2 shows that DoD Phase I awards have generally averaged just under $90,000 since 1997. The increase from 1994–1997 resulted from changes in SBA guidelines after the 1992 SBIR reauthorization.

### 3.2.2 Phase I—New Winners

The share of Phase I awards going to new winners—firms that have not previously participated in the DoD SBIR program—is an important measure of the

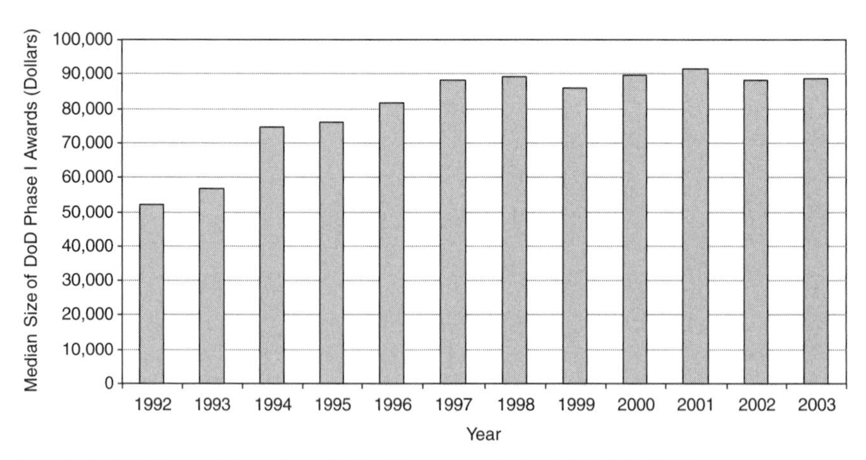


**FIGURE 3-2** Phase I Awards at DoD: Mean award size, FY1992–2003.
SOURCE: DoD awards database.

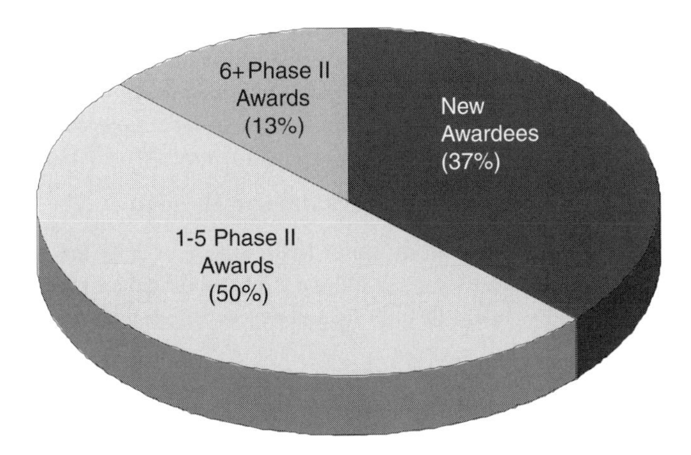


**FIGURE 3-3** Prior SBIR participation/success of the pool of Phase I award winners, FY2005.
SOURCE: Michael Caccuitto, DoD SBIR/STTR Program Administrator, and Carol Van Wyk, DoD CPP Coordinator, Presentation to SBTC SBIR in Rapid Transition Conference, September 27, 2006, Washington, DC.

openness of the program. In this context, "openness" means the extent to which the SBIR program remains open to new entrants and has not been "captured" by a limited set of winning companies with well-established connections to DoD.

DoD has provided data covering FY2005 (as of March 2006) that show that 37 percent of awards went to firms that had not previously won a DoD SBIR award. An additional 50 percent of awards were given to companies with five or less Phase II awards. Only 13 percent of Phase I awards went to companies with more than five Phase II awards. (See Figure 3-3.)

Data from the Navy also suggest that the program is open to outside firms without an SBIR track record. For example, about half of all Phase I contracts from NAVAIR go to companies which have never won an SBIR at NAVAIR before; about 40–45 percent of Phase II contracts go to "new" firms as well.[1]

Figures for NAVAIR alone are naturally higher than for DoD overall, as some of the "new" winners at NAVAIR previously have won SBIR awards elsewhere within DoD, and would therefore not be classified as "new" winners for the department as a whole.

Fundamentally, the evidence is clear that DoD SBIR programs are systematically including large numbers of new entrants. SBIR awards have thus not become the preserve of a small group of multiple winners. While some companies

[1]Carol Van Wyk, NAVAIR SBIR Program Manager, presentation to PMA-209, September 2005.

have won a large number of awards, there are structural characteristics of DoD that tend to encourage staff to work with companies that have performed well in the past. Overall, the SBIR program is remarkably open to new entrants even as some companies are able to repeatedly win in open competitions for awards.

### 3.2.3 Phase I—The States and Regions

One of the persistent questions about SBIR concerns how the awards are distributed across the states. Like other federal R&D funding distributed by merit, SBIR funding tends to cluster in high-tech states and high-tech regions within those states.

DoD Phase I SBIR awards go disproportionately to states with well-established traditions of science and engineering (see Table 3-2). The top five award-winning states received 53.8 percent of all DoD Phase I awards between 1992 and 2005. California and Massachusetts together account for 37.42 percent of all Phase I awards between 1992 and 2005.[2]

Concentration at the top is mirrored in the limited number of awards given to companies in low-award states (see Table 3-3). The bottom 15 states accounted for 1.85 percent of Phase I awards over the same time period, and 10 states averaged less than three awards per year.

This concentration of awards is not unique to the SBIR program. The GAO pointed out in its 1999 study of the SBIR program that, according to the SBA, one-third of the states received 85 percent of all SBIR awards, but also found that the distribution of SBIR awards tends to mirror the distribution of R&D funds in general.[3] The same 1999 GAO study also noted concern about the concentration of awards, not only by company (see below), but also by geographic location. With regard to geographic distribution, the GAO report noted that "Companies in a small number of states, especially California and Massachusetts, have submitted the most proposals and won the majority of awards, although the distribution of awards generally follows the pattern of distribution of non-SBIR expenditures for R&D, venture capital investments, and academic research."[4]

The study notes further that the data on the "proposal-to-award ratios show that proposals from companies in states with historically lesser amounts of federal research funding won awards at almost the same rate as proposals from companies in other states" (i.e. those receiving fewer awards).[5] This suggests that rates

---

[2]As a comparison, California and Massachusetts accounted for 36.5 percent of Phase I awards at NIH.

[3]The SBA study mentioned in the report (no citation given) referred to SBIR awards from FY1983 through FY1986. U.S. Government Accounting Office, "Federal Research: Evaluation of Small Business Innovation Research Can Be Strengthened," GAO/RCED-99-114, Washington, DC: U.S. Government Accounting Office, June 1999, p. 17.

[4]Ibid., p. 21. See also pp. 26-27.

[5]Ibid, p. 27.

**TABLE 3-2** Phase I Awards 1992–2005 by State

| State | Number of Phase I Awards | | | | | | | | | | | | | | Total |
|---|---|---|---|---|---|---|---|---|---|---|---|---|---|---|---|
| | 1992 | 1993 | 1994 | 1995 | 1996 | 1997 | 1998 | 1999 | 2000 | 2001 | 2002 | 2003 | 2004 | 2005 | |
| AK | | | | 1 | 1 | 1 | | | | | 1 | 1 | 1 | 1 | 7 |
| AL | 14 | 20 | 24 | 32 | 24 | 31 | 21 | 38 | 21 | 32 | 45 | 59 | 63 | 89 | 513 |
| AR | | 1 | | 1 | | | | 2 | 1 | 3 | 1 | 3 | 1 | 2 | 15 |
| AZ | 14 | 16 | 14 | 23 | 41 | 41 | 31 | 31 | 38 | 32 | 40 | 44 | 46 | 50 | 461 |
| CA | 262 | 336 | 370 | 294 | 332 | 344 | 275 | 321 | 254 | 302 | 500 | 439 | 436 | 464 | 4,929 |
| CO | 30 | 43 | 47 | 38 | 47 | 68 | 52 | 57 | 76 | 59 | 114 | 101 | 99 | 97 | 928 |
| CT | 35 | 33 | 55 | 27 | 30 | 30 | 24 | 22 | 14 | 15 | 35 | 27 | 26 | 19 | 392 |
| DC | 2 | | 2 | 3 | 4 | 3 | 2 | 8 | 4 | 3 | 6 | 3 | 3 | 6 | 49 |
| DE | 6 | 3 | 7 | 3 | 3 | 3 | 4 | 3 | 2 | 3 | 1 | 6 | 6 | 6 | 56 |
| FL | 31 | 30 | 35 | 36 | 36 | 44 | 35 | 37 | 38 | 38 | 51 | 46 | 60 | 65 | 582 |
| GA | 6 | 11 | 10 | 7 | 13 | 17 | 16 | 17 | 15 | 13 | 20 | 19 | 17 | 21 | 202 |
| HI | | | 4 | 3 | 3 | 4 | 3 | 8 | 3 | | 6 | 7 | 4 | 14 | 59 |
| IA | | 1 | | 1 | | | 3 | 1 | | 3 | 4 | 1 | 5 | 2 | 21 |
| ID | | 1 | 4 | | 2 | | 1 | | 2 | 5 | 7 | 3 | 3 | 8 | 36 |
| IL | 12 | 15 | 4 | 11 | 14 | 10 | 12 | 17 | 15 | 9 | 23 | 17 | 29 | 26 | 214 |
| IN | 5 | 4 | 1 | 4 | 5 | 3 | 8 | 4 | 2 | 3 | 14 | 9 | 8 | 24 | 94 |
| KS | 4 | 3 | 2 | 3 | 2 | 5 | 3 | 3 | 3 | 4 | 1 | 6 | 1 | 5 | 45 |
| KY | 1 | 1 | 3 | 3 | 1 | 2 | 2 | 1 | | | 2 | 2 | 1 | 1 | 20 |
| LA | 4 | 5 | 1 | 5 | 6 | 2 | 3 | 2 | 1 | 3 | 5 | 8 | 7 | 5 | 57 |

*continued*

**TABLE 3-2** Continued

| State | Number of Phase I Awards | | | | | | | | | | | | | | Total |
|---|---|---|---|---|---|---|---|---|---|---|---|---|---|---|---|
| | 1992 | 1993 | 1994 | 1995 | 1996 | 1997 | 1998 | 1999 | 2000 | 2001 | 2002 | 2003 | 2004 | 2005 | |
| MA | 181 | 219 | 199 | 194 | 204 | 247 | 189 | 198 | 182 | 176 | 317 | 310 | 298 | 298 | 3,212 |
| MD | 37 | 65 | 59 | 48 | 63 | 73 | 48 | 84 | 53 | 68 | 118 | 101 | 108 | 122 | 1,047 |
| ME | 6 | 5 | 3 | 3 | | 1 | 1 | 4 | 1 | 1 | 11 | 9 | 8 | 10 | 63 |
| MI | 14 | 13 | 20 | 26 | 20 | 37 | 27 | 19 | 23 | 18 | 54 | 41 | 40 | 68 | 420 |
| MN | 13 | 19 | 25 | 19 | 18 | 26 | 15 | 25 | 20 | 18 | 27 | 23 | 20 | 26 | 294 |
| MO | 4 | 4 | 6 | 7 | 3 | 10 | 8 | 4 | 5 | 4 | 12 | 6 | 7 | 12 | 92 |
| MS | | 2 | | | 2 | | | 3 | 1 | 3 | 4 | 4 | 5 | 2 | 26 |
| MT | 1 | 2 | 2 | | 3 | | 2 | 4 | 3 | 4 | 6 | 4 | 8 | 8 | 47 |
| NC | 9 | 10 | 13 | 10 | 10 | 8 | 10 | 9 | 13 | 15 | 9 | 20 | 26 | 16 | 178 |
| ND | 2 | | | 1 | | 1 | 1 | 2 | | 1 | 1 | | | | 9 |
| NE | 3 | 1 | 1 | | | 1 | | 2 | 3 | 3 | 3 | 1 | | | 18 |
| NH | 8 | 15 | 14 | 17 | 10 | 12 | 15 | 20 | 14 | 18 | 27 | 29 | 22 | 27 | 248 |
| NJ | 40 | 42 | 55 | 41 | 44 | 51 | 44 | 44 | 41 | 36 | 63 | 68 | 53 | 65 | 687 |
| NM | 37 | 40 | 24 | 38 | 37 | 27 | 32 | 33 | 26 | 27 | 35 | 35 | 35 | 30 | 456 |
| NV | 3 | 2 | 1 | 6 | 4 | 5 | 1 | 6 | 2 | 5 | 14 | 12 | 4 | 4 | 69 |
| NY | 36 | 53 | 64 | 50 | 57 | 54 | 50 | 52 | 40 | 56 | 78 | 89 | 82 | 108 | 869 |
| OH | 38 | 51 | 50 | 56 | 45 | 63 | 64 | 58 | 63 | 56 | 92 | 87 | 95 | 111 | 929 |
| OK | 5 | 2 | 3 | 2 | 6 | 3 | 4 | 3 | 1 | 5 | 6 | 7 | 21 | 16 | 84 |
| OR | 5 | 7 | 10 | 13 | 8 | 12 | 6 | 5 | 4 | 11 | 13 | 14 | 15 | 17 | 140 |
| PA | 44 | 38 | 52 | 44 | 53 | 61 | 48 | 46 | 48 | 56 | 74 | 80 | 92 | 101 | 837 |

| | | | | | | | | | | | | | | |
|---|---|---|---|---|---|---|---|---|---|---|---|---|---|---|
| RI | 9 | 6 | 4 | 8 | 7 | 1 | 5 | 3 | 6 | 2 | 6 | 10 | 2 | 4 | 73 |
| SC | | | 2 | 2 | 3 | 4 | 1 | 1 | 7 | 5 | 12 | 12 | 3 | 4 | 56 |
| SD | | | | | | 1 | | | 2 | 1 | 1 | 1 | | 1 | 7 |
| TN | 7 | 10 | 15 | 9 | 9 | 8 | 13 | 11 | 10 | 9 | 9 | 11 | 11 | 10 | 142 |
| TX | 26 | 45 | 42 | 55 | 53 | 55 | 54 | 38 | 41 | 51 | 80 | 95 | 98 | 115 | 848 |
| UT | 13 | 8 | 9 | 10 | 16 | 4 | 11 | 14 | 2 | 8 | 14 | 12 | 7 | 23 | 151 |
| VA | 75 | 88 | 86 | 79 | 105 | 113 | 112 | 94 | 95 | 109 | 151 | 170 | 144 | 186 | 1,607 |
| VT | 2 | 3 | 2 | 1 | 4 | 6 | 3 | 3 | 1 | 1 | 4 | 5 | 4 | 4 | 43 |
| WA | 16 | 26 | 17 | 23 | 16 | 22 | 17 | 25 | 17 | 14 | 35 | 34 | 32 | 32 | 326 |
| WI | 5 | 4 | 9 | 5 | 4 | 12 | 5 | 8 | 5 | 2 | 3 | 5 | 8 | 8 | 83 |
| WV | | | | | 3 | | 4 | 2 | 1 | | 7 | 15 | 10 | 8 | 50 |
| WY | | | | | 1 | | 1 | 1 | 1 | | | 2 | 1 | 3 | 10 |
| Total | 1,065 | 1,303 | 1,370 | 1,262 | 1,372 | 1,526 | 1,286 | 1,393 | 1,220 | 1,310 | 2,162 | 2,113 | 2,075 | 2,344 | 21,801 |

SOURCE: Department of Defense.

**TABLE 3-3** Phase I Awards to the 15 Lowest Award-receiving States, FY1992–2005

| | Number of Phase I Awards | | | | | | | | | | | | | | |
|---|---|---|---|---|---|---|---|---|---|---|---|---|---|---|---|
| State | 1992 | 1993 | 1994 | 1995 | 1996 | 1997 | 1998 | 1999 | 2000 | 2001 | 2002 | 2003 | 2004 | 2005 | Total |
| AK | | | | 1 | 1 | 1 | | | | | 1 | 1 | 1 | 1 | 7 |
| SD | | | | | | 1 | | | 2 | 1 | 1 | 1 | | 1 | 7 |
| ND | 2 | | | 1 | | 1 | 1 | 2 | | 1 | 1 | | | | 9 |
| WY | | | | | 1 | | 1 | 1 | 1 | | | 2 | 1 | 3 | 10 |
| AR | | 1 | | 1 | | | | 2 | 1 | 3 | 1 | 3 | 1 | 2 | 15 |
| NE | 3 | 1 | 1 | | | 1 | | 2 | 3 | 3 | 3 | 1 | | | 18 |
| KY | 1 | 1 | 3 | 3 | 1 | 2 | 2 | 1 | | | 2 | 2 | 1 | 1 | 20 |
| IA | | 1 | | 1 | | | 3 | 1 | | 3 | 4 | 1 | 5 | 2 | 21 |
| MS | | 2 | | | 2 | | | 3 | 1 | 3 | 4 | 4 | 5 | 2 | 26 |
| ID | | 1 | 4 | | 2 | | 1 | | 2 | 5 | 7 | 3 | 3 | 8 | 36 |
| VT | 2 | 3 | 2 | 1 | 4 | 6 | 3 | 3 | 1 | 1 | 4 | 5 | 4 | 4 | 43 |
| KS | 4 | 3 | 2 | 3 | 2 | 5 | 3 | 3 | 3 | 4 | 1 | 6 | 1 | 5 | 45 |
| MT | 1 | 2 | 2 | | 3 | | 2 | 4 | 3 | 4 | 6 | 4 | 8 | 8 | 47 |
| DC | 2 | | 2 | 3 | 4 | 3 | 2 | 8 | 4 | 3 | 6 | 3 | 3 | 6 | 49 |
| WV | | | | | 3 | | 4 | 2 | 1 | | 7 | 15 | 10 | 8 | 50 |

SOURCE: DoD awards database.

**TABLE 3-4** Phase I Awards—Top Zip Codes

| Zip Code | Number of Phase I Awards |
|---|---|
| 92121 | 353 |
| 01801 | 324 |
| 90501 | 314 |
| 01803 | 257 |
| 02154 | 253 |
| 35805 | 203 |
| 01824 | 193 |
| 80301 | 192 |
| 02451 | 191 |
| 01810 | 190 |
| 85706 | 177 |
| 02138 | 163 |
| 94043 | 144 |
| 90505 | 141 |
| 24060 | 137 |
| 93117 | 135 |
| 87109 | 132 |
| 77840 | 131 |
| 01730 | 129 |
| 03755 | 129 |

SOURCE: DoD awards database.

of application are a major determinant of success in winning awards from the program.

Awards are also distributed unequally within states. The top 20 winning zip codes account for 17.8 percent of Phase I awards overall (see Table 3-4). This is a lower degree of concentration than at NIH, but in both cases, the data illustrate that the SBIR awards, like other innovation activity, tend to be concentrated in relatively small geographic areas. These clusters of innovation are, in effect, the relevant unit of measure. Even states with high numbers of awards find that they are not distributed across the state but instead are concentrated in these innovation clusters. Moreover, other sources of early-stage funding such as venture capital tend to be concentrated as well, and normally in the same areas.[6]

[6]For example, venture capital investment is widely recognized to be concentrated in California with some 47 percent of national venture funding, yet 35 percent of the nation's VC investments are in Silicon Valley, just under 7 percent in Los Angeles/Orange County, 4.6 percent in San Diego, while the rest of California receives 0.5 percent of the $7.6 billion invested there in 2005. See the presentation "The Private Equity Continuum" by Steve Weiss, Executive Committee Chair of Coachella Valley Angel Network, citing PricewaterhouseCoopers Money Tree data at the Executive Seminar on Angel Funding, University of California at Riverside, December 8-9, 2006, Palm Springs, California.

### 3.2.4 Commercialization and Multiple-award Winners

In the first eight years of the program, a number of companies were successful in winning multiple awards. Many of the projects funded by these awards were not commercialized. There are several reasons for this. In part, the low commercialization rates reflect the uncertainties inherent in the funding of relatively early-stage technology development. It may also reflect imperfect alignment between solicitations and the needs of procurement agencies and the complexities and long lead times of the procurement process. Perhaps most important, it may reflect the lower emphasis on commercialization in the early years of the program than is now the case.[7] Lastly, it reflects the different goals of participating companies documented in previous NRC research.[8] Understanding these different goals is important in this context.

Companies that participate in the program, like the agencies themselves, often have multiple objectives.

- Firms approach the SBIR award process at different stages of development and with different objectives. Some firms are developing technology concepts; some firms see their vocation as contract research organizations; others actively seek to develop commercial products, either for public agencies or for the marketplace.[9]
- Investigator-led firms, limited in size and focused on a single concept may seek multiple awards as they advance research on a promising technology.[10]

---

[7]As a Creare representative, Nabil Elkouh, points out, in the early years of the program, small companies had not figured out how to use it, nor had the departments figured out how to run the program, and the award process was less competitive than it is today. Emphasis on commercialization was minimal. Program managers defined topics that represented an interesting technical challenge. See the case study of Creare, Inc., August 2005, in National Research Council, *An Assessment of the SBIR Program at the Department of Energy*, Charles W. Wessner, ed., Washington, DC: The National Academies Press, 2008.

[8]National Research Council, *The Small Business Innovation Research Program: An Assessment of the Department of the Defense Fast Track Initiative*, Charles W. Wessner, ed., Washington, DC: National Academy Press, 2000.

[9]See Reid Cramer, "Patterns of Firm Participation in the Small Business Innovation Research Program in the Southwestern and Mountain States," in National Research Council, *The Small Business Innovation Research Program: An Assessment of the Department of Defense Fast Track Initiative*, op. cit., p. 151. The author describes the incremental nature of technical advance, which sometimes necessitates several awards. See also John T. Scott, "An Assessment of the Small Business Innovation Research Program in New England: Fast Track Compared with Non-Fast Track," in National Research Council, *The Small Business Innovation Research Program: An Assessment of the Department of Defense Fast Track Initiative*, op. cit., p. 109, for a discussion of Foster-Miller, Inc.

[10]Ibid. The mirror image of this approach is the program manager that makes several awards for similar technologies among different companies. In fact, it is not uncommon to have multiple awardees on the same Phase I topic. For an example, see the Navy's SBIR Web site selections page for their FY-06.1 awardees, available at <*http://www.navysbir.com/06_1selections.html*>. This page not only shows several awardees for each topic, but if you click on the "Details" link, you can see the differences in companies' approaches to the topics.

- For firms that carry out research as a core activity, success is often measured in multiple contract awards. Some firms, mainly at DoD, have won large numbers of awards over the life of the program. Yet, even with many awards, there is nothing intrinsically wrong with a process that provides high-quality research at a lower cost than might otherwise be available to the department.[11] Inexpensive exploration of new technological approaches can be valuable, particularly if they limit expenditure on technological dead ends. For research oriented firms, the key issue is the quality of the research and its alignment with service and agency needs.[12]
- In some cases, firms respond to an agency solicitation and "solve" the problem, provide the needed data, or propose a solution that can then be adopted by the agency with no further "commercialization" revenues for the firm.[13]
- Those firms that seek to develop commercial products may, in an initial phase, seek multiple awards to rapidly develop a technology. For the high-growth firms, this period is limited in time, before private investment becomes the principal source of funding.[14]

---

[11]To secure additional awards, a small company has to resubmit its proposal through the regular review process. These awards are relatively small in amount—the normal Phase I and Phase II awards would total $850,000. As a point of comparison, the top three U.S. prime contractors in 2004 garnered over $86 billion in defense revenues.

[12]For example, Foster-Miller, a multiple-award winner, developed robots for use in Iraq to identify roadside improvised explosive devices. Creare is also won a large number of awards and tends to focus engineering problem solving rather than commercialization. Nonetheless the firm has 21 patents resulting from SBIR-funded work, has published dozens of papers, and licensed a variety of technologies. These technologies include high-torque threaded fasteners, a breast cancer surgery aid, corrosion preventative coverings, an electronic regulator for firefighters, and mass vaccination devices (pending). Products and services developed at Creare include thermal-fluid modeling and testing, miniature vacuum pumps, fluid dynamics simulation software, network software for data exchange, and the NCS Cryocooler used on the Hubble Space Telescope to restore the operation of the telescope's near-infrared imaging device. See the case study of Creare, Inc., August 2005, in National Research Council, *An Assessment of the SBIR Program at the Department of Energy*, op. cit.

[13]There are cases where a small business successfully completes the requirements and objectives of a Phase II contract, meeting the needs of the customer, without gaining additional commercialization revenues. For example, Aptima, Inc., a multiple-award winner, designed an instructional system to improve boat handling safety by teaching the use of strategies that mitigate shock during challenging wave conditions. A secondary goal was to demonstrate how an innovative learning environment could establish robust skill levels while compressing learning time. Phase I of the project developed a training module, and in Phase II, instructional material, including computer animation, videos, images, and interviews were developed. The concept and the supporting materials were adopted as part of the introductory courses for Special Operations helmsmen with the goal of reducing injuries and increasing mission effectiveness. Michael Paley, Aptima, Inc., personal communication, September 30, 2006.

[14]For a discussion of Martek as an example, see Maryann P. Feldman, "Role of the Department of Defense in Building Biotech Expertise." in National Research Council, *The Small Business Innovation Research Program: An Assessment of the Department of Defense Fast Track Initiative*, op. cit., pp. 266-268. See also Reid Cramer, "Patterns of Firm Participation in the Small Business Innovation Research Program in the Southwestern and Mountain States," in National Research Council, *The*

In short, the participating firms, like the services and agencies, use the program in a variety of ways. Some are start-ups, some are well-established firms, all have differing strategies and objectives, and many are new entrants. Some are strong performers with regard to the commercialization metric, while others make valuable, if less commercially oriented, contributions.[15]

(The DoD commercialization database provides the best data on overall outcomes from awards to FAWs, not least because it is specifically designed to do so.[16])

#### 3.2.4.1 Background to the Multiple-award Winner Issue

As the 1992 SBIR reauthorization approached, there was some concern on the part of the Small Business Committee that "large, multiple-award winners might hurt the program. . . ."[17] This concern was included in the second GAO evaluation of the program, required by the 1982 legislation, and released in March 1992.[18] The GAO evaluators found very preliminary evidence that in the 1984–1987 period, SBIR companies receiving five or more awards (deemed "frequent winners") had a somewhat lower commercialization record than companies

---

*Small Business Innovation Research Program: An Assessment of the Department of Defense Fast Track Initiative*, op. cit., pp. 146-147, who discusses several firms that realized commercial success after several awards.

[15]One of the earliest (1992) GAO studies on SBIR found a positive record on commercialization. The study noted that "even though many of the SBIR projects have not yet had sufficient time to achieve their full commercial potential, the program is showing success in Phase III activity," with the majority of this activity occurring in the private sector, a goal of the program. U.S. Government Accounting Office, *Federal Research: Small Business Innovation Research Shows Success but Can Be Strengthened*, GAO/RCED-92-37, March 1992, p. 4.

[16]The DoD database does not contain information on all companies or all awards. According to BRTRC, which manages the database, the data collected from the agencies on Phase II awards made from 1992 to 2001 identified 2,257 firms that had received at least one Phase II, but were not in the DoD database, and were therefore not included in Table 3-5. Of these 2,257, only six had received 15 or more Phase II during the ten years for which BRTRC received award data. Although inclusion of pre-1992 and post-2001 awards would have increased that number, it seems reasonable to conclude that the firms in the DoD data represent a large majority of the multiple winners.

[17]House Report (REPT. 102-554) Part I (Committee on Small Business), *The Small Business Research and Development Enhancement Act of 1992*, p. 17.

[18]The GAO analysis was carried out between August 1990 and August 1991. At the time, the GAO cautioned that the group examined consisted of the Phase II awardees from the first four years in which Phase II awards were made, the GAO analysts chose the earlier recipients "because studies by experts on technology development concluded that five to nine years are needed for a company to progress from a concept to a commercial product." U.S. Government Accounting Office, *Federal Research: Small Business Innovation Research Shows Success but Can Be Strengthened*, GAO/RCED-92-37, op. cit. p. 17. They note further that "even with this early group of Phase II recipients, additional time is required for projects to mature." They add that "about ten percent of the projects responding to our survey had not even completed Phase II," adding that "our findings therefore represent an early interpretation of the trends in Phase III."

receiving fewer than five awards."[19] The preliminary nature of this analysis was not fully appreciated at the time, despite the GAO's qualifications concerning the limited time between, for example, the 1987 awards and the study's analysis in 1990–1991.

Notwithstanding the tentative nature of the findings, the problem of "SBIR mills" was established and the focus shifted to efforts to enhance commercialization. To this end, Section 9(e)(4) of the Small Business Act (15 U.S.C. 638(e)(4)) was amended in 1992 to require that agencies consider "the small business concern's record of successfully commercializing SBIR or other research" when making Phase II awards. Thus Section 9(e)(4) of the Small Business Act (15 U.S.C. 638(e)(4)) was amended as follows—

> (B) a second phase, to further develop proposals which meet particular program needs, in which awards shall be made based on the scientific and technical merit and feasibility of the proposals, as evidenced by the first phase, considering, among other things, the proposal's commercial potential, as evidenced by—(i) the small business concern's record of successfully commercializing SBIR or other research; (ii) the existence of second phase funding commitments from private sector or non-SBIR funding sources; (iii) the existence of third phase, follow-on commitments for the subject of the research; and (iv) the presence of other indicators of the commercial potential of the idea.

#### 3.2.4.2 The Creation of the Commercialization Achievement Index

At DoD, this requirement led to development of the "Commercialization Achievement Index (CAI)," which normalized reported sales and further investment resulting from the Phase II based on the length of time since the award and allowed a numerical comparative evaluation of a firm's success in commercialization to firms with comparable SBIR experience.

The CAI is one component of the Company Commercialization Report (CCR), which is electronically included in every proposal. The CCR also captures other indicators of the commercial potential, such as firm growth, IPO resulting from SBIR, number of patents received as a result of SBIR, firm revenue and the percent of that revenue that is SBIR. Evaluation of the proposal's commercial potential is based on the commercialization plan, which would include any funding commitments for Phase II for or Phase III from private sector or non-SBIR sources, and on the CCR. The CCR also allows the firm to describe commercialization that may have small sales dollars, but major impact, such as in health care or cost savings.

Firm winning more than four or five awards (the number changed in 2005) are now required to complete a CCR with every application for further DoD SBIR awards. As the report requires information about all awards, not just those

[19]House Report (REPT. 102-554) Part I (Committee on Small Business), *The Small Business Research and Development Enhancement Act of 1992*, op. cit., p. 17.

at DoD, it includes some data on awards at other agencies, for firms which also apply at DoD.

#### 3.2.4.3 Commercialization Outcomes from DoD Commercialization Data

Table 3-5 shows commercialization data for firms by number of Phase II awards.[20]

#### 3.2.4.4 Reconceptualizing Multiple-award Winners

Criticisms of multiple-award winners seem in general to be misplaced. They result from an overly negative reading of limited data, focused on one element of the program, often using a highly simplified, essentially linear conception of commercialization.

A more comprehensive assessment of the role of companies with multiple awards reveals multiple dimensions.

1. **Evolution in Company Revenue.** Data from the NRC Phase II Survey shows that larger companies tend to rely less on SBIR as a source of company revenue. This is supported by case research: At Radiation Monitoring, for example, SBIR is now only 16 percent of total firm revenues.[21]
2. **Graduation.** The companies evolve over time. Some of the large Phase II winners have "graduated" from the program either by growing beyond the 500 employee limit or by being acquired. In the case of Foster-Miller, a particularly strong award winner, the company was acquired by a foreign-owned firm.
3. **Meeting Agency Needs.** Case studies show that some of the biggest award winners have successfully commercialized, and have also in other ways met the needs of sponsoring agencies. Some are effective at contract research. Contract research is often a valuable contribution in its own right. DoD staff indicate that SBIR fills multiple needs, many of which do not show up in sales data. For example, agency staff suggest that SBIR awards permit efficient probes of the technological frontier, conducted in a short time frame, with a very limited budget. These awards can effectively explore new technological approaches, saving time and resources; and some companies succeed in providing viable alternatives to program managers.[22]
4. **Company Creation.** Some frequent winners frequently spin off

[20]Although the database covers all agencies, some agencies are underrepresented owing to the focus on DoD-oriented firms. NIH awardees, for example, account for only 7 percent of entries in the database.

[21]Michael Squillante, Vice President, Radiation Monitoring, private communication, June 2004.

[22]See National Research Council, *SBIR and the Phase III Challenge of Commercialization*, Charles W. Wessner, ed., Washington, DC: The National Academies Press, 2007.

**TABLE 3-5** DoD Commercialization Data on Multiple-award Winners, All Agencies, 1992–2005

| Number of Phase II SBIR Awards per Firm[a] | Number of Firms | Number of Projects in CCR Database | Number of Projects with Award Years prior to 2004 | Average Commercialization of Projects with Award Years prior to 2004 ($) | Average Sales of Projects with Award Years prior to 2004 ($) | Average Funding of Projects with Award Years prior to 2004 ($) |
|---|---|---|---|---|---|---|
| ≥125 projects | 5 | 941 | 823 | 2,067,719 | 1,384,571 | 683,148 |
| ≥75 and <110 (no firms had between 111 and 124 projects) | 5 | 485 | 411 | 1,117,325 | 526,623 | 590,703 |
| ≥50 and <75 | 17 | 1,067 | 945 | 4,103,125 | 3,586,611 | 516,514 |
| ≥25 and <50 | 77 | 2,692 | 2,330 | 1,710,140 | 1,048,787 | 661,354 |
| ≥15 and <25 | 101 | 1,858 | 1,535 | 1,375,061 | 863,310 | 511,750 |
| >0 and <15 | 2,715 | 8,101 | 6,243 | 1,300,886 | 751,418 | 549,468 |

[a]Awards are all Phase II from any agency, not just DoD awards.
SOURCE: Derived from DoD CCR Commercialization Database.

companies—like Optical Sciences, Creare, and Luna. Creating new firms is a valuable contribution of the program especially with regard to the defense industrial base. The creation of these firms creates new opportunities for defense contractors, greater competition, and permits more rapid development of new defense solutions.

5. **Flexibility and Speed.** Some FAWs have provided the highly efficient and flexible capabilities needed to solve pressing problems rapidly. For example, Foster-Miller, Inc., responded to needs in Iraq by developing and the manufacturing add-on armor for Humvees.[23]

### 3.2.4.5 Conclusions

The data and analysis above suggest three core conclusions:

1. While some companies win a substantial number of awards, perhaps not unlike leading universities, there does not appear to be a widespread problem inherent to the program at DoD (or at other agencies). The most recent data suggest these companies commercialize on average more than companies with fewer awards.
2. Analysis of other dimensions of the program also strongly suggest that frequent winners provide powerful benefits: Given that our analysis of selection procedures suggests that in general these are both fair and competitive, the presumption must be that this limited number of companies are winning awards because they meet the needs of the agency, as expressed in published solicitations. More broadly, it is too narrow an approach to evaluate company performance solely on the basis of commercialization: The SBIR program is designed to meet other equally important congressional objectives as well.
3. The current focus on commercialization records is a valuable stimulus. DoD is currently meeting congressional requirements in this area by maintaining the CAI and requiring completion of the CCR, and by including commercialization information with Phase II applications. Efforts to further enhance reporting and analysis are recommended elsewhere.
4. DoD has implemented what might be called the "enhanced surveillance model for FAWs—requiring closer scrutiny of the commercialization efforts in the course of the selection process. While elements of this process will undoubtedly be adjusted and fine-tuned in light of ongoing experience, the fact is that DoD is already taking steps to ensure that "research for the sake of research" is not encouraged.

---

[23]Foster-Miller's LAST® Armor, which uses Velcro-backed tiles to protect transport vehicles, helicopters and fixed wing aircraft from enemy fire, was developed on two Phase I SBIRs and a DARPA Broad Agency Announcement. The technology has helped improve the safety of combat soldiers and fliers in Bosnia and Operation Desert Storm. Access at *<http://www.dodsbir.net/SuccessStories/fostermiller.htm>*.

Given that SBIR awards meet multiple agency needs and multiple congressional objectives, it is difficult to see how the program might be enhanced by the imposition of an arbitrary limit on the number of applications per year. The evidence supports the conclusion that the department does not have a general problem with multiple-award winners.

If, over time, agencies see issues emerging in this area, they might consider adopting some version of the DoD "enhanced surveillance" model, in which multiple winners are subject to enhanced scrutiny in the context of the award process.

### 3.2.5 Phase I Awards—By Company

Some companies are very successful in winning Phase I awards at DoD. The most successful applicant between FY1992 and FY2005 won 361 Phase I awards (and is no longer a small business). The top 20 Phase I winners among the 7,113 companies that received at least one Phase I award from DoD over this period accounted for 11.2 percent of all Phase I awards (compared to 8.9 percent at NIH).

Twenty-seven companies received at least 50 awards from DoD during this 14-year period, and ten received more than 100. Two received more than 300. (See Table 3-6.) These data indicate a considerably greater degree of concentra-

**TABLE 3-6** Multiple-award Winning Companies at DoD FY1992–2005

| Firm Name | Number of Phase I Awards |
|---|---|
| FOSTER-MILLER, INC. | 361 |
| PHYSICAL OPTICS CORP. | 316 |
| PHYSICAL SCIENCES, INC. | 170 |
| MISSION RESEARCH CORP. | 126 |
| ALPHATECH, INC. | 117 |
| CREARE, INC. | 129 |
| CHARLES RIVER ANALYTICS, INC. | 112 |
| CFD RESEARCH CORP. | 107 |
| TRITON SYSTEMS, INC. | 125 |
| COHERENT TECHNOLOGIES, INC. | 101 |
| TECHNOLOGY SERVICE CORP. | 90 |
| CYBERNET SYSTEMS CORP. | 95 |
| SCIENTIFIC SYSTEMS CO., INC. | 91 |
| DIGITAL SYSTEM RESOURCES, INC. | 48 |
| STOTTLER HENKE ASSOC., INC. | 84 |
| TEXAS RESEARCH INSTITUTE AUSTIN, INC. | 77 |
| ORINCON CORP. | 97 |
| METROLASER, INC. | 66 |
| SYSTEMS & PROCESS ENGINEERING CO. | 69 |
| TOYON RESEARCH CORP. | 65 |

SOURCE: DoD awards database.

tion of awards among the top winning companies than at other agencies, including NIH, which has the second largest SBIR program.

It might also be observed that a number of the companies listed in Table 3-6 have grown and are now large firms, no longer eligible for SBIR. Of course, this successful growth is a desirable result of their prior SBIR work.

On the other side of the spectrum, 95 percent of SBIR awardees received less than 10 awards, and 74 percent received no more than two.

### 3.2.6 Phase I Awards—Demographics

Data from the DoD awards database indicate that the percentage of DoD SBIR awards going to woman- and socially and economically disadvantaged small business concerns has hovered around 20 percent for years (see Figure 3-4). Within that 20 percent, the percentage going to woman-owned firms has slowly increased, while the share going to minority-owned firms has fallen steadily from a peak of 14 percent in 1999 to 9.3 percent in 2005. However, the actual numbers of awards to both have increased during this period as the overall number of SBIR awards has expanded with DoD research funding.

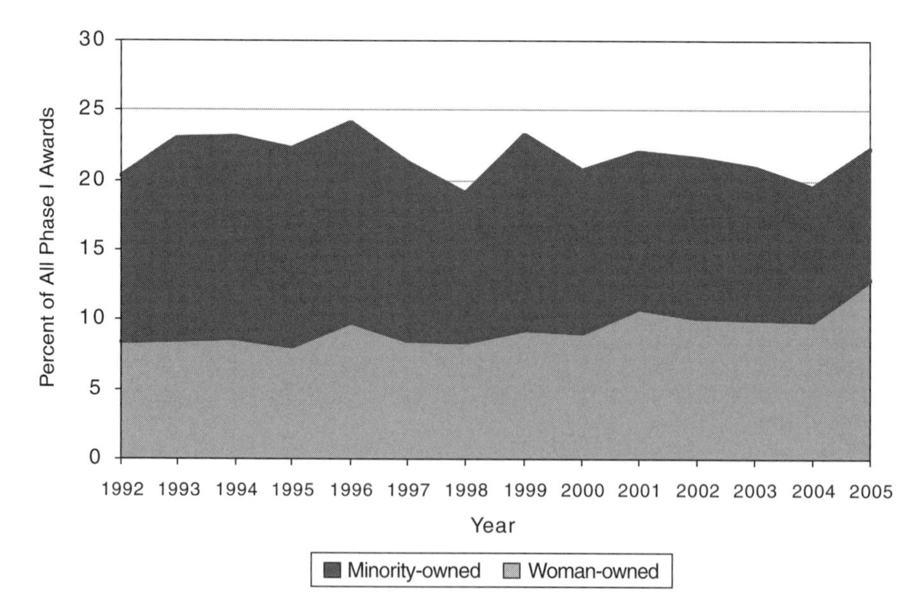


**FIGURE 3-4** Phase I awards to woman- and socially and economically disadvantaged small business concerns.
SOURCE: DoD awards database.

### 3.2.7 Phase I Awards—By Agency and Component

The substantial size differences between the various components of DoD mean that different components award different numbers of contracts. The percentage of Phase I contracts awarded by each component is displayed in Table 3-7.

The data show that the three largest components—Army, Air Force, and Navy—account for a fraction under 70 percent of all DoD SBIR awards between 1992 and 2005. This dominance varies substantially, however, ranging from a high of 83 percent in 1995 to a low of 59 percent in 2002 (as shown in Figure 3-5).

### 3.2.8 Phase I Awards—Size of Awards

None of the DoD components has experimented with oversized Phase I awards in the same way as NIH. In general, awards are kept slightly below the SBA guideline maximum of $100,000. Some components hold back up to $30,000 of a possible Phase I award as an "option" which can be released as bridge funding between Phase I and Phase II after a Phase II contract has been awarded but before the contract is in place.

Overall, less than 0.15 percent of all Phase I awards were made for more than $150,000, although it is worth noting that in recent years some large Phase I contracts have been awarded, as shown in Table 3-8.

DoD staff have suggested that these extra-large awards—and similar extra-large Phase II awards—have resulted from the addition of non-SBIR funding to existing SBIR awards. This technique is a permissible and apparently not uncommon event at DoD and is considered by many to be a very desirable additional incentive and success measure. Ideally, the award data should indicate such additional funding.

## 3.3 PHASE II AWARDS

As R&D funding for DoD has increased, the number of Phase II contracts awarded has increased. The trendline in Figure 3-6 reflects growth in the number of Phase II contracts awarded, from about 400 in 1992 to about 1,000 in 2005. The substantial jump in numbers awarded in 2003 partly reflects the 2002 increase in Phase I awards.

One strategic question for all SBIR agencies is the balance between Phase I and Phase II funding. Too many Phase I awards might leave insufficient funding to provide for the critical Phase II research that can result in technologies that the agencies will use, or that can be commercialized. Too few Phase I awards, and agencies find they have starved the "pipeline," and must subsequently award Phase II funds to projects that may not deserve it. This balancing act is captured by the percentage of total SBIR funds that are allocated to Phase II, described in Figure 3-7. DoD allocates about 75–80 percent of funding to Phase II awards.

**TABLE 3-7** Phase I Awards, by Agency and Component

| | Number of Phase I Awards | | | | | | | | | | | | | |
|---|---|---|---|---|---|---|---|---|---|---|---|---|---|---|
| Year | AF | ARMY | BMDO | CBD | DARPA | DSWA | DTRA | MDA | NAVY | NGA | NIMA | OSD | SOCOM | Total |
| 1992 | 230 | 224 | 0 | 0 | 169 | 24 | 0 | 209 | 209 | 0 | 0 | 0 | 0 | 1,065 |
| 1993 | 470 | 288 | 0 | 0 | 70 | 19 | 0 | 147 | 306 | 0 | 0 | 0 | 3 | 1,303 |
| 1994 | 390 | 175 | 0 | 0 | 328 | 9 | 0 | 161 | 290 | 0 | 0 | 0 | 17 | 1,370 |
| 1995 | 427 | 252 | 0 | 0 | 31 | 11 | 0 | 116 | 373 | 0 | 0 | 42 | 10 | 1,262 |
| 1996 | 310 | 247 | 0 | 0 | 237 | 15 | 0 | 140 | 361 | 0 | 0 | 56 | 6 | 1,372 |
| 1997 | 431 | 291 | 0 | 0 | 146 | 19 | 0 | 189 | 380 | 0 | 0 | 59 | 11 | 1,526 |
| 1998 | 471 | 200 | 167 | 20 | 44 | 0 | 9 | 11 | 229 | 0 | 3 | 115 | 17 | 1,286 |
| 1999 | 416 | 197 | 170 | 20 | 119 | 0 | 26 | 0 | 406 | 0 | 2 | 29 | 8 | 1,393 |
| 2000 | 367 | 198 | 230 | 26 | 82 | 0 | 20 | 0 | 216 | 0 | 6 | 56 | 19 | 1,220 |
| 2001 | 373 | 281 | 161 | 18 | 82 | 0 | 24 | 0 | 241 | 0 | 2 | 108 | 20 | 1,310 |
| 2002 | 427 | 282 | 2 | 34 | 120 | 0 | 23 | 520 | 578 | 0 | 4 | 128 | 44 | 2,162 |
| 2003 | 449 | 352 | 0 | 25 | 90 | 0 | 7 | 454 | 550 | 0 | 2 | 156 | 28 | 2,113 |
| 2004 | 527 | 356 | 0 | 27 | 155 | 0 | 0 | 315 | 585 | 2 | 0 | 83 | 25 | 2,075 |
| 2005 | 608 | 705 | 0 | 21 | 74 | 0 | 40 | 240 | 466 | 2 | 0 | 163 | 25 | 2,344 |
| Total | 5,896 | 4,048 | 730 | 191 | 1,747 | 97 | 149 | 2,502 | 5,190 | 4 | 19 | 995 | 233 | 21,801 |

SOURCE: DoD awards database.

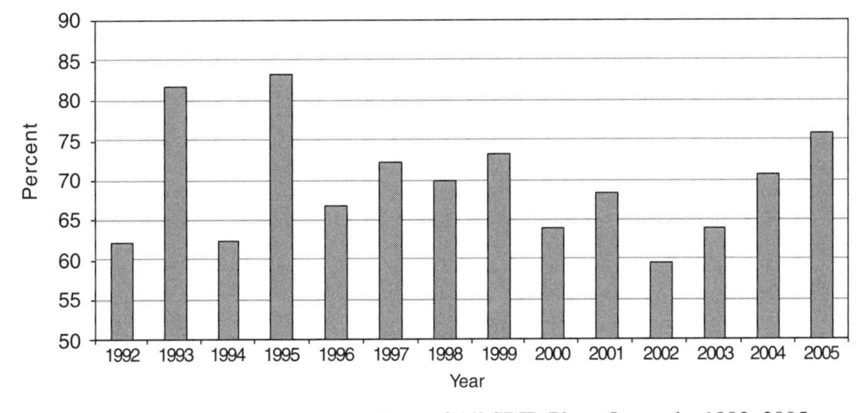


**FIGURE 3-5** Army/Air Force/Navy Share of All SBIR Phase I awards, 1992–2005.
SOURCE: DoD awards database.

### 3.3.1 Phase II—Average Size of Award

As with Phase I, the data show that DoD Phase II awards are closely aligned with the SBA guidelines. The median size of award rose when the guidelines were increased after the 1992 reauthorization, but has remained at slightly under $750,000 in nominal terms since 1997.

The DoD awards database does not distinguish clearly the source of funding on a contract. As a result, the database includes contracts where substantial additional funds were added from non-SBIR sources to an SBIR contract. As a

**TABLE 3-8** Maximum Size of Phase I Award, 1992–2005

| Year | Maximum Phase I Award Size ($) |
|---|---|
| 1992 | 150,000 |
| 1993 | 238,729 |
| 1994 | 118,086 |
| 1995 | 129,770 |
| 1996 | 150,803 |
| 1997 | 153,675 |
| 1998 | 163,805 |
| 1999 | 198,216 |
| 2000 | 179,968 |
| 2001 | 262,540 |
| 2002 | 303,996 |
| 2003 | 448,796 |
| 2004 | 597,999 |
| 2005 | 163,050 |

SOURCE: DoD awards database.

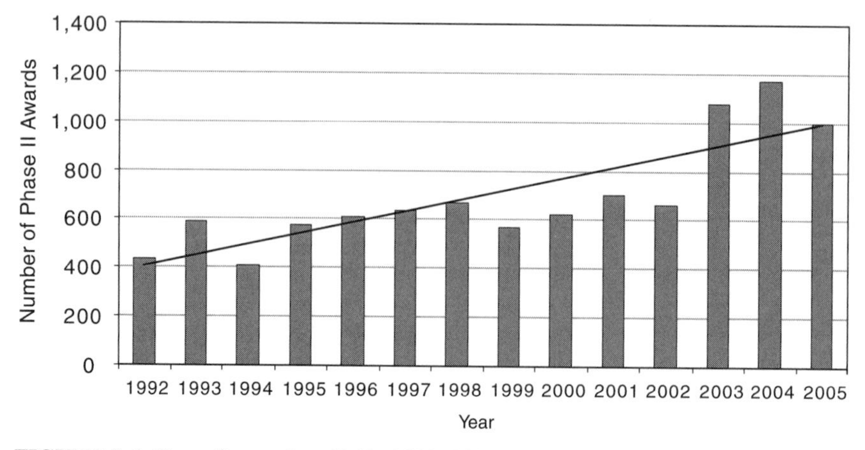


**FIGURE 3-6** Phase II awards at DoD, 1992–2005.
SOURCE: DoD awards database.

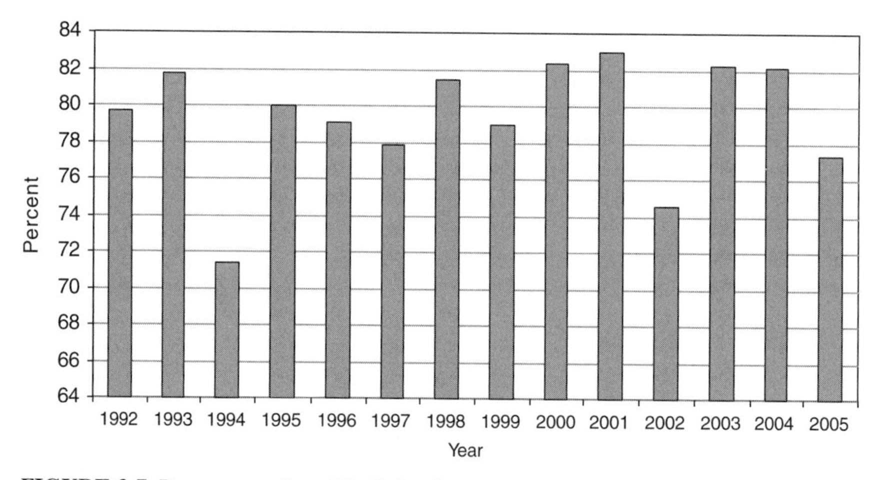


**FIGURE 3-7** Percentage of total DoD funding spent on Phase II, 1992–2005.
SOURCE: DoD awards database.

result, the awards database indicates some significant extra-large awards (see Figure 3-9).

The extent to which these awards are actually oversized SBIR awards rather than SBIR contracts supplemented with non-SBIR funds cannot be determined conclusively from the DoD awards database.[24]

[24]Component-level data is, according to the data contractor BRTRC, likely to be more detailed in this regard but is not maintained centrally and has not been used for this analysis.

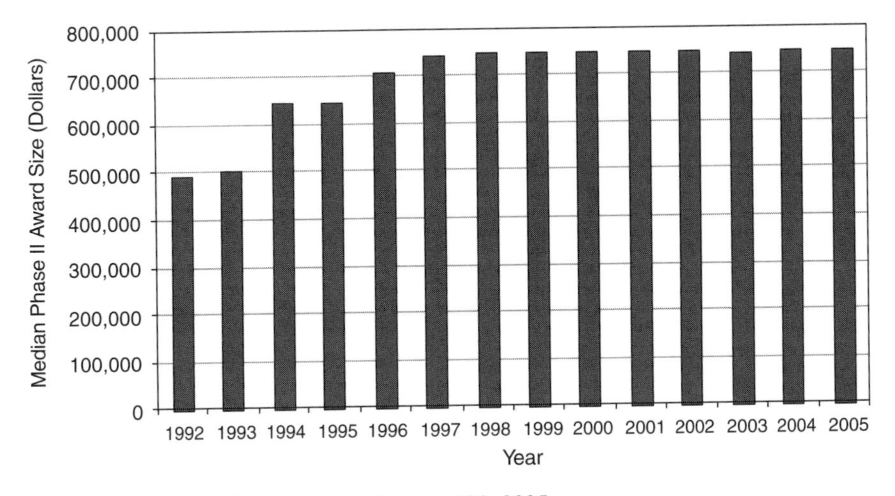


**FIGURE 3-8** Phase II median award size, 1992–2005.
SOURCE: DoD awards database.

### 3.3.2 Phase II Awards—By Company

As with Phase I, some companies have received numerous Phase II awards. The companies receiving many Phase I awards are often also successful in applying for multiple Phase II awards, as on average, 42 percent of Phase I winning proposals receive Phase II awards.

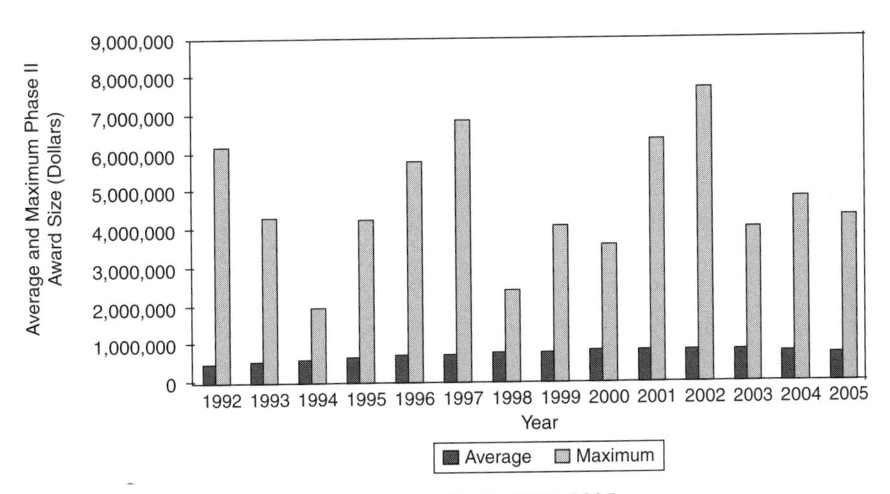


**FIGURE 3-9** Oversized Phase II awards at DoD, 1992–2005.
SOURCE: DoD awards database.

**TABLE 3-9** Phase II Multiple-award Winners 1992–2005 at DoD

| Firm Name | Number of Phase I Awards | Number of Phase II Awards | Phase I/ Phase II Conversion Rate |
|---|---|---|---|
| FOSTER-MILLER, INC. | 361 | 140 | 38.8 |
| PHYSICAL OPTICS CORP. | 316 | 117 | 37.0 |
| PHYSICAL SCIENCES, INC. | 170 | 75 | 44.1 |
| MISSION RESEARCH CORP. | 126 | 69 | 54.8 |
| ALPHATECH, INC. | 117 | 68 | 58.1 |
| CREARE, INC. | 129 | 60 | 46.5 |
| CHARLES RIVER ANALYTICS, INC. | 112 | 60 | 53.6 |
| CFD RESEARCH CORP. | 107 | 56 | 52.3 |
| TRITON SYSTEMS, INC. | 125 | 55 | 44.0 |
| COHERENT TECHNOLOGIES, INC. | 101 | 53 | 52.5 |
| TECHNOLOGY SERVICE CORP. | 90 | 42 | 46.7 |
| CYBERNET SYSTEMS CORP. | 95 | 41 | 43.2 |
| SCIENTIFIC SYSTEMS CO., INC. | 91 | 38 | 41.8 |
| DIGITAL SYSTEM RESOURCES, INC. | 48 | 36 | 75.0 |
| STOTTLER HENKE ASSOC., INC. | 84 | 36 | 42.9 |
| TEXAS RESEARCH INSTITUTE AUSTIN, INC. | 77 | 36 | 46.8 |
| ORINCON CORP. | 97 | 36 | 37.1 |
| METROLASER, INC. | 66 | 35 | 53.0 |
| SYSTEMS & PROCESS ENGINEERING CO. | 69 | 35 | 50.7 |
| TOYON RESEARCH CORP. | 65 | 34 | 52.3 |
| Total and Average (conversion rate) | 2,446 | 1,122 | 48.6 |

SOURCE: DoD awards database.

Table 3-9 shows the top Phase II award winners. Note that these results are estimates only.[25]

Together, the top 20 winners account for 11.5 percent of all Phase II awards made at DoD from FY1992 to FY2005. This compares with 11.1 percent at NIH. It is also worth noting that some of the top 20 winners are no longer eligible. For example, Foster-Miller, Inc., has been purchased by a foreign-owned corporation; Alphatech, Inc., Digital System Resources, Inc., and Triton Systems, Inc., have each been acquired and are now part of companies which have more than 500 employees.

### 3.3.3 Phase II Awards—By State

As would be expected with merit-based R&D awards, the geographical distribution of Phase II awards approximates but does not equal the distribution for Phase I awards. As can be seen in Table 3-10, the states with many Phase I

[25]Because companies change names, and in some cases tax ID numbers, a precise count would require a manual examination of all records.

**TABLE 3-10** Phase II Awards—By State

| State | Number of Phase II Awards | | | | | | | | | | | | | | Total |
|---|---|---|---|---|---|---|---|---|---|---|---|---|---|---|---|
| | 1992 | 1993 | 1994 | 1995 | 1996 | 1997 | 1998 | 1999 | 2000 | 2001 | 2002 | 2003 | 2004 | 2005 | |
| AK | | | | | | | | | 1 | | | 1 | | | 2 |
| AL | 11 | 13 | 3 | 7 | 16 | 21 | 16 | 6 | 18 | 8 | 15 | 32 | 31 | 30 | 227 |
| AR | | | | | | | | | 2 | | 1 | | 3 | 2 | 8 |
| AZ | 1 | 5 | 8 | 5 | 7 | 11 | 15 | 10 | 11 | 16 | 14 | 16 | 17 | 19 | 155 |
| CA | 100 | 130 | 114 | 132 | 147 | 155 | 160 | 103 | 133 | 163 | 160 | 236 | 254 | 211 | 2,198 |
| CO | 11 | 22 | 13 | 22 | 19 | 25 | 31 | 33 | 29 | 39 | 32 | 54 | 57 | 48 | 435 |
| CT | 20 | 19 | 5 | 18 | 20 | 15 | 10 | 7 | 7 | 12 | 6 | 18 | 17 | 17 | 191 |
| DC | | 2 | 1 | | | | 1 | 2 | 1 | 1 | 2 | 2 | 1 | | 13 |
| DE | 1 | 4 | 1 | 3 | 2 | 2 | | | 2 | 2 | 1 | | 1 | 2 | 21 |
| FL | 4 | 17 | 4 | 16 | 17 | 13 | 19 | 18 | 15 | 18 | 14 | 33 | 26 | 26 | 240 |
| GA | 3 | 1 | 2 | 5 | 1 | 8 | 8 | 10 | 8 | 9 | 6 | 12 | 9 | 10 | 92 |
| HI | 3 | | | 1 | 3 | | | | 3 | 1 | | 2 | 3 | 4 | 20 |
| IA | | | | | | 1 | | 1 | | | 3 | 2 | 1 | 1 | 9 |
| ID | | | | 1 | | 1 | | | | 1 | 4 | 2 | 1 | 3 | 13 |
| IL | 7 | 6 | 6 | 2 | 2 | 2 | 4 | 3 | 10 | 8 | 3 | 12 | 13 | 16 | 94 |
| IN | | 1 | 3 | 1 | 2 | 3 | | 4 | 1 | 1 | 2 | 7 | 6 | 5 | 36 |
| KS | 1 | 2 | 2 | | 1 | 2 | 4 | 2 | | 1 | 3 | 1 | 3 | | 22 |
| KY | | 1 | | | 2 | | 1 | 1 | 1 | | | | 1 | | 7 |
| LA | 1 | 1 | 1 | | 1 | | 2 | | 2 | 2 | 1 | 2 | 3 | 2 | 18 |

*continued*

**TABLE 3-10** Continued

| State | Number of Phase II Awards | | | | | | | | | | | | | | Total |
|---|---|---|---|---|---|---|---|---|---|---|---|---|---|---|---|
| | 1992 | 1993 | 1994 | 1995 | 1996 | 1997 | 1998 | 1999 | 2000 | 2001 | 2002 | 2003 | 2004 | 2005 | |
| MA | 67 | 102 | 66 | 87 | 92 | 91 | 106 | 87 | 86 | 100 | 98 | 152 | 159 | 119 | 1,412 |
| MD | 19 | 19 | 20 | 25 | 24 | 26 | 24 | 23 | 35 | 29 | 25 | 52 | 56 | 54 | 431 |
| ME | | 5 | | | 2 | | | | 2 | 2 | | 3 | 9 | 3 | 26 |
| MI | 12 | 5 | 6 | 10 | 7 | 14 | 22 | 9 | 11 | 15 | 9 | 30 | 24 | 20 | 194 |
| MN | 5 | 9 | 3 | 6 | 14 | 9 | 4 | 7 | 12 | 9 | 12 | 18 | 10 | 9 | 127 |
| MO | 2 | 1 | 1 | 1 | | 2 | 6 | 1 | 5 | 1 | 2 | 1 | 6 | 1 | 30 |
| MS | | | | 1 | | | | | 1 | | 1 | 1 | 3 | 4 | 11 |
| MT | | 1 | 1 | 3 | | | 1 | 2 | 2 | 2 | 3 | 3 | 2 | 5 | 25 |
| NC | 3 | 4 | 7 | 5 | 2 | 7 | 5 | 6 | 4 | 5 | 4 | 5 | 12 | 10 | 79 |
| ND | | 1 | | 1 | 1 | | | | 2 | 1 | | | | | 6 |
| NE | 2 | 2 | | | | | 1 | | 2 | 2 | 1 | | | 1 | 11 |
| NH | 8 | 11 | 2 | 7 | 11 | 7 | 5 | 7 | 10 | 7 | 8 | 11 | 22 | 8 | 124 |
| NJ | 12 | 25 | 19 | 26 | 24 | 17 | 19 | 21 | 21 | 20 | 21 | 28 | 32 | 24 | 309 |
| NM | 20 | 14 | 10 | 15 | 7 | 16 | 15 | 19 | 14 | 11 | 10 | 13 | 17 | 12 | 203 |
| NV | 3 | 3 | 1 | 1 | 2 | 2 | 4 | 1 | 2 | 3 | 6 | 6 | 11 | 3 | 48 |
| NY | 24 | 26 | 16 | 29 | 21 | 29 | 23 | 25 | 23 | 21 | 20 | 44 | 49 | 50 | 400 |
| OH | 14 | 13 | 14 | 31 | 27 | 19 | 36 | 25 | 31 | 39 | 29 | 57 | 53 | 59 | 447 |
| OK | 1 | 1 | | 2 | | 3 | 2 | 3 | 1 | 1 | 3 | 2 | 4 | 6 | 29 |
| OR | 2 | 2 | 3 | 4 | 7 | 4 | 4 | 4 | 1 | 5 | 4 | 6 | 8 | 9 | 63 |
| PA | 15 | 27 | 16 | 32 | 16 | 32 | 29 | 31 | 22 | 28 | 30 | 49 | 47 | 39 | 413 |

| | | | | | | | | | | | | | | | |
|---|---|---|---|---|---|---|---|---|---|---|---|---|---|---|---|
| RI | 2 | 4 | 2 | | 2 | 3 | | 2 | 2 | 3 | 2 | 4 | 5 | 1 | 32 |
| SC | | | | | 1 | 1 | | | 2 | 4 | 1 | 5 | 1 | 4 | 19 |
| SD | | | | | | | 1 | | | | | | | 1 | 2 |
| TN | 3 | 3 | 4 | 5 | 8 | 4 | 4 | 3 | 5 | 7 | 3 | 1 | 9 | 10 | 69 |
| TX | 10 | 19 | 9 | 16 | 24 | 14 | 22 | 25 | 25 | 26 | 22 | 40 | 50 | 44 | 346 |
| UT | 4 | 12 | 4 | 4 | 6 | 5 | 4 | 4 | 4 | 4 | 4 | 10 | 5 | 6 | 76 |
| VA | 29 | 48 | 30 | 37 | 48 | 56 | 48 | 47 | 42 | 60 | 63 | 73 | 83 | 77 | 741 |
| VT | | 1 | | | 2 | 1 | 2 | 2 | 2 | | | 3 | 5 | 1 | 19 |
| WA | 8 | 7 | 7 | 11 | 9 | 11 | 11 | 8 | 9 | 11 | 12 | 18 | 29 | 14 | 165 |
| WI | 5 | 2 | 2 | 3 | 4 | 4 | 3 | 4 | 2 | 2 | 1 | 5 | 6 | 3 | 46 |
| WV | | | | | | 1 | | 2 | 2 | 1 | | 5 | 9 | 5 | 25 |
| WY | | | | | | 1 | | | | 1 | | 1 | | | 3 |
| Total | 433 | 591 | 406 | 575 | 611 | 638 | 672 | 568 | 626 | 702 | 661 | 1,078 | 1,173 | 998 | 9,732 |

SOURCE: DoD awards database.

award-winners tended to get the most Phase II awards. Not surprisingly, states with few Phase I awards had few Phase II awards.

Still, states do vary substantially in the degree to which their companies successfully convert Phase I awards into Phase II. Table 3-11 shows the percentage share of Phase II awards between 1992 and 2005, by state, expressed as a percentage of the Phase I awards between 1992 and 2005, by state. This metric indicates which states appear to be particularly successful at converting Phase I awards into Phase II awards.

The data show that the top 10 states on this metric had companies that converted Phase I into Phase II at a rate of 50 percent or better; the ten lowest receiving states all converted at rates of less than 35 percent. This suggests avenues for state-level research. It is possible that enthusiastic outreach efforts at the state level—perhaps by state S&T or economic development agencies—have encouraged firms to submit Phase I proposals that in the end have not justified Phase II funding. This may not necessarily be a good strategy for either the firm or the state. On the other hand, states can perhaps help companies learn to develop a more successful approach to Phase II. These data may also be impacted by sample size. None of the 15 states with the most Phase II awards are on either list.

The number of "low award" states—those with 10 or fewer Phase II awards per year—has fallen substantially between 1992 and 2005, from 28 to 16. This may be partly explained by the substantial increase that took place during this period in the number of awards. Nonetheless, it is clear that companies from areas traditionally not regarded as S&T hubs do have opportunities to win Phase II wards at DoD, an advantage of the program given the required concentration of early-stage capital.

**TABLE 3-11** Phase II Awards—Conversion Rates for Phase IIs by State, 1992-2005, Expressed as a Percent of Phase Is

| High Conversion | | Low Conversion | |
|---|---|---|---|
| NV | 69.6 | OK | 34.5 |
| ND | 66.7 | SC | 33.9 |
| NE | 61.1 | HI | 33.9 |
| WI | 55.4 | AZ | 33.6 |
| AR | 53.3 | MO | 32.6 |
| MT | 53.2 | LA | 31.6 |
| WA | 50.6 | WY | 30.0 |
| UT | 50.3 | AK | 28.6 |
| NH | 50.0 | SD | 28.6 |
| WV | 50.0 | DC | 26.5 |

SOURCE: DoD awards database.

Naturally, Phase II awards are further concentrated within states. However, the zip code with the largest number of Phase II awards received only 1.6 percent of Phase I awards, and 1.5 percent of Phase IIs. Overall, the top 10 zip codes accounted for 11.2 percent of both Phase I and Phase II awards. This contrasts with NIH, where the top zip code accounted for 19.9 percent of Phase I awards, and the top 10 zip codes for 13.6 percent. Science and engineering talent in the disciplines relevant to DoD appear to be more widely distributed than that in the life sciences.[26]

### 3.3.4 Phase II—Awards by Component

Like Phase I, Phase II awards are concentrated in the major components of DoD—Army, Navy, Air Force, MDA, and DARPA (see Table 3-12).

As shown by Figure 3-10, Army, Navy, Air Force, and MDA account for 83 percent of Phase II awards on average since FY2000: The remaining 17 percent is largely accounted for by DARPA.

These percentages vary somewhat over time, although that has stabilized at about 85 percent since 2002 (see Figure 3-11).

## 3.4 WOMAN- AND MINORITY-OWNED FIRMS

One of the stated objectives of the SBIR program is to expand opportunities for women and minorities in the federal S&T contracting process. One way to measure program performance in this area is to review the share of awards being made to woman- and minority-owned firms.

While Phase I awards to woman-owned firms have continued to increase as a percentage of all Phase I awards, the percentage of Phase I awards being made to minority-owned firms has declined quite substantially since the mid-1990s. The percentage fell below 10 percent for the first time in 2004.

DoD data suggest that the decline in Phase I award shares for minority-owned firms is reflected in Phase II, although there was in fact an uptick in the percentage of awards to minority-owned firms in FY2005. (See Figure 3-12.)

These data also indicate that both woman- and minority-owned firms are converting Phase I awards into Phase II at a rate very close to that of all award winners. On average, their share of all Phase II awards is 0.3 percent higher than their share of Phase I awards. This suggests that the overall quality of Phase I awards from woman- and minority-owned firms is comparable to that of all firms, in that these awards appear equally deserving of the substantially greater investment required from the agency at Phase II.

Further analysis of applications data is required to determine whether the declining Phase I awards rate for minority-owned firms reflects a declining share

[26] Data from DoD awards database and NIH IMPAC database respectively.

**TABLE 3-12** Phase II Awards, by Agency and Component

| | Number of Phase II Awards | | | | | | | | | | | |
|---|---|---|---|---|---|---|---|---|---|---|---|---|
| Year | AF | ARMY | MDA/ BMDO | NAVY | CBD | DARPA | DSWA/ DTRA | NGA | NIMA | OSD | SOCOM | Total |
| 1992 | 110 | 172 | 46 | 52 | 0 | 41 | 12 | 0 | 0 | 0 | 0 | 433 |
| 1993 | 199 | 123 | 60 | 120 | 0 | 83 | 6 | 0 | 0 | 0 | 0 | 591 |
| 1994 | 152 | 78 | 23 | 107 | 0 | 43 | 3 | 0 | 0 | 0 | 0 | 406 |
| 1995 | 191 | 131 | 34 | 127 | 0 | 84 | 6 | 0 | 0 | 0 | 2 | 575 |
| 1996 | 217 | 100 | 36 | 156 | 0 | 78 | 3 | 0 | 0 | 16 | 5 | 611 |
| 1997 | 221 | 113 | 51 | 164 | 0 | 57 | 6 | 0 | 0 | 23 | 3 | 638 |
| 1998 | 243 | 111 | 69 | 136 | 0 | 71 | 7 | 0 | 0 | 33 | 2 | 672 |
| 1999 | 212 | 105 | 43 | 107 | 6 | 44 | 7 | 0 | 0 | 37 | 7 | 568 |
| 2000 | 195 | 112 | 56 | 186 | 8 | 45 | 5 | 1 | 0 | 13 | 5 | 626 |
| 2001 | 221 | 180 | 59 | 136 | 7 | 53 | 6 | 3 | 0 | 29 | 8 | 702 |
| 2002 | 233 | 124 | 28 | 170 | 11 | 47 | 6 | 1 | 0 | 36 | 5 | 661 |
| 2003 | 234 | 323 | 184 | 193 | 10 | 66 | 3 | 1 | 0 | 50 | 14 | 1,078 |
| 2004 | 317 | 273 | 211 | 212 | 11 | 75 | 3 | 2 | 0 | 60 | 9 | 1,173 |
| 2005 | 339 | 123 | 102 | 290 | 7 | 80 | 4 | 1 | 0 | 38 | 14 | 998 |
| Total | 3,084 | 2,068 | 1,002 | 2,156 | 60 | 867 | 77 | 9 | 0 | 335 | 74 | 9,732 |

SOURCE: DoD awards database.

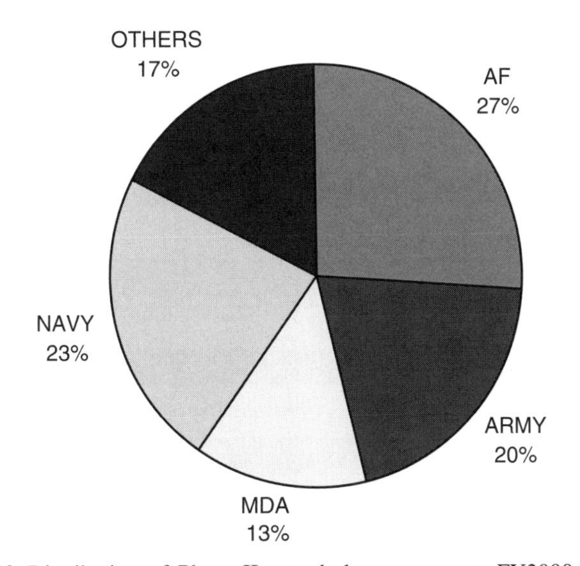


**FIGURE 3-10** Distribution of Phase II awards by component, FY2000–2005 (annual average percent).
SOURCE: DoD awards database.

of applications, the rejection rate that is increasingly greater than that for all other applicants, or whether the rate of increase in awards is growing faster than number of minority firms.

Finally, a note on data. NRC research has determined that the DoD applications database is a poor source of information on the woman/minority status of the approximately 15,000 entries for a given year. The data come directly from the proposals, but firms are sometimes inaccurate in what they enter for ownership status. In FY2005 we identified 53 firms that listed minority or woman ownership on some, but not all of the proposals they submitted. Looking across years, firms were identified that showed woman ownership some years, then no status, then woman ownership again. One firm that had about ten proposals annually listed itself as minority-owned, then several years of no special ownership, then woman-owned. After awards are made and moved to a separate database table, DoD works to correct some obvious errors in the demographic status.

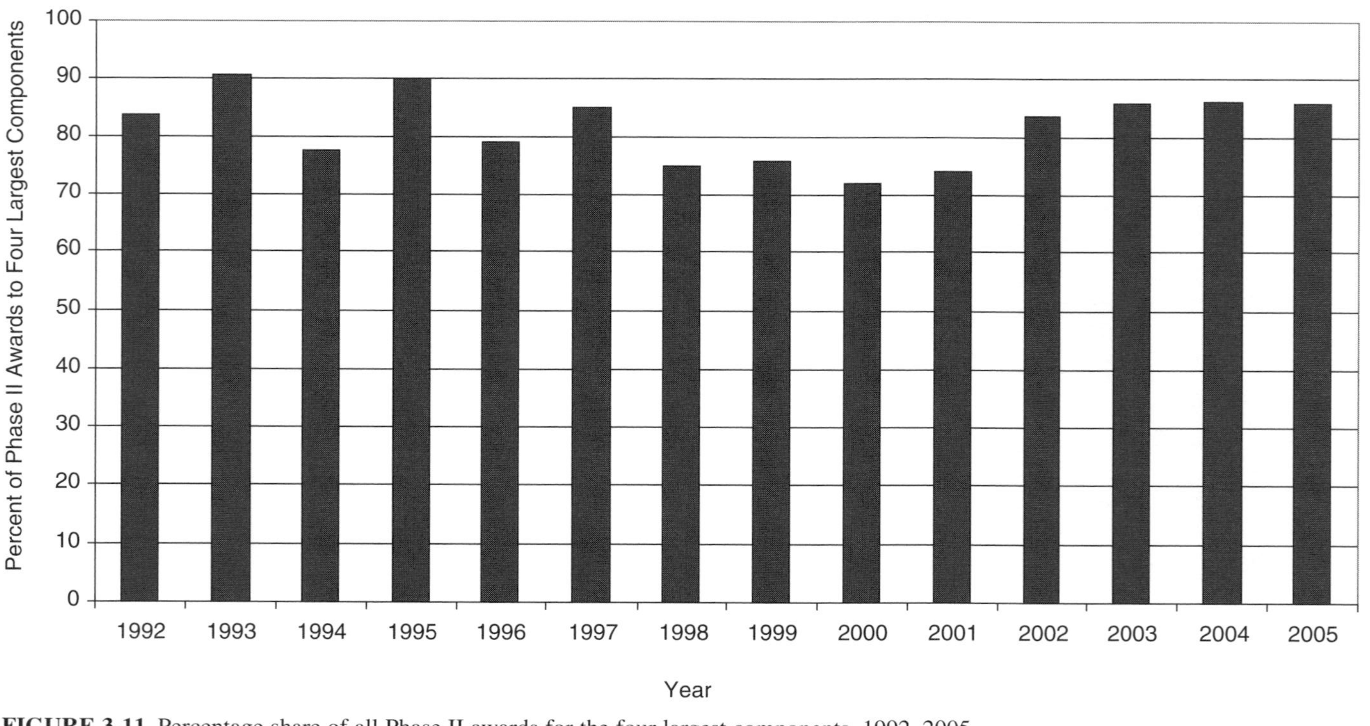


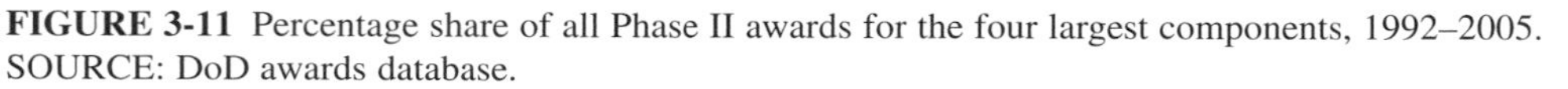

**FIGURE 3-11** Percentage share of all Phase II awards for the four largest components, 1992–2005.
SOURCE: DoD awards database.

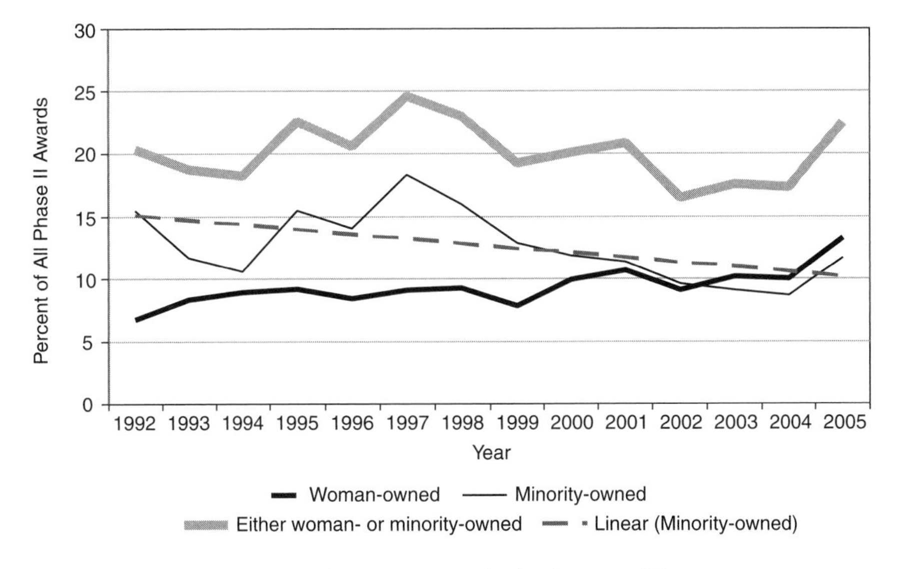


**FIGURE 3-12** Phase II awards to woman- and minority-owned firms.
SOURCE: DoD awards database.

# 4

# Outcomes

## 4.1 INTRODUCTION

Identifying the specific outcomes resulting from an early-stage R&D program such as SBIR is challenging.[1] The long lag between input (funding) and output (possible products and services), combined with the frequent need for multiple inputs for successful technology development, make definitive assessments of the link between a single input and a complex output difficult. In addition there are very substantial data collection problems, as awardees and agencies cannot consistently capture outcomes for all supported projects. Many early-stage research projects generate little that is tangible in the form of products and services while a few projects can generate very large returns. The large skew means that anything short of an all-inclusive analysis risks missing important contributions from the program.

While keeping these caveats in mind, this chapter seeks to provide as broad an assessment of outcomes from the DoD SBIR program as possible. It will focus on whether SBIR is meeting its four congressional objectives. These are "(1) to stimulate technological innovation; (2) to use small business to meet federal research and development needs; (3) to foster and encourage participation by minority and disadvantaged persons in technological innovation; and (4)

[1]For a summary of the challenges of tracking specific SBIR outcomes, see National Research Council, *SBIR: Program Diversity and Assessment Challenges*, Charles W. Wessner, ed., Washington, DC: The National Academies Press, 2004, pp. 32-35. Data in this chapter are derived from the NRC Phase II Survey unless otherwise specified.

to increase private sector commercialization derived from federal research and development."[2]

## 4.2 COMMERCIALIZATION

### 4.2.1 Background

Bringing new technologies developed under the research supported by SBIR awards to the marketplace has been a central objective of the SBIR program since its inception. The program's initiation in the early 1980s in part reflected a concern that American investment in research was not being transformed adequately into products that could generate greater wealth, more employment, and increased competitiveness. Directing a portion of federal investment in R&D to small businesses was thus seen as a new means of meeting the mission needs of federal agencies while increasing the participation of small business and thereby the proportion of innovation that would be commercially relevant.[3]

Congressional and Executive Branch interest in the commercialization of SBIR research has increased over the life of the program.

A 1992 GAO study[4] focused on commercialization in the wake of congressional expansion of the SBIR program in 1986.[5] The 1992 reauthorization specifically "emphasize[d] the program's goal of increasing private sector commercialization of technology developed through federal research and development[6] and noted the need to "emphasize the program's goal of increasing private sector commercialization of technology developed through federal research and development." The 1992 reauthorization also changed the order in which the

[2]The Small Business Innovation Development Act (PL 97-219).

[3]A growing body of evidence, starting in the late 1970s and accelerating in the 1980s indicates that small businesses were assuming an increasingly important role in both innovation and job creation. See, for example, J. O. Flender and R. S. Morse, *The Role of New Technical Enterprise in the U.S. Economy*, Cambridge, MA: MIT Development Foundation, 1975, and David L. Birch, "Who Creates Jobs?" *The Public Interest*, 65:3-14, 1981. Evidence about the role of small businesses in the U.S. economy gained new credibility with the empirical analysis by Zoltan Acs and David Audretsch of the U.S. Small Business Innovation Data Base, which confirmed the increased importance of small firms in generating technological innovations and their growing contribution to the U.S. economy. See Zoltan Acs and David Audretsch, "Innovation in Large and Small Firms: An Empirical Analysis," *The American Economic Review* 78(4):678-690, Sept 1988. See also Zoltan Acs and David Audretsch, *Innovation and Small Firms*, Cambridge, MA: The MIT Press, 1990.

[4]U.S. Government Accounting Office, *Federal Research: Small Business Innovation Research Shows Success but Can Be Strengthened*, GAO/RCED-92-37, Washington, DC: U.S. Government Accounting Office, March 1992.

[5]PL 99-443, October 6, 1986.

[6]PL 102-564 October, 28, 1992.

program's objectives are described, moving commercialization to the top of the list.[7]

The term "commercialization" means, "reaching the market," which some agency managers interpret as "first sale"—that is the first sale of a product in the market place, whether to public or private sector clients. This definition, however misses significant components of commercialization that do not result in a discrete sale. It also fails to provide any guidance on how to evaluate the scale of commercialization, an important element in assessing the degree to which SBIR programs successfully encourage commercialization. The metrics for assessing commercialization can also be elusive. It's not straightforward, for example, to calculate the full value of an "enabling technology" that can be used across industries. Also elusive is the value of materials that enable a commercial service.[8]

In light of the difficulties in measuring commercialization effectively, the Navy SBIR program manager has suggested that a firm's success in securing Phase III funding from an agency be substituted for the current weight accorded commercialization in the Commercialization Achievement Index (CAI) measure used during the proposal selection competition (the CAI is discussed in more detail in Chapter 6). Given the earlier noted variations in the S&T needs, sizes, and institutional arrangements of the services and units, applying this performance measure consistently is not always self evident.

In fact, efforts to identify Phase III results may have been unduly limited. The initial 1982 SBIR legislation noted that Phase III is not time-bound and can come long after the end of the Phase II; Phase III can include private sector sales. The law indicates that commercialization "may also involve non-SBIR, government-funded production contracts with a federal agency. . . ."[9]

Moreover, Phase III funding comes via a wide variety of mechanisms. Firms receive modifications to add Phase III federal R&D funding to Phase II SBIR contracts, they have won production contracts or R&D contracts competitively, sold to prime contractors, received additional private sector funding, and sold products commercially. All of these are Phase III activities in accordance with the legislation and with the SBA policy directive. Consequently, a narrow definition of Phase III, as a noncompetitively awarded further R&D or production

---

[7]These changes are described by R. Archibald and D. Finifter in "Evaluation of the Department of Defense Small Business Innovation Research Program and the Fast Track Initiative: A Balanced Approach" in National Research Council, *The Small Business Innovation Research Program: An Assessment of the Department of Defense Fast Track Initiative*, Washington, DC: National Academy Press, 2000.

[8]For a discussion of this and related methodological challenges, see, National Research Council, *Assessment of the Small Business Innovation Research Program—Project Methodology*, Washington, DC: The National Academies Press, 2004, accessed at <*http://books.nap.edu/catalog.php?record_id=11097#toc*>.

[9]U.S. Government Accounting Office, *Federal Research: Small Business Innovation Research Shows Success but Can Be Strengthened*, GAO/RCED-92-37, op. cit., p. 14.

contract, and one that is only captured if properly entered in the DD350 report is too limited as an approach.

In addition, DD350 documentation of Phase III funding does not occur until at least 1 year following completion of any Phase II enhancement awards, the form itself is often not filled out completely or appropriately, as data tests run by BRTRC indicate.[10] DoD staff indicate that Navy makes a considerable effort to ensure that its DD350 forms fully capture SBIR Phase III activities to the maximum extent possible. Other agencies do not.[11]

In sum, while they do provide one important measure of commercialization and one that could be used more effectively, under current circumstances, the DD350 reports may not provide sufficiently comprehensive or accurate data on which to make definitive determinations about the success of DoD SBIR commercialization. Indeed, the multiple goals of the SBIR program mean that multiple measures are appropriate for evaluation.

### 4.2.2 Proposed Commercialization Indicators and Benchmarks

This report uses three sets of indicators to quantitatively assess commercialization success:

1. **Sales and licensing revenues** ("sales" hereafter, unless otherwise noted). Revenues flowing into a company from the commercial marketplace constitute the most obvious measure of commercial success. They are also an important indicator of uptake for the product or service. Sales indicate that the result of a project has been sufficiently positive to convince buyers that the product or service is the best available solution.

Yet if there is general agreement that sales are a key benchmark, there is no such agreement on what constitutes "success." Companies, naturally enough, focus on projects that contribute to the bottom line—that are profitable. Agency staff provide a much wider range of views. Some view any sales a substantial success for a program focused on such an early stage of the product and development cycle, while others seem more ambitious.[12] Some senior executives in the private sector viewed only projects that generated cumulative revenues at $100 million or more as a complete commercial success.[13]

Rather than seeking to identify a single sales benchmark for "success," it therefore seems more sensible to simply assess outcomes against a range of

[10] Peter Cahill, BRTRC, private communication, December 1, 2006.

[11] Michael Caccuitto, DoD SBIR/STTR Program Administrator, Interview, November 28, 2006.

[12] Interviews with SBIR program coordinators at DoD, NIH, NSF, and DoE.

[13] Pete Linsert, CEO, Martek, Inc., Meeting of the Committee for Capitalizing on Science, Technology, and Innovation: An Assessment of the Small Business Innovation Research Program, June 5, 2005.

benchmarks reflecting these diverse views, with each marking the transition to a greater level of commercial success:

a. *Reaching the market*—A finished product or service has made it to the marketplace.
b. *Reaching $1 million in added cumulative sales* (beyond SBIR Phases I and II)—The approximate combined amount of standard DoD Phase I and Phase II awards.
c. *Reaching $5 million in cumulative sales*—A modest commercial success that may imply that a company has broken even on a project.
d. *Reaching $50 million in cumulative sales*—A full commercial success.

2. **Phase III activities within DoD.** As noted above, Phase III activities within DoD are a primary form of commercialization for DoD SBIR projects. These activities are considered in Section 4.3 and Chapter 5.

3. **R&D investments and research contracts.** Further R&D investments and contracts are good evidence that the project has been successful in some significant sense. These investments and contracts may include partnerships, further grants and awards, or government contracts. The benchmarks for success at each of these levels should be the same as those above, namely:

a. Any R&D additional funding.
b. Additional funding of $1 million or more.
c. Additional funding of $5 million or more.
d. Funding of $50 million or more.

4. **Sale of equity.** This is a less clear-cut indicator of commercial success. but it is unlikely investors or competitors would buy equity in a company that had not shown its ability to produce something of significant value. Key metrics include:

a. Equity investment in the company by independent third party.
b. Sale or merger of the entire company.

### 4.2.3 Sales and Licensing Revenues from DoD SBIR Awards

The most basic of all questions on commercialization is whether a project produced a good or service that reached the marketplace. Figure 4-1 shows the status of surveyed projects. It shows that only a fairly small percentage of SBIR projects have been discontinued with no hope of ever generating sales (26 percent), though it is likely that a significant portion of the projects currently in development will also fail to achieve significant commercial success.

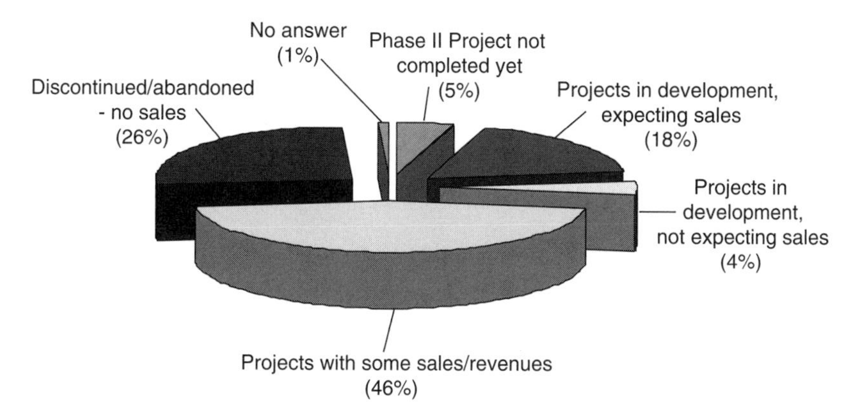


**FIGURE 4-1** Status of surveyed projects.
NOTE: The NRC deployed two surveys to the population of Phase II recipients as part of the research conducted for this project. The NRC Phase II Survey focused on individual projects. In addition, the NRC Firm Survey was sent to every firm receiving a Phase II award between 1992 and 2001, and focused on firm-level questions. Unless otherwise stated, all references here to the NRC Survey are to the project survey.
SOURCE: NRC Phase II Survey.

These data indicate that 46 percent of surveyed projects reported some revenues from their project; a further 18 percent were still in development and expected sales, and 5 percent had not yet completed Phase II.

### 4.2.3.1 Sales Ranges

Early-stage technology projects are inherently risky. As a result, there is a very skewed distribution of results. Many projects generate no commercial results at all, and relatively few of those that do reach the market have substantial commercial successes (see Figure 4-2).

The data suggest that at DoD—as at other agencies—the overwhelming majority of sales are concentrated in the $0–$1 million range. Ten percent of reporting projects generated at least $5 million in revenues, while more than 65 percent of respondents with sales reported total sales of less than $1 million revenues (as of May 2005, the date of the survey). As a result of this very skewed distribution, the mean amount of sales for all companies that reported sales was $2,894,834, while the median was $500,000.

**Underreporting of Sales Results.** The average total sales for older projects is much higher than for recent ones. In fact, the average sales for the 176 reporting DoD projects awarded Phase II contracts from 1992 to 1994 was $2.78 million, whereas the average sales for the 415 reporting DoD projects awarded Phase II

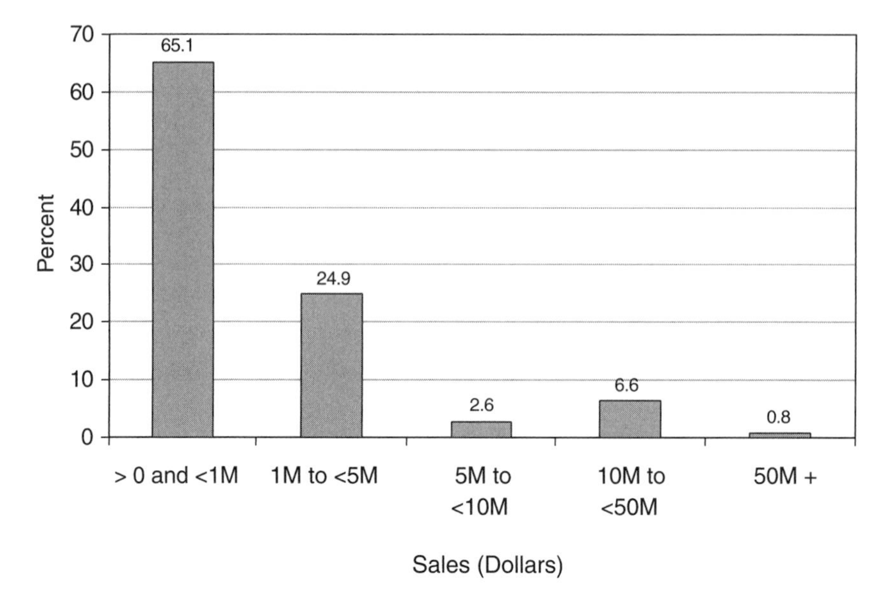


**FIGURE 4-2** Sales by sales range.
SOURCE: NRC Phase II Survey.

contracts between 1999 to 2001 was only $982,000. This difference in part reflects the number of recent awardees whose products have yet to be commercially introduced or fully exploited commercially. Reported aggregate and average sales data up through the survey date of May 2005 are therefore only a partial estimate of the total commercial impact of the 920 awards covered by the NRC Phase II Survey.[14] According to former senior DoD staff, average major DoD weapons system R&D cycle is approximately 12 years (before production)—so SBIR

[14]Using the trendline shown in Figure 4-3, we find that the best fit generates average sales of approximately $5.5 million per project for those with awards in 1992, declining to averages sales of $1 million for those in 2001. Note that these data cover only firms reporting some sales. The trendline gives us a means of estimating the eventual sales generated by each project, using simplified assumptions (notably, that all sales end by May 2005 (the date of the NRC Phase II Survey), and that commercialization remains constant across time (in fact, it is likely to have increased as agencies have increased their focus on supporting projects with better prospects of commercial success). Other assumptions tend to reduce the size of the estimated revenues.

Using these estimates to project forward, we find that by the time commercialization of all projects is completed (i.e., ten years after the last project funded, or 2011), total revenues generated by the projects reporting revenues so far is $2.13 billion.

These data suggest that the reported revenues as of May 2005 may understate eventual total revenues as of 2011 by as much as 50 percent.

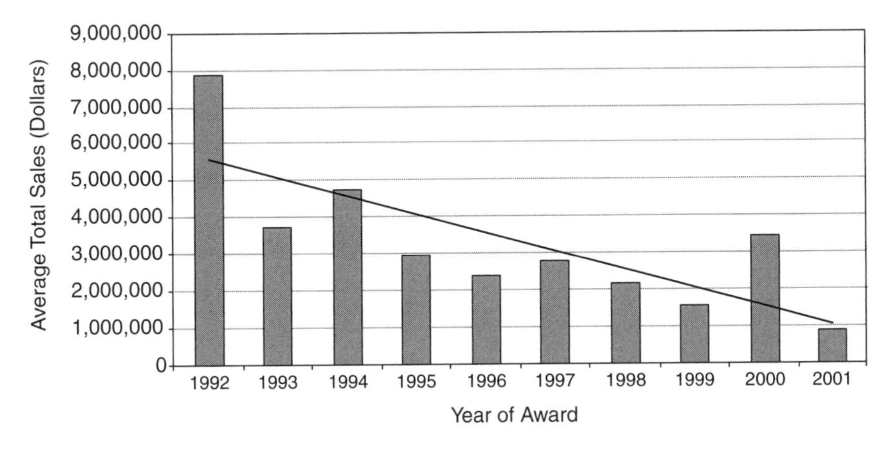


**FIGURE 4-3** Average sales, by year of award, plus trendline.
SOURCE: NRC Phase II Survey.

**TABLE 4-1** Reported and Projected Revenues for Companies that Reported Sales as of May 2005

| | Amount ($) |
|---|---|
| | 1,094.0 |
| Reported Sales (Millions of Dollars) | 1,040.5 |
| Total Sales (Millions of Dollars) | 2,134.5 |
| Average Sales per Project with Sales ($) | 5,646,787 |
| Average Sales per Project—All Projects ($) | 2,151,699 |

SOURCE: NRC Phase II Survey.

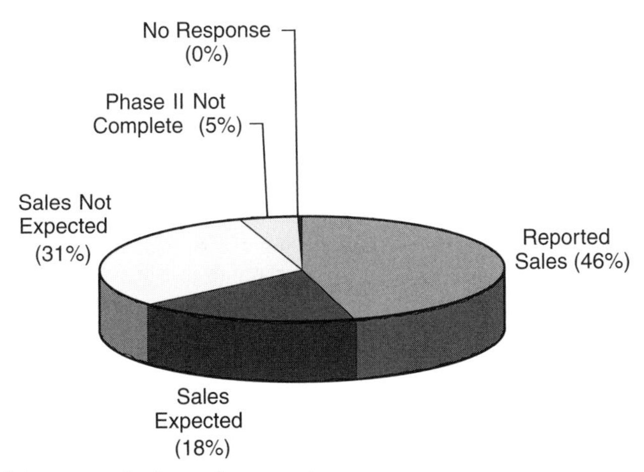


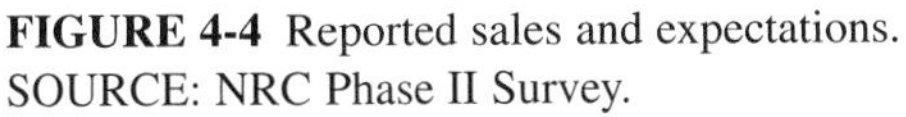
**FIGURE 4-4** Reported sales and expectations.
SOURCE: NRC Phase II Survey.

**TABLE 4-2** For Companies Anticipating Sales, Year of Expected First Sale

| Year of Expected Sales | Number of Projects | Percent of Projects |
|---|---|---|
| 2005 | 37 | 22.0 |
| 2006 | 70 | 41.7 |
| 2007 | 31 | 18.5 |
| 2008 | 18 | 10.7 |
| 2009 | 5 | 3.0 |
| 2010 | 7 | 4.2 |
| Total | 168 | |

NOTE: Survey Date: 2005.
SOURCE: NRC Phase II Survey.

products for defense sales would tend to have a long time lag.[15] Thus, Figure 4-3 should not be taken to mean that sales are declining; it largely reflects the extent of the lag in DoD-oriented sales.

**Sales Concentration.** Total revenues from sales are highly concentrated in the very few projects that have generated at least $5 million in cumulative revenues. Just under 75 percent of all cumulative sales were accounted for by the 38 projects (out of 920 overall) that reported at least $5 million in sales. This very high concentration confirms the view that from the perspective of sales, the SBIR program at DoD generates a few major winners, rather than a more widely dispersed range of more modest successes. This is similar to commercial outcomes from early-stage R&D programs.

### 4.2.3.2 Sales Expectations

About a quarter of projects reported that they expected sales in the future. Of those companies not yet reporting sales on their projects, about 67 percent still expect them.

Most of the respondents that had not yet received sales expected sales to come in the very near future, as shown by Table 4-2. Of those expected sales, 80 percent anticipated that their first sale would occur within 3 years.

The data in Figure 4-5 show, for projects that have received sales, the time that elapsed between the Phase II award and the first sales. Survey responses indicate that 87.6 percent of first sales occurred within 4 years of the award date. This relatively short time from award to first sale is supported by the comments of John Williams, Navy SBIR Manager.[16] However, it must be stressed that

[15]Dr. Jacques Gansler, former Under Secretary of Defense for Acquisition, Technology and Logistics, January 29, 2007.

[16]John Williams, U.S. Navy SBIR program manager, presentation at SBTC SBIR in Rapid Transition Conference, Washington, DC, September 27, 2006.

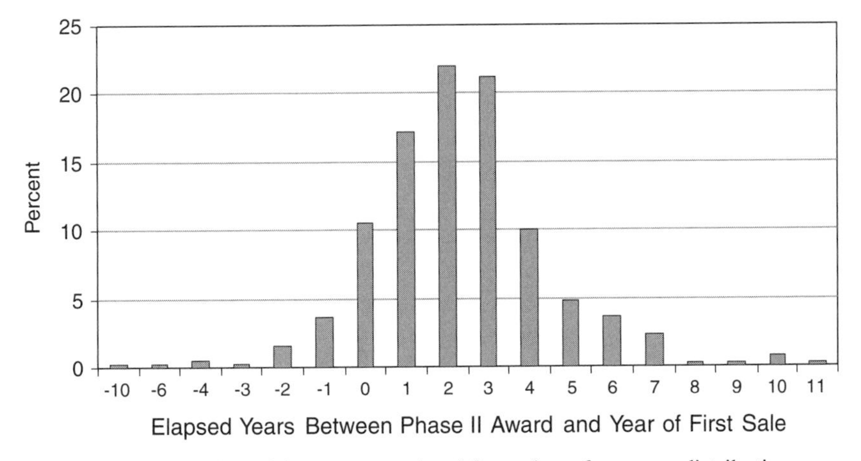


**FIGURE 4-5** Time elapsed between award and first sales—frequency distribution.
SOURCE: NRC Phase II Survey.

interviews and cases strongly support the view that the bulk of sales will occur some years after the date of first sale. The latter is therefore best seen as a leading indicator for sales.

These data from the NRC's May 2005 Phase II Survey help us to evaluate claims of companies that they will generate sales in the future. About 25 percent of all DoD survey respondents made this claim. However, the likelihood of commercialization diminishes substantially with time elapsed since the award. Projects still expecting sales are clustered toward the 2000–2001 timeframe, but even here the likely window of opportunity for success appears to be closing rapidly. *The median time to first sale is before the end of the second year after the award.*

This analysis suggests that though a considerable number of companies anticipate sales in the future, the actual likelihood of this occurring is relatively low. By the start of the 9th year after the Phase II award is made, 99 percent of projects that will eventually report sales have done so; the likelihood a project without sales reporting sales after the 8th year is less than 1 percent. Similarly, projects not reporting sales by the start of year five have a 13.8 percent chance of eventually reporting sales (to put it yet another way, by the end of the fourth year after the award year, 86.2 percent of projects that eventually make sales will have started to do so).[17]

Table 4-3 helps us to determine likely revenues for companies that expect

[17]This is not to say that all sales will have been completed by the end of that year, only that it is quite reasonable to apply these percentages to the more recent awards, as a way of estimating eventual sales data.

**TABLE 4-3** Total Projected Sales for Companies Without Sales That Still Expect Them

| Year of Award | Projects Expecting Sales | Historical Success Percentage | Projected Total Revenues ($) |
|---|---|---|---|
| 1992 | 3 | 0.0 | 0 |
| 1993 | 4 | 0.0 | 0 |
| 1994 | 4 | 0.3 | 58,730 |
| 1995 | 8 | 0.8 | 352,381 |
| 1996 | 6 | 0.3 | 88,095 |
| 1997 | 13 | 0.3 | 190,873 |
| 1998 | 18 | 2.4 | 2,378,571 |
| 1999 | 19 | 3.7 | 3,905,556 |
| 2000 | 32 | 4.8 | 8,457,143 |
| 2001 | 61 | 10.1 | 34,034,127 |
| Total | | | 49,465,476 |

NOTE: Projected sales are calculated by multiplying the number of companies reporting that they still expect sales, by the percentage likelihood that a company in that award year will in fact generate sales, by the average total sales for all companies that did record more than $0 in sales.
SOURCE: NRC Phase II Survey.

them, based on historical records. The total projected revenues for these companies is relatively low—about $50 million, or less than 2.5 percent of all projected sales. We can therefore conclude that while ongoing revenues from companies with some sales in hand will be relatively large (see above), the revenues expected from companies that had not yet reported some sales as of May 2005 are likely to be of limited significance.

These findings suggest that while the product cycle for the entire defense industry may be long—a well-known characteristic of major defense systems—the first sales cycle for most SBIR-related products is actually relatively short. Most successful project start receiving initial sales revenues within 4 years of the award, while large sales tend to come considerably later due to the defense procurement cycle.[18]

**Sales and Projected Sales: Conclusions.** The NRC Phase II Survey provided the following summary data regarding sales and projected sales:

- 378 out of a total of 920 respondents ( 41.1 percent) report sales greater than $0.
- The average reported sales is $1.3 million for all projects (n=920), and $3.2 million for those reporting sales greater than $0 (n=378).[19]

[18]This paragraph concerns only first sales. The bulk of sales occur at some unknown period after the first sale.

[19]When projected out to 2011, estimated average sales per project are $5.6 million for projects already reporting some sales, and $2.2 million for all projects.

**TABLE 4-4** Customer Base

| Customer | Percent of Total Sales |
|---|---|
| Domestic private sector | 21 |
| Department of Defense | 38 |
| Prime contractors for DoD or NASA | 12 |
| NASA | 1 |
| Other federal agencies | 1 |
| State or local governments | 1 |
| Export markets | 11 |
| Other | 16 |
| Total | 100 |

SOURCE: NRC Phase II Survey.

- Finally, additional sales from projects with no reported sales as of May 2005 are likely to be of limited importance—less than $50 million in total.[20]

### 4.2.3.3 Sales by Sector

The NRC Phase II Survey asked respondents to identify the customer base for their products. The responses are summarized in Table 4-4.

While the fact that half of sales went either to DoD or DoD/NASA prime contractors is not surprising, the balance between the two is somewhat at odds with comments made by many interviewees and speakers at the NRC Phase III conference. Those comments indicated that it was very hard for SBIR firms to sell directly to DoD and that sales had to be mediated through the primes. The data above suggest that this is much less the case than conventional wisdom would suggest, as more than one-third of sales went directly to DoD, and these sales constitute the largest single sector market for DoD SBIR recipients. However, it is also possible that the question was asked with insufficient precision, or that some of these were limited sales—e.g., prototypes to DARPA.

Phase II projects lead to several forms of new products and processes, with some new technologies having multiple characteristics. Allowing for more than one response, the most prevalent form reported was "hardware" (60 percent), which may occur as a final product, component, or intermediate product. "Hardware" was followed by "software" (32 percent) and "process technology" (23 percent). Of note is that some reported outputs occur in the forms of "new or improved service capability" (18 percent) and "research tool" (15 percent).[21]

[20]However, given the highly skewed nature of sales outcomes, it is entirely possible that one of the companies that does reach the market after May 2005 will turn out to be a major success, but there is no way of predicting whether that will be the case.

[21]Other federal agencies report a significantly higher percent of research tools and educational materials (26 percent and 13 percent respectively for the other four agencies in aggregate).

**BOX 4-1**
**Multiple Sources of Bias in Survey Response**

Large innovation surveys involve multiple sources of bias that can skew the results in both directions. Some common survey biases are noted below. These biases were tested for and responded to in the NRC surveys.[a]

- **Successful and more recently funded firms are more likely to respond.** Research by Link and Scott demonstrates that the probability of obtaining research project information by survey decreases for less recently funded projects and it increased the greater the award amount.[b] Nearly 40 percent of respondents in the NRC Phase II Survey began Phase I efforts after 1998, partly because the number of Phase I awards increased, starting in the mid-1990s, and partly because winners from more distant years are harder to reach. They are harder to reach as time goes on because small businesses regularly cease operations, are acquired, merge, or lose staff with knowledge of SBIR awards.
- **Success is self-reported.** Self-reporting can be a source of bias, although the dimensions and direction of that bias are not necessarily clear. In any case, policy analysis has a long history of relying on self-reported performance measures to represent market-based performance measures. Participants in such retrospectively analyses are believed to be able to consider a broader set of allocation options, thus making the evaluation more realistic than data based on third-party observation.[c] In short, company founders and/or principal investigators are in many cases simply the best source of information available.
- **Survey sampled projects at firms with multiple awards.** Projects from firms with multiple awards were underrepresented in the sample, because they could not be expected to complete a questionnaire for each of dozens or even hundreds of awards.
- **Failed firms are difficult to contact.** Survey experts point to an "asymmetry" in their ability to include failed firms for follow-up surveys in cases where the firms no longer exist.[d] It is worth noting that one cannot necessarily infer that the SBIR project failed; what is known is only that the firm no longer exists.
- **Not all successful projects are captured.** For similar reasons, the NRC Phase II Survey could not include ongoing results from successful projects in firms that merged or were acquired before and/or after commercialization of the project's technology. The survey also did not capture projects of firms that did not respond to the NRC invitation to participate in the assessment.
- **Some firms may not want to fully acknowledge SBIR contribution to project success.** Some firms may be unwilling to acknowledge that they received important benefits from participating in public programs for a variety of reasons. For example, some may understandably attribute success exclusively to their own efforts.
- **Commercialization lag.** While the NRC Phase II Survey broke new ground in data collection, the amount of sales made—and indeed the number of projects that generate sales—are inevitably undercounted in a snapshot survey taken at a single point in time. Based on successive data sets collected from NIH SBIR award recipients, it is estimated that total sales from all responding projects will likely be on the order of 50 percent greater than can be captured in a single survey.[e] This underscores the importance of follow-on research based on the now-established survey methodology.

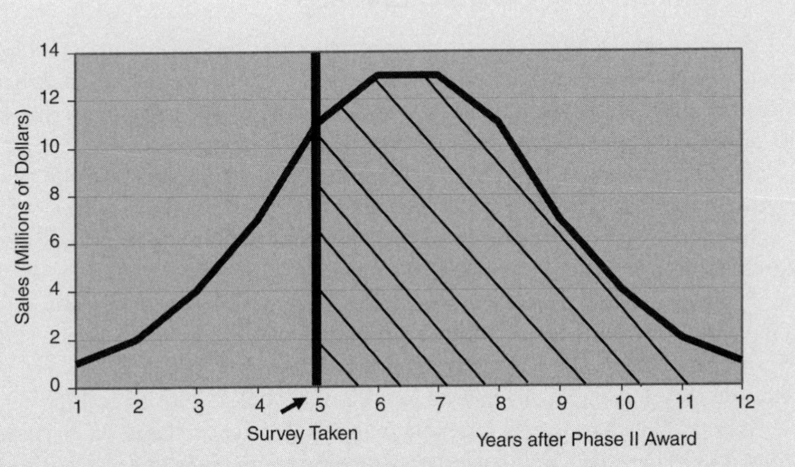


FIGURE B-4-1 Survey bias due to commercialization lag.

These sources of bias provide a context for understanding the response rates to the NRC Phase I and Phase II Surveys conducted for this study. For the NRC Phase II Survey for DoD, of the 2,191 firms that could be contacted out of a sample size of 3,055, 920 responded, representing a 42 percent response rate. The NRC Phase I Survey captured 9 percent of the 13,103 awards made by DoD between 1992 to 2001. See appendixes B and C for additional information on the surveys.

---

[a]For a technical explanation of the sample approaches and issues related to the NRC surveys, see Appendix B.

[b]Albert N. Link and John T. Scott, *Evaluating Public Research Institutions: The U.S. Advanced Technology Program's Intramural Research Initiative*, London: Routledge, 2005.

[c]While economic theory is formulated on what is called "revealed preferences," meaning individuals and firms reveal how they value scarce resources by how they allocate those resources within a market framework, quite often expressed preferences are a better source of information especially from an evaluation perspective. Strict adherence to a revealed preference paradigm could lead to misguided policy conclusions because the paradigm assumes that all policy choices are known and understood at the time that an individual or firm reveals its preferences and that all relevant markets for such preferences are operational. See (1) Gregory G. Dess and Donald W. Beard, "Dimensions of Organizational Task Environments," *Administrative Science Quarterly*, 29: 52-73, 1984. (2) Albert N. Link and John T. Scott, *Public Accountability: Evaluating Technology-Based Institutions*, Norwell, MA: Kluwer Academic Publishers, 1998.

[d]Albert N. Link and John T. Scott, *Evaluating Public Research Institutions: The U.S. Advanced Technology Program's Intramural Research Initiative*, op. cit.

[e]Data from NIH indicates that a subsequent survey taken two years later would reveal very substantial increases in both the percentage of firms reaching the market, and in the amount of sales per project. See National Research Council, *An Assessment of the SBIR Program at the National Institutes of Health*, Charles W. Wessner, ed., Washington, DC: The National Academies Press, 2009.

#### 4.2.3.4 Sales by Size of Company (Employees)

Another important question is whether the size of a SBIR contract recipient seems to significantly affect commercialization. It would seem, for instance, that companies that are extremely small would have more difficulty managing both the research and marketing functions needed for commercial success.

The data shown in Table 4-5 indicates that very small companies do tend to have less commercialization success (as measured by cumulative sales).

Further analysis suggests that companies larger than 25 employees report significantly better sales outcomes, as shown by Table 4-6.

Companies with more than 25 employees seem to consistently outperform companies with less than 25 employees, in terms of projects that generate at least $5 million in sales. The former account for 31.5 percent of all responding projects, but 74.5 percent of all projects reporting sales of at least $5 million.

One possible hypothesis for explaining this difference, based on discussions at the NRC Phase III Conference and with case study companies, might be that acquisitions officers are more comfortable engaging with larger and presumably more stable companies that will likely have a longer track record (all other things being equal).

#### 4.2.3.5 Sales by Licensees

Licensing revenues are an important source of commercialization activity for SBIR companies. Indeed, interviews with staff and awardees suggest that in some

**TABLE 4-5** Cumulative Project Sales by Company Size at Time of Survey

| | Cumulative Project Sales (Number of Projects) | | | | | | |
|---|---|---|---|---|---|---|---|
| Number of Employees | <$100K | $100K to < $1M | $1M to <$5M | $5M to <$50M | >$50M | Total | Percent |
| 0–5 | 19 | 33 | 6 | 2 | | 60 | 15.8 |
| 6–10 | 12 | 24 | 9 | 6 | | 51 | 13.5 |
| 11–15 | 12 | 21 | 10 | 3 | | 46 | 12.1 |
| 16–25 | 8 | 19 | 24 | 2 | | 53 | 14.0 |
| 26–50 | 7 | 21 | 17 | 17 | | 62 | 16.4 |
| 51–100 | 6 | 12 | 20 | 8 | | 46 | 12.1 |
| 101–250 | 6 | 20 | 8 | 5 | | 39 | 10.3 |
| 251–500 | 1 | 5 | 6 | 5 | 2 | 17 | 5.0 |
| 500+ | 0 | 2 | 0 | 1 | | 3 | 0.8 |
| Missing | | | | | | | |
| Total | 71 | 157 | 100 | 49 | 2 | 377 | |
| Percent | 18.7% | 41.4% | 26.4% | 12.9% | 0.5% | | 100.0% |

SOURCE: NRC Phase II Survey.

**TABLE 4-6** Sales Outcomes by Size of Company at Time of Survey

| | Sales Outcomes (Percent Distribution) | | | | | | |
|---|---|---|---|---|---|---|---|
| Number of Employees | <$100K | $100K to < $1M | $1M to <$5M | $5M to <$50M | >$50M | Percent of responses | Percent |
| 0–5 | 31.7 | 55.0 | 10.0 | 3.3 | 0.0 | 15.8 | 100.0 |
| 6–10 | 23.5 | 47.1 | 17.6 | 11.8 | 0.0 | 13.5 | 100.0 |
| 11–15 | 26.1 | 45.7 | 21.7 | 6.5 | 0.0 | 12.1 | 100.0 |
| 16–25 | 15.1 | 35.8 | 45.3 | 3.8 | 0.0 | 14.0 | 100.0 |
| 26–50 | 11.3 | 33.9 | 27.4 | 27.4 | 0.0 | 16.4 | 100.0 |
| 51–100 | 13.0 | 26.1 | 43.5 | 17.4 | 0.0 | 12.1 | 100.0 |
| 101–250 | 15.4 | 51.3 | 20.5 | 12.8 | 0.0 | 10.3 | 100.0 |
| 251–500 | 5.3 | 26.3 | 31.6 | 26.3 | 10.5 | 5.0 | 100.0 |
| 500+ | 0.0 | 66.7 | 0.0 | 33.3 | 0.0 | 0.8 | 100.0 |
| All Responses | 18.7 | 41.4 | 26.4 | 12.9 | 0.5 | 100.0 | 100.0 |

SOURCE: NRC Phase II Survey.

important cases, SBIR has been commercialized primarily not by the awardee company, but by a licensee.

The NRC Phase II Survey asked respondents to estimate sales by licensees. Table 4-7 shows what they reported.

As with direct sales data, these responses suggest both that a large majority of licensee sales are less than $1 million, and that there are a few very substantial licensing streams. However, this is an area that seems to be underreported in the survey. This may be due to a lack of information, as indicated by the considerable number of firms reporting sales by licensees, but no first date for those sales. Total sales reported for licensees were $124.9 million. Of this, $88 million (70.5 percent) came from the 6 projects reporting more than $10 million in licensee sales.

Case analysis also suggests that for some technologies, licensing may be the only realistic method of commercializing a product. For many research compa-

**TABLE 4-7** Sales by Licensees

| | Count | Percent |
|---|---|---|
| >0 and <$1M | 17 | 51.5 |
| $1M to <$5M | 10 | 30.3 |
| $5M to <$10M | 2 | 6.1 |
| $10M to <$50M | 4 | 12.1 |
| $50M+ | 0 | 0.0 |
| Total | 33 | 100.0 |

SOURCE: NRC Phase II Survey.

nies, the decision to begin manufacturing is seen as a high-risk venture that is often rejected in favor of a licensing-based strategy.

### 4.2.3.6 Employment Effects

#### *4.2.3.6.1 Employment Effects*

Employment analysis provides another method for identifying commercial success; it also offers another indication of support for small business.

The median size of company receiving SBIR awards at DoD is relatively small—far lower than the 500 employee limit imposed by the SBA (as shown in Table 4-8).

Just over 41 percent of respondents had 15 or fewer employees, while about 15 percent had more than 100. Three firms exceeded the 500-employee SBA limit for participation in the SBIR program (having grown since receiving the award). The median size of SBIR awardees at the time of the award was 10 employees; 62 percent of respondents had 15 employees or fewer.

Although employment is not a direct indicator for commercialization, it is clear that the two are related. Commercialization tends to require additional staff, and that additional staff requires additional revenues to support it. Those revenues, in turn, normally have to come from successfully commercializing a product.

#### *4.2.3.6.2 Employment Gains*

The NRC Phase II Survey sought detailed information about the number of employees SBIR awardees had at the time of the relevant award, the number of employees the awardees had at the time of the survey, and the estimated direct impact of the award on employment. Overall, it showed that the average employment gain at each responding firm since the date of the SBIR was 29 full-time

**TABLE 4-8** Distribution of Companies, by Employees, at Time of the Survey

| Number of Employees | Number of Firms | Percent of Firms |
|---|---|---|
| 0–5 | 60 | 15.9 |
| 6–10 | 51 | 13.5 |
| 11–15 | 46 | 12.2 |
| 16–25 | 53 | 14.1 |
| 26–50 | 62 | 16.4 |
| 51–100 | 46 | 12.2 |
| 101–250 | 39 | 10.3 |
| 251–500 | 17 | 4.5 |
| 500+ | 3 | 0.8 |
| Total | 377 | 100.0 |

SOURCE: NRC Phase II Survey.

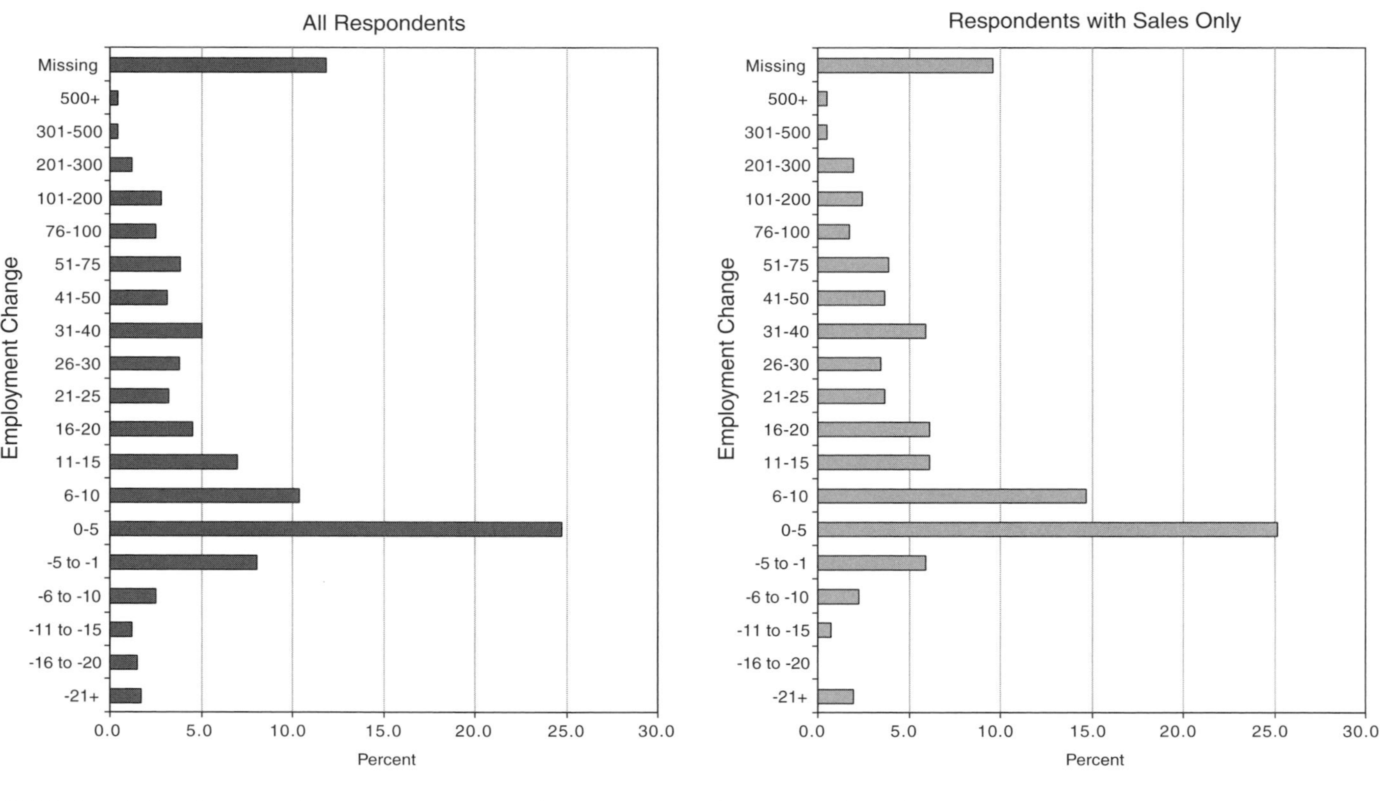


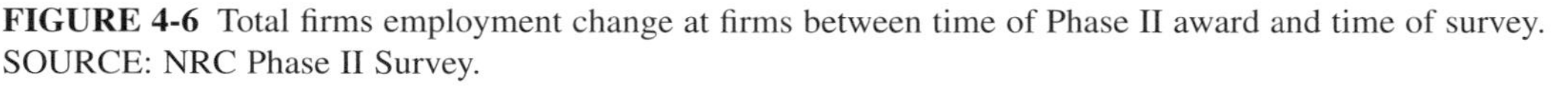
**FIGURE 4-6** Total firms employment change at firms between time of Phase II award and time of survey.
SOURCE: NRC Phase II Survey.

equivalent (FTE) employees. In addition, on average, respondents estimated that, specifically as a result of the SBIR project, their firms were able to hire 2.5 FTE employees, and to retain 2.3 more.[22]

The NRC Phase II Survey results do track with information from case studies, where several interviewees noted that a Phase II award typically funded the addition of slightly more than one full-time researcher for two years plus overhead[23] but firms then grew on future sales and third-party funding.

Respondent firms report modest employment growth as a result of the specific SBIR award addressed in the survey. Forty-eight percent of respondents reported that they had added no employees as a result of receiving the Phase II award; 42 percent reported adding between one to five additional employees to work on the project. Five percent reported adding between 6–20 employees, and only two percent reported adding more than 20.[24]

Eight percent of firms were true "start-ups" at the time of the surveyed award, having no employees.[25] At the time of their responses to the survey, only one percent of the firms remained in this category. At the time they submitted the Phase II proposal, 67 percent of SBIR awardees had 20 employees or fewer. In 2005, when they completed the survey, the firms had grown, and this percentage had dropped to 48 percent. At the upper end of the distribution, whereas 13 percent of respondents had between 21–50 employees at the time of their proposal submission and only eight percent had more than 100 employees, these firms had also grown, and these percentages had increased to 22 percent and 18 percent respectively by 2005.

### 4.2.4 Additional Investment, Funding, and Other Partnerships

Post-SBIR investment in a company is a powerful validation that its work is of value. About 53 percent of DoD respondents said that they had received some additional funding related to the surveyed project, other than further SBIR awards. As with sales, the distribution of funding is highly skewed, with a few companies receiving most of the additional investment (see Table 4-9).

The average investment was about $850,000 ($1.6 million per project if only those that received additional funding are counted).

---

[22]NRC Phase II Survey, Questions 16b and 16c.

[23]Interview with Josephina Card, Sociometrics, Inc.

[24]See the remarks by Joshua Lerner of the Harvard Business School in National Research Council, *The Small Business Innovation Research Program: Challenges and Opportunities*, Charles W. Wessner, ed., Washington, DC: National Academy Press, 1999, p. 23.

[25]Other than the founder, who may not have been drawing full-time equivalent income from the award.

**TABLE 4-9** Further Investments in SBIR Projects

| | Number of Responses | Investment ($) |
|---|---|---|
| $50M+ | 2 | 106,700,000 |
| $5M to <$50M | 32 | 324,151,193 |
| $1M to <$5M | 96 | 202,819,919 |
| $100K to <$1M | 241 | 90,112,919 |
| <$100K | 80 | 3,291,603 |
| Total Investments | 451 | 727,075,634 |
| Average | | 1,612,141 |
| No additional investment | 402 | 0 |
| Average (all responses) | | 852,375 |

SOURCE: NRC Phase II Survey, DoD awards database.

#### 4.2.4.1 Sources of Investment Funding

Conventional wisdom is that the bulk of funding for near-market development comes from venture capitalists and angel investors. However, data from the NRC Phase II Survey—shown in Table 4-10—do not validate the conventional view.

About one-quarter of all further investments come from non-SBIR federal funds, while only 3.8 percent are from U.S. venture capital companies. However, the picture does change somewhat when we consider the amount of funding, rather than the number of investments (see Figure 4-7 and Table 4-11).

**TABLE 4-10** Sources of Investment Funding (NRC)

| Source of Investment | Number of Investments | Percent |
|---|---|---|
| Non-SBIR Federal Funds | 205 | 25.8 |
| Private Investment from U.S. Venture Capital | 30 | 3.8 |
| Your Own Company | 274 | 34.4 |
| Private Investment from Other Private Equity | 55 | 6.9 |
| Private Investment from Other Domestic Private Company | 105 | 13.2 |
| Private Investment from Foreign Investment | 19 | 2.4 |
| Personal Funds | 65 | 8.2 |
| State or Local Government | 31 | 3.9 |
| College or Universities | 12 | 1.5 |
| Total | 796 | 100.0 |

SOURCE: NRC Phase II Survey.

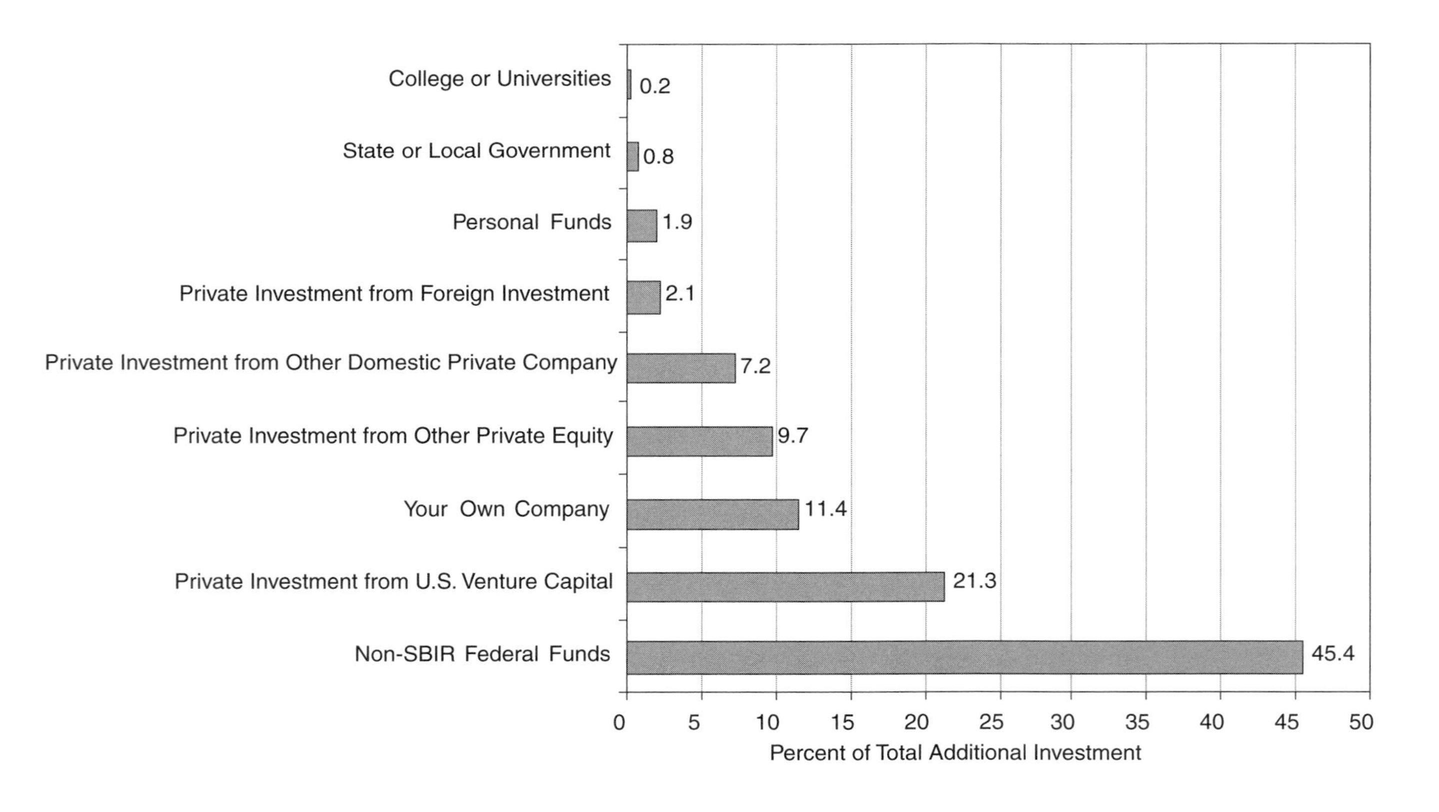


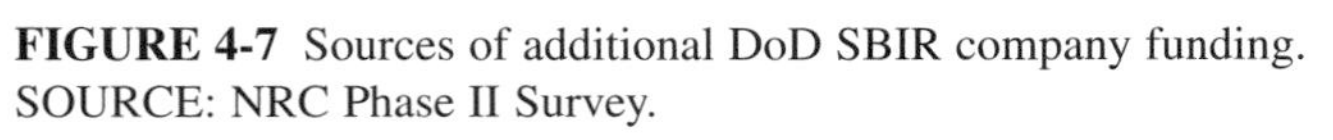

**FIGURE 4-7** Sources of additional DoD SBIR company funding.
SOURCE: NRC Phase II Survey.

**TABLE 4-11** Average Additional Funding by Category, for Firms That Received Additional Investment in Each Category.

| Source of Investment | Number of Investments Reported | Average Investment ($) |
|---|---|---|
| Non-SBIR Federal Funds | 205 | 1,621,339 |
| Private Investment from U.S. Venture Capital | 30 | 5,192,267 |
| Your Own Company | 274 | 305,257 |
| Private Investment from Other Private Equity | 55 | 1,292,124 |
| Private Investment from Other Domestic Private Company | 105 | 500,981 |
| Private Investment from Foreign Investment | 19 | 810,788 |
| Personal Funds | 65 | 215,597 |
| State or Local Government | 31 | 178,608 |
| Colleges or Universities | 12 | 138,939 |

SOURCE: NRC Phase II Survey. See also Table App-A-37.

While venture funding did not provide the largest total amount of additional support, it did offer the largest average support per project funded, at $5.2 million per project. This is in line with the growing size of VC-funded deals, which in 2007 averaged at $8 million per project.[26]

However, venture funding supported only 30 projects—less than 4 percent of the 920 responses. Of course, the relatively small number of VC investments overall means that SBIR-funded companies may still account for a significant share of all VC investments.[27]

Self-finance continues to be the source of additional funding for many companies. More than one-third of respondents with additional funding indicated that additional funds came from their own company, although this accounted for only 1 percent of total funding received. Just under 10 percent reported funding from other private equity sources, which seems in most cases likely to mean angel funding. The widespread use of private investment funding (probably angel

[26]According to the 2007 First Quarter Capital Report by Dow Jones VentureOne and Ernst & Young, LLP, "The median deal size reached $8 million, up from $7 million in the first quarter of 2006 and making it the highest quarterly median round size since the fourth quarter of 2000." Ernst and Young, "U.S. Venture Capital Investment Increases to 8 percent to $6.96 Billion in First Quarter of 2007," April 23, 2007. Accessed at <*http://www.ey.com/global/content.nsf/US/Media_-_Release_-_04-23-07DC*> on May 24, 2007.

[27]In fact, venture capital firms often use SBIR as a signal of quality in making investment decisions. According to John Cottrell of L3 Communications, 12 percent of U.S. VC investment has involved firms with SBIR funding. John Cottrell, L3 Communications. Presentation at SBTC SBIR in Rapid Transition Conference, Washington, DC, September 27, 2006.

funding) is not surprising, and the average amount invested is $1.2 million per funded project.

Investments from state and local government and academic sources played a limited role in terms of numbers and amount of funding provided (less than 5.5 percent of all investments in projects with additional funding). These sources also provided relatively low amounts of funding per project.

#### 4.2.4.2 SBIR Impact on Further Investment

The NRC Phase II Survey also sought additional information about the impact of the SBIR program on company efforts to attract third-party funding. Some case study companies mentioned that SBIR awards can have a "halo effect," acting as a form of validation for external inventors.[28]

Case study interviews provided mixed views on this perception. Some interviewees strongly supported the view that SBIR helps to attract investment, while others claimed that the halo effect was weaker than commonly thought—views that may reflect the individual firm or university experience.

Forty-six percent of DoD SBIR respondents did not attract any outside funding, and venture funding only accounted for 3.8 percent of investments. This suggests that an SBIR award is in itself no guarantee of further external funding, or (possibly) that outside funding was not needed.

### 4.2.5 Additional SBIR Funding

Aside from third-party investment, the federal government in many cases makes further investments via the SBIR programs itself. The NRC Phase II Survey attempted to determine how many additional SBIR awards followed each initial award.

The data shown in Table 4-12 indicate that 43.5 percent of respondent projects had at least one related SBIR award. These data suggest that while SBIR awards are to some extent concentrated, this effect is not overwhelming. 56.5 percent of respondents report no additional related SBIR awards at all, although small businesses reported that it takes multiple awards (often complementary) to build a product for effective sale.

[28]E.g., Neurocrine, Illumina. See National Research Council, *An Assessment of the SBIR Program at the National Institutes of Health*, Charles W. Wessner, ed., Washington, DC: The National Academies Press, 2009, Appendix D. See also Maryann Feldman "Assessing the ATP: Halo Effects and Added Value" in National Research Council, *The Advanced Technology Program: Assessing Outcomes*, Washington, DC: National Academy Press, 2001.

**TABLE 4-12** Related SBIR Awards

| Number of Phase II Awards | Number of Companies | Percent |
|---|---|---|
| 28 | 1 | 0.1 |
| 9 | 1 | 0.1 |
| 7 | 4 | 0.5 |
| 6 | 6 | 0.7 |
| 5 | 12 | 1.5 |
| 4 | 23 | 2.8 |
| 3 | 29 | 3.6 |
| 2 | 106 | 13.0 |
| 1 | 172 | 21.2 |
| 0 | 459 | 56.5 |

SOURCE: NRC Phase II Survey.

### 4.2.6 Sales of Equity and Other Corporate-level Activities

The NRC assessment explored the different types of activities ongoing or completed among surveyed companies.

The data in Table 4-13 show that marketing-related activities were most widespread, with licensing agreements related to 33.4 percent of projects, and marketing/distribution agreements to 22.5 percent. Agreements likely to involve the direct transfer of equity were much less common. Only 3.6 percent of respondents reported finalized or ongoing mergers, while only 6.3 percent reported a sale of the company. Note, however, that the question asked specifically for outcomes that were the "result of the technology developed during this project"[29]—a very tight description.

Activities with foreign partners were, unsurprisingly, substantially lower than similar activities with U.S. partners. Again, marketing-related activities were most widespread.

In addition, the NRC Firm Survey[30] determined that three firms (with SBIR awards at DoD) had had initial public offerings, and that a further three planned such offerings for 2005/2006. Seventy-five out of 445 companies at all agencies had established one or more spin-off companies (16.9 percent).[31]

The diversity of these outcomes underscores the challenge of early-stage

[29]NRC Phase II Survey, Question 12.

[30]The NRC Firm Survey did not assign companies (as opposed to projects) to specific agencies, as many had received awards from more than one agency.

[31]The NRC Firm Survey was sent to senior executives at all firms receiving Phase II awards between 1992 and 2001.

**TABLE 4-13** Equity-related Activities

| Activities | U.S. Companies/Investors | | | Foreign Companies/Investors | | | All Companies/Investors | | |
|---|---|---|---|---|---|---|---|---|---|
| | Finalized (%) | Ongoing (%) | Total (%) | Finalized (%) | Ongoing (%) | Total (%) | Surveys | Total | %t |
| Licensing Agreement(s) | 16 | 16 | 32 | 3 | 5 | 8 | 619 | 207 | 33.4 |
| Sale of Company | 1 | 5 | 6 | 0 | 1 | 1 | 619 | 39 | 6.3 |
| Partial Sale of Company | 1 | 4 | 5 | 0 | 1 | 1 | 619 | 33 | 5.3 |
| Sale of Technology Rights | 4 | 10 | 14 | 1 | 3 | 4 | 619 | 95 | 15.3 |
| Company Merger | 0 | 3 | 3 | 0 | 1 | 1 | 619 | 22 | 3.6 |
| Joint Venture Agreement | 4 | 8 | 12 | 1 | 2 | 3 | 619 | 83 | 13.4 |
| Marketing/Distribution Agreement(s) | 11 | 9 | 20 | 5 | 4 | 9 | 619 | 139 | 22.5 |
| Manufacturing Agreement(s) | 3 | 9 | 12 | 3 | 2 | 5 | 619 | 94 | 15.2 |
| R&D Agreement(s) | 14 | 14 | 28 | 3 | 3 | 6 | 619 | 175 | 28.3 |
| Customer Alliance(s) | 14 | 14 | 28 | 5 | 3 | 7 | 619 | 173 | 27.9 |
| Other | 2 | 2 | 4 | 0 | 1 | 1 | 619 | 24 | 3.9 |

SOURCE: NRC Phase II Survey.

technology development and the multiple paths followed by small companies as they partner, license, are acquired, and/or attract additional capital.

Some companies have made a practice of acquiring SBIR firms. Titan Corp. purchased 12 SBIR-funded companies, and was then acquired itself by L3 Communications. L3 purchased a further 14 SBIR-funded companies, meaning that it has bought a total of 27 SBIR-funded companies. GE and Invitrogen are other examples of companies that have made a number of acquisitions among SBIR-funded companies.[32]

Arguably this is a powerful validation of the value expected by some SBIR awards and the opportunity represented by the technologies developed through the program. It also underscores the difficulties small companies face in developing a product set independently.

### 4.2.7 Initiatives to Improve Commercialization Outcomes

Prior to the 1992 reauthorization of SBIR, DoD agencies focused on using SBIR to meet DoD research and development goals. Growth in SBIR funding in late 1980s coincided with an overall reduction in funding for 6.2 (applied) research.[33] Consequently, many DoD laboratories began using SBIR as a substitute for what would have been Broad Agency Announcements (BAA) to get their research program accomplished. Topics which in prior years would have been BAA topics were made into SBIR topics. Reportedly, for many topics prior to 1992, topic authors and selection panels paid little attention to the commercial potential of either a topic or an award. Thus, SBIR was a direct and subordinate aspect of DoD's R&D program: it was used when DoD needed to investigate a technology or wanted a prototype built, but lacked either the funds or the budgetary flexibility to take these steps.[34]

This was the background against which the increased push for commercialization should be viewed. Consequently, a number of efforts have been made to improve commercialization outcomes from the SBIR program.

#### 4.2.7.1 DoD Initiatives

During the period after 1992, DoD's topic generation focus gradually changed, with increasingly greater emphasis on topics that could result in sales to DoD or within the commercial marketplace. Several influences contributed to this changed emphasis. In 1994, the Principal Deputy Under Secretary of Defense for Acquisitions chartered a Process Action Team (PAT) to develop a comprehensive

---

[32]John Cottrell, L3 Communications, Presentation at SBTC SBIR in Rapid Transition Conference, Washington, DC, September 27, 2006.

[33]DoD uses a coding system to describe kinds of R&D, ranging from 6.1 (basic research) to 6.7 (testing and deployment). 6.2 is the DoD code for applied research.

[34]Peter Cahill, BRTRC, private communication, December 1, 2006.

set of recommendations to improve SBIR contracting and funding processes; increase the commercialization of SBIR research in both military and private sector markets; and expand program outreach, particularly to socially and economically disadvantaged small businesses and woman-owned businesses.

These recommendations resulted in a number of changes for the program. Over the next few years, DoD undertook a wide range of initiatives, including:

- Creation of the Fast Track program, which provided benefits for companies that could show third-party investment in their projects.[35]
- Efforts to reduce award processing time.
- Establishment of pre-release procedures for each solicitation, allowing potential applicants to talk with topic authors.
- An increase in outreach activities toward woman and minority business owners and PIs.
- New efforts to train small business owners in commercialization (see discussion of the Navy outreach and training program below).
- New requirements that firms identify commercialization strategy in proposals (partly reflecting new congressional mandates in this area).
- Implementation of the Commercialization Achievement Index (CAI), which created a commercialization metric for multiple Phase II award winners.

For a period in the mid- to late 1990s, the Director of Defense Research and Engineering (DDR&E) strongly pushed dual-use technology which could find markets in both the commercial and military sectors. The DoD Office of Small and Disadvantaged Business Utilization (SADBU) SBIR Program Manager also emphasized the importance of the commercial sector, on the assumption that commercially attractive SBIR topics and awards would widen the downstream market for the technology as well as attract private sector investment, thus reducing the cost of the technology and making it more affordable for DoD. However, at the service and component level, the primary emphasis continued to be on what was good for agency mission needs as defined at the technical monitor level. If commercial sales resulted, that was a just a bonus.[36]

#### 4.2.7.2 The Fast Track Initiative

In 1996, DoD SBIR Solicitation 96.1 established, on a two-year pilot basis, a "fast track" SBIR process for companies which during their Phase I projects identified independent third-party investors that would match Phase II SBIR funding. Fast Track projects received (1) interim SBIR funding between Phases

[35]See below for a description of this program.

[36]Peter Cahill, BRTRC, private communication, December 1, 2006.

I and II, (2) priority for Phase II funding, and (3) an expedited Phase II selection decision and award.

As early as 1992, DoD's Ballistic Missile Defense Organization (BMDO) had begun to reward applications whose technologies demonstrated commercial private sector interest in the form of investment potential. This BMDO initiative, called "co-investment," was effectively an informal "fast track" program. Under this approach, the evaluation process for Phase II proposals gave preference to applicants who could demonstrate that they would commit internal funding to the research or that they had financial or in-kind commitments from third parties to bring the technology to market in Phase III. With that commitment, applicants received essentially continuous funding from Phase I to Phase II.

In October 1995, DoD launched a broader Fast Track initiative to attract new firms and encourage commercialization of SBIR-funded technologies throughout the Department. With this initiative, DoD sought to improve commercialization through preferential evaluation and efforts to close the funding gap that could develop between Phase I and Phase II grants. The Fast Track program addressed this gap by providing expedited review and essentially continuous funding from Phase I to Phase II as long as applying firms could demonstrate that they had obtained third-party financing for their technology. In this context, third-party financing could mean that another company or government agency had agreed to invest in or purchase the SBIR firm's technology; it could also mean a venture capital commitment to invest in the firm or that other private capital is available.

The expedited decision-making process for the Phase II award is justified from the agency's perspective because outside funding validates the commercial promise of the technology. More broadly, the Fast Track program sought to address the need to shorten government decision cycles in order to interact more effectively with small firms focused on rapidly evolving technologies.

Based on commissioned case studies, surveys, and empirical research, the Moore Committee's 2000 report[37] suggested that the Fast Track initiative was meeting its goals of encouraging commercialization and attracting new firms to the program.[38] Consequently, the Committee recommended that Fast Track be continued and expanded where appropriate.

[37]National Research Council, *The Small Business Innovation Research Program: An Assessment of the Department of Defense Fast Track Initiative*, op. cit.

[38]As always, there are caveats and limitations to this research. The first limitation concerns the relatively short time that the Fast Track program had been in place. This necessarily limited the Committee's ability to assess the impact of the program. The case studies and surveys constituted what was clearly the largest independent assessment of the SBIR program at the Department of Defense, the study was nonetheless constrained by the limitations of the case-study approach and the size of the survey sample. The study nevertheless represents the largest external review of the Fast Track program and SBIR undertaken until the current review.

#### 4.2.7.3 The Commercialization Achievement Index (CAI)[39]

The efforts to increase the focus on and the success of commercialization led to the introduction of the Commercialization Achievement Index (CAI) in 1999. Companies with five or more Phase II awards were required to submit data about commercial outcomes as part of the application process for further awards. Following review of its SBIR and STTR programs as part of OMB's Program Assessment Rating Tool (PART) process, DoD modified the CAI reporting requirements in its FY2006 SBIR Program Solicitation. The new requirements (Section 3.5d) specify that:

a. Firms with four or more completed Phase II projects will receive a CAI score. Formerly, a CAI score was assigned only to firms with five or more completed Phase II projects.

b. Firms with a CAI at the 10th percentile or below may receive no more than half of the evaluation points available for commercial potential criteria. Formerly this provision applied to firms with a CAI at the 5th percentile or below.

c. DoD will now comprehensively examine the company commercialization review (CCR) data supplied by all firms participating in the program. Formerly, the review consisted mainly of periodic cross-checks.

The projected effect of these changes is to (a) slightly increase the number of firms whose new proposals are subject to criteria relating to past commercialization results, and (b) reduce the evaluation scores of more firms whose prior Phase II projects have not produced desirable levels of commercialization relative to other awardees.

**CAI—A positive development?** Efforts to find metrics for commercialization should be applauded. Eventual insertion of SBIR-developed products and services into DoD acquisition programs is an important objective of the program. Finding ways to ensure that money is targeted to firms with a good commercialization records ensures a positive return to the program (although it also implies multiple awards to a single firm).

However, the case histories suggest that developing such metrics is challenging. The length of time needed for the development of a commercially viable technology or for market demand to match an emerging technology may be longer than allowed for in the CAI. Thus the CAI may not be a sufficiently accurate measure of the commercial impact of SBIR-funded technologies at DoD for its use to be expanded further with the current metrics.

[39]The Company Commercialization Report (CCR) was first required in DoD SBIR Solicitation 93.2 as a result of the 1992 SBIR reauthorization. The CCR was redefined and formatted as Appendix E to any proposal in DoD SBIR Solicitation 96.1. The CCR became an electronic submission with an embedded evaluation of a firm's history of commercializing prior Phase II awards, the Commercialization Achievement Index (CAI), in DoD SBIR Solicitation 99.2.

It is also worth pointing out that CAI is a limited tool focused exclusively on commercialization. SBIR's other objectives are not addressed. Moreover, use of the CAI omits several important economic benefits of the SBIR program that contribute to the viability, growth, and profitability of small firms. These components of value are discussed in other sections of this chapter.

Thus, while CAI is undoubtedly a useful tool, it is important that DoD—and the selection process in particular—remain aware of its limitations.

### 4.2.8 Commercialization: Conclusions

As noted in the introduction to this section, there is no single metric for identifying commercial success. Instead, multiple indicators, and multiple metrics within those indicators, are needed to develop a broad assessment of commercialization within the DoD SBIR program.

This assessment supports the view that there has been considerable effort to bring SBIR projects at DoD to the market, with some substantial success. Even though the number of spectacular commercial successes (as is typical in most early-stage research efforts) has been few, the overall commercialization effort is substantial.[40]

Products are coming to market and significant licensing and marketing efforts are underway for many projects. Approximately 40 percent of projects generate products or services that eventually reach the marketplace (an unusually high percentage). These data all paint a picture of a program where the commercialization objective is well understood by award recipients and by the agency.

Overall, integrating survey findings and case study narratives suggests that firm experiences under the DoD SBIR program do not appear to differ markedly from private sector firms that invest in their own R&D. Christensen and Raynor, for example, note that recent surveys of private sector R&D report: "Over 60 percent of new product efforts are scuttled before they ever reach the market, and of the 40 percent that do see the light of day, 40 percent fail to become profitable and are withdrawn from the market."[41]

This approximate benchmark puts an even more positive light on DoD SBIR commercialization outcomes, since private sector R&D is typically weighted far more to product development than to the science and technology activities (6.1–6.3) within the DoD SBIR portfolio; thus private sector R&D should be expected to produce comparatively higher success rates.

However, as we shall see in Chapters 5 and 6, there are still considerable barriers to SBIR awardees achieving success within the DoD acquisition system itself—which represents an area of significant potential improvement.

---

[40]The distortion is echoed in venture and angel investments, both of which are often downstream of SBIR, which normally makes earlier-stage investments.

[41]C. Christensen and M. Raynor, *Innovator's Solution*, Boston, MA: Harvard Business School, 2003, p. 73.

## 4.3 AGENCY MISSION

One of the core legislative objectives of the SBIR program is that it contributes to each agency's mission. DoD's SBIR program has nurtured many technological innovations that have made significant contributions to DoD's mission capabilities.[42] SBIR-spawned innovations have contributed to enhanced combat capabilities, and provided technological solutions to meet sudden, unexpected military threats.

Detailed analysis of Phase III at DoD is provided in Chapter 5.

### 4.3.1 Unique Benefits of SBIR at DoD

#### 4.3.1.1 Enhanced Flexibility and Innovation

SBIR-funded projects have proven to be of especial value in generating technological approaches to new, unexpected problems that have arisen in ongoing military engagements in Afghanistan and Iraq. The high degree of flexibility characteristic of small firms means that SBIR has provided DoD with an increased number of suppliers capable of quickly responding on short notice[43] to unanticipated battlefield situations (such as the use of improvised explosive devices [IEDs] in Iraq).

DoD has used the SBIR program to move quickly from identification of a DoD need to issuance of Phase I and then Phase II awards, and then to rapid deployment of operational equipment to meet pressing needs. Among the success stories contained in the case studies are the use of unmanned aerial vehicles for collecting over the horizon intelligence (Advanced Ceramics), the development of hand-held language translators (Marine Acoustics/Voxtex), the invention of radio detection and explosive devices to combat improvised explosive devices (First RF), and the production of an automated ammunitions sorter (Cybernet).

According to DoD SBIR program officials, SBIR has also proven an important and successful means of attracting the interest of small, high-technology firms to address specific R&D and operational needs where the potential market is too small to attract the interest of large defense contractors or venture-backed firms.

DoD officials point to a number of other benefits from the SBIR program:

- It has increased the number of potential suppliers for new technologies, and also created new opportunities for these firms to partner together in new undertakings.

---

[42]Dr. Charles Holland, Deputy Under Secretary for Science and Technology, Department of Defense, "Meeting Mission Needs," in National Research Council, *SBIR and the Phase III Challenge of Commercialization*, Charles W. Wessner, ed., Washington, DC: The National Academies Press, 2007.

[43]Dr. Mike McGrath, Deputy Assistant Secretary for RDT&E, U.S. Navy.

- It has helped DoD personnel learn about a set of high-tech firms with whom they would not otherwise ordinarily have contact.[44]
- It allows DoD to draw on the creativity of the small- and medium-sized firms that comprise the SBIR community, and of the commitment of these firms to serve the needs of Services and agencies.
- The SBIR program also has served as a filter to determine if firms have the technical, management, and financial capabilities to become reliable suppliers to DoD, whether on additional Phase III awards or for consideration in subsequent procurement competitions.

Survey responses from 347 SBIR technical monitors or Technical Points of Contact (TPOC) indicate that they perceive of DoD SBIR projects' research to be of high quality—on average close to the quality estimated for non-SBIR awards.[45] On a ten-point scale, where 10 represented the best research ever produced in the research unit/office in which the TPOC was located, SBIR awards received a mean score of 6.95. This average score was slightly below the mean score of 7.27 for non-SBIR research projects.[46] This difference may be explained by outliers among the surveyed group.

### 4.3.1.2 Usefulness

SBIR projects were also found to have affected the way in which the DoD unit/office conducted research or supported research in other contracts. Fifty-three percent of TPOC respondents indicated that the specific SBIR project referred to in the survey produced results that were useful to them and which they had followed up on in other research.

Another indicator of the relative cost-effectiveness of the knowledge or informational contribution of SBIR projects to the design of DoD's research program was that one-third of TPOC respondents noted that the SBIR project had more benefits for the agency's mission than the average dollar spent on other re-

[44]Firms express this outcome as follows: Had it not been for SBIR, their business with DoD services or agencies would not have developed. Services likely would have stayed with their pre-existing sources of supply. Program managers are too busy with multiple contracts to search out or respond attentively to new sources of technology. Their orientation is to hire a contractor to solve problems, not necessarily to seek out the most technologically innovative performer. The SBIR program requires that they become involved with small firms, to look at technical options, and to allow for increased competition in the selection of R&D performers.

[45]NRC Program Manager Survey in National Research Council, *An Assessment of the SBIR Program at the National Aeronautics and Space Administration*, Charles W. Wessner, ed., Washington, DC: The National Academies Press, 2009—contains a full description of the survey methodology and findings.

[46]The difference in mean scores was statistically significant (at the .01 level), but was attributable to the considerably higher percentage of SBIR than non-SBIR projects that received low scores (e.g., 3 or below).

search contracts that the TPOC sponsored. Forty percent of respondents reported the same level of benefits per dollar spent; 27 percent reported fewer benefits per dollar spent. This at a minimum indicates that TPOC respondents saw SBIR dollars as generating research of quality equal to that of funding spent through other mechanisms—an important result in an environment where SBIR has often been seen as an unwanted congressional imposition on research managers' discretion with research funding.

#### 4.3.1.3 Low-cost Technological Probes

SBIR projects also provide DoD project managers with relatively low-cost explorations of novel scientific and technological approaches, the outcomes of which serve to increase the effectiveness of DoD's laboratories and R&D facilities. While SBIR awardees overheads are normally not lower than universities (though this varies with company size) SBIR firms normally do have a much lower overhead than laboratories or large corporations.

#### 4.3.1.4 Cost Savings

A number of DoD SBIR projects, for example RLW's technique for real-time monitoring of ship maintenance and SAVI's radio frequency detection system for monitoring shipments, have also been directed specifically at cost reduction (see case studies).

### 4.3.2 Assessment of SBIR's Contributions to DoD Missions

#### 4.3.2.1 Previous Assessment Efforts

There is currently no comprehensive formal measurement of DoD SBIR performance. The Navy considers the Phase III reported in the DD350 its primary tool for such purposes, emphasizing once again the stress laid on active insertion into weapons programs by the services as the core metric of success from the agency perspective. However, the other services have not emphasized the DD350 to the same degree, and there are substantial interservice differences in the extent to which the DD350 reporting system actually captures program outcomes.[47]

Of special concern to DoD SBIR program officials is that the CCR does not fully capture the program's mission benefits and that undue use of commercialization measures will undercut core processes of technological development. They note that the closer a technology effort is to basic research, the less likely it is that its impacts will be captured in the PART system. Latent in these concerns is the

[47] These limitations are emphasized by the DoD's primary data subcontractor on SBIR. Pete Cahill, BRTRC, private communication, December 1, 2006.

implication that program managers and laboratory personnel will rebalance their R&D portfolios to perform well according to PART and GPRA measures. They will select short-term product development topics that will improve their individual or unit annual performance reports but will diminish, over time, the rate at which agency R&D contributes to significantly enhanced agency performance.

DoD did make one prior effort to more fully analyze the performance of its SBIR program. In 1996–1997, the Director of Defense Research and Engineering (DDR&E) in conjunction with the DoD Small and Disadvantaged Business Utilization Office (SADBU) SBIR Program Manager, contracted BRTRC, a technology research firm, to study the commercialization of DoD SBIR awards. The DDR&E chose not to publish the results of this internal study. The survey's responses, however, were shared with GAO. GAO used and referenced these results in two reports to Congress.[48]

DoD's SBIR monitoring and reporting activities currently emphasize reporting on compliance. These data are suitable for documenting attainment of selected program objectives, such as support for woman- and minority-owned firms, but not to those related to mission support.

Conceptual and empirical difficulties clearly exist in documenting these contributions. Mission support (or "benefits") and "sales revenue" are not identical. Thus, for example, it is hard to exactly measure the "benefit" of the SBIR-supported technology developed by Ophir Corp., which made stealth bombers more "stealthy." Is the benefit (a) the $27.5 million in the firm's sales, (b) an estimate for the increase in aircraft survivability (and, ultimately, lives saved), (c) the enhanced force capability, or (d) the sum of all of the above?

These complexities are not unique to the Ophir project. They are relevant to many SBIR projects that successfully promoted an agency mission. In most cases, direct measurement of agency mission impacts is simply not quantifiable—as NIH found when seeking to identify the public health impacts of SBIR awards. To give another example, how do you determine the contribution (or value) to mission support of Vista Controls Corporation's electronic computing card in the ballistic computer of the M-1 tank? How do we answer the question, "how many lives and how much equipment are saved by getting off the first shot and hitting a target before it can hit you?"

Sometimes educated guesses can be made. Pentagon logisticians estimated that the SaviTag (an SBIR-developed bar-coding device for managing the logistics of military and nonmilitary material) could have saved $2 billion dollars had it been used in Desert Shield. The technology was used when U.S. forces were deployed in Bosnia and it is now in widespread use in Iraq.[49] While the survey

[48]U.S. General Accounting Office, *Observations on the Small Business Innovation Research Program*, GAO/RCED-98-132, Washington, DC: U.S. General Accounting Office, April 1998, and U.S. General Accounting Office, *Evaluation of Small Business Innovation Research Can Be Strengthened*, GAO/RCED-99-114, Washington, DC: U.S. General Accounting Office, June 1999.

[49]Available at <*http://www.savi.com/products/SaviTag_654.pdf*>.

and other data make it relatively easy to compute Savi's sales, this amount has little relationship to the cost savings the technology provided to the department, the alternative uses of resources it enabled and the efficiency gains and savings permitted, especially during crises when resources are constrained.

#### 4.3.2.2 Data Issues

The quality of the data available to DoD to assess the impacts of its SBIR program is also uneven. The DoD commercialization database contains information on Phase II awards for all agencies, not just DoD. Projects awarded prior to 1999 are reported/updated only if the firm is applying for further SBIR. For awards since 1999, updates on the project are required one year after the start of Phase II, at the completion of Phase II, and subsequently when the contractor submits a new SBIR or STTR proposal to DoD. Firms that do not submit a new proposal to DoD are asked (but not required) to provide updates on an annual basis after the completion of Phase II.[50]

If a company failed and ceased to exist, was very successful and outgrew SBIR, was sold to a larger firm, or became disenchanted with SBIR or DoD prior to 1999, no information on that firm will exist in the commercialization database. Over 30 percent of the Phase II awards made by DoD prior to 2000 have no information in the commercialization database. For more recent years, the data is more complete; for the period 1992–2002, over 82 percent of DoD Phase II awardees have entries in the commercialization database. The voluntary nature of reporting after the Phase II contract is finished clearly limits the completeness of the data.

"Success stories," another frequently employed indicator of SBIR's contribution to DoD's mission objectives-success stories, also have limitations. DoD's compilation of "success stories" is somewhat uneven. In practice, DoD accepts all submissions offered by firms, topic authors, and program managers. This approach cannot be completely comprehensive or representative, being shaped more by the initiatives and motivations of individuals rather than from any agencywide attempt at systematic coverage. There also are no visible quality control checks to insure the accuracy of specific statements. Moreover, since the content of success stories tends to focus on technological performance and sales,

[50]The consequence of collecting data on pre-1999 awards only if a firm submits a new proposal can be seen in the following example. In terms of reported sales, the seventh most commercially successful award reported in CCR at $92 million was the Air Force's award to SPEAKR Engineering. This award Phase II award however was made in 1983. SPEAKR did not submit an SBIR proposal in 1999, 2000, 2001, or 2002. Thus, there is no record of this $92 million in the commercialization database until the second solicitation of 2003, when SPEAKR submitted a new Phase I proposal. At the time of this submission, their proposal also reported three other DoD Phase II awards (one of which had $14 million in sales).

it is unlikely that many of the other benefits of the DoD SBIR program are fully accounted for, e.g., firm knowledge effects.

These challenges frequently arise in assessments of the economic and non-economic impacts of other public sector investments. But answers to these questions, couched variously in terms of benefit-cost, cost-effectiveness, or rate-of-return analysis, exist for related R&D programs at several federal agencies. In addition, an extensive, long-standing theoretical and empirical literature exists on assessment of the impacts of technologies comparable to those developed under DoD's SBIR awards.[51]

The distinctive aspect of the current state of affairs in assessing the contribution of the DoD's SBIR program to missions is therefore not the inherent complexity of the challenges faced in producing acceptable estimates (although, admittedly, it is difficult), but the absence of such studies.

Interviews with DoD officials indicate that the primary reason for this absence is the lack of internal budget, staff, and expertise to systemically monitor, document, and assess the contribution of the SBIR program to its mission objectives. DoD program managers note that they are legislatively required to use all appropriated SBIR funds for awards to firms, and are explicitly prohibited (by Title 15, Chapter 14A) from using these funds for in-house activities such as program assessment. In the absence of nonprogram resources, DoD finds itself without the resources to maintain an adequate database or analytical studies to respond to questions about program management and program outcomes. When the department does initiate changes, it is not well positioned to systematically assess their impact.

Notwithstanding these constraints, the Department of Defense has made a significant effort to evaluate its SBIR program. For many years, the department has relied on an outside consulting firm (BRTRC) to maintain a database and, in the past, to carry out some analysis of the program. The department has also led the way among SBIR agencies in external assessment, commissioning exploratory workshops and then a major review of the Fast Track program by the National Academies in 1999–2000. More recently, Defense was an early contributor to the current study and has commissioned additional work with the Academies and with the RAND Corporation. Despite the real resource constraints described above, the department leads all other agencies in its efforts to better

[51]Notable examples include several studies of the economic benefits of health related research: S. C. Silverstein, H. H. Garrison, and S. J. Heinig, "A Few Basic Economic Facts about Research in the Medical and Related Life Sciences," FASEB 9:833-840, 1995; D. Cutler, *Your Money or Your Life*, New York: Oxford University Press, 2005; agricultural research: R. Evenson, P. Waggoner, and P. Ruttan "Economic Benefits from Research: An Example from Agriculture," *Science*, 205(14 September):1101-1107, 1979, and R. Ruegg and I. Feller, *A Toolkit for Evaluating Public R&D Investment—Models, Methods, and Findings from ATP's First Decade*, NIST GCR 03-857, Gaithersburg, MD: National Institute of Standards and Technology, 2003.

understand the program's operations and accomplishments as well as the impact of new policy initiatives.

That said, program managers recognize that current resources do not allow them to collect and maintain the data needed to better understand the program, nor to provide an effective tracking system and internal analysis that would facilitate external analysis of the program.[52]

## 4.4 SUPPORT FOR SMALL BUSINESS AND FOR MINORITY- AND WOMAN-OWNED BUSINESSES

This section provides an analysis of the ways in which this support helps different kinds of companies, and the overall impact of SBIR on small business activities at DoD. At one level, SBIR obviously provides support for small business, in that it provides funding only to businesses with no more than 500 employees—the SBA definition of a small business.

### 4.4.1 Small Business Shares of DoD Funding

Data from DoD show that for all contracts (not just SBIR), small business direct awards have for the past 15 years been approximately equal to the subcontracts to small business reported by the prime contractors (see Figure 4-8).

### 4.4.2 Project-level Impacts

A basic question concerning any "intervention" is whether the observed outcome would have occurred in the absence of the intervention. In terms of the SBIR program, this question generally relates to whether or not an R&D project would have been undertaken by a firm in the absence of an award, and if so, what affect the award had on the size, scope, or final characteristics of the project. After all, if most SBIR projects would have gone forward even without the SBIR contract, then the program's impact would appear to be limited.

#### 4.4.2.1 Project Initiation and Beyond

Figure 4-10 reports firm responses to the question of whether they would have undertaken the specific project in question in the absence of the SBIR award.[53] Over one-third (37 percent) of responses stated that the project would "definitely not" have been undertaken, and another third (33 percent) said "prob-

---

[52]As noted in the introduction, a major challenge for the Academies' assessment of the program was the absence of the necessary data with regard to outcomes and program impact.

[53]This question was asked of the 618 respondents who had completed their Phase II project and then continued to develop the project.

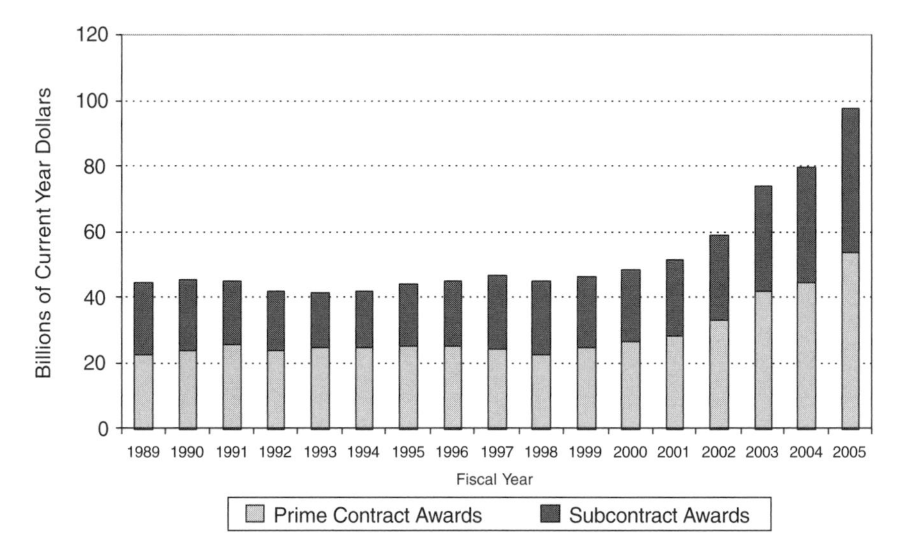

**FIGURE 4-8** DoD direct and subcontract awards to small businesses.
NOTE: Subcontract Awards category includes only reportable subcontracts. A significant portion of small business awards are for a variety of necessary services, e.g., facilities maintenance, but are not high-tech research and product development.
SOURCE: Presentation by Dr. James Finley, Deputy Under Secretary for Acquisition and Technology, DoD. SBTC Conference on SBIR in Transition, September 27, 2006. Washington, DC.

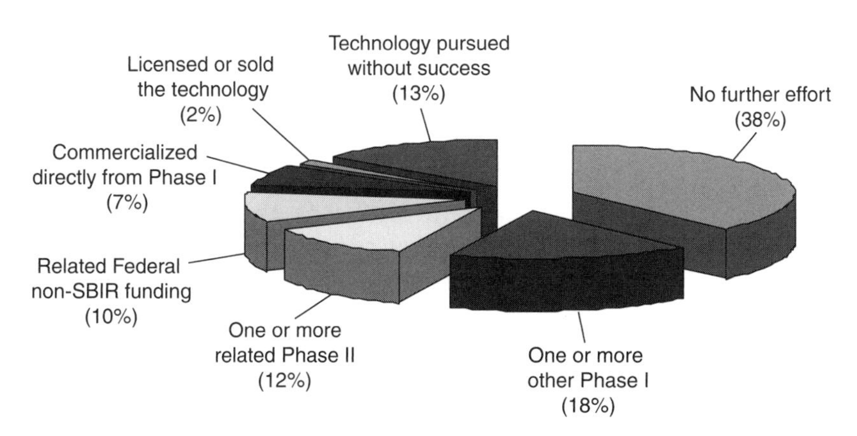


**FIGURE 4-9** Outcomes for firms not successful in receiving Phase II.
SOURCE: NRC Phase I Survey.

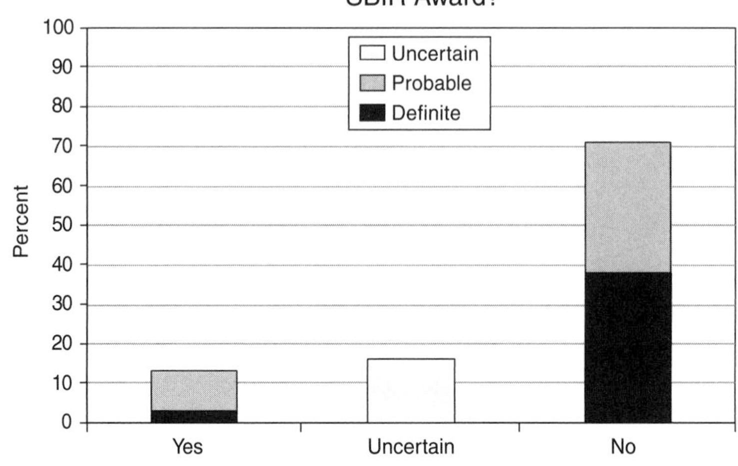


**FIGURE 4-10** Project-level impacts—Phase II recipients.
SOURCE: NRC Phase II Survey.

ably not," Only 13 percent of responses indicated that the project "definitely" or "probably" would have gone forward in the absence of SBIR.

These responses strongly imply that the SBIR contract had a substantial impact in terms of the firm's decision to carry out the proposed project. The NRC also surveyed companies that did not receive a Phase II award. Almost half did not pursue the technology.

Case studies suggest that the "definitely not" respondents can be divided into two groups. Some SBIR recipients claim that without the SBIR award the firm itself would either not have been formed or would not have survived for long. SBIR provided incentives and opportunities for the founders to leave their existing occupation and start a firm, and the initial working capital essential for pursuing the proposed research.

For existing firms with ongoing R&D or production activities, the primary value of the SBIR program was that SBIR topics indicate a well-funded potential market for the firm's technology.[54] In effect, technology-intensive firms may be seen as constantly searching for potential applications for their core technologies. At any point in time, there are more technology paths to pursue than resources allow. SBIR topics focus a firm's attention on a specific technological and market

[54]N. Rosenberg, "The Direction of Technological Change: Inducement Mechanisms and Focusing Devices," *Economic Development and Cultural Change* 18:1-24, 1969.

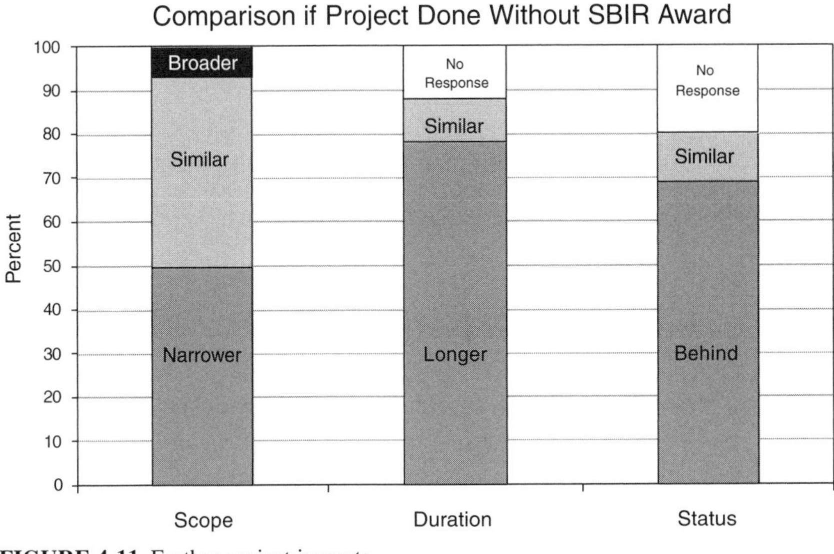


**FIGURE 4-11** Further project impacts.
SOURCE: NRC Phase II Survey.

objective. And when the firm's application is successful, an SBIR award provides the capital, contacts, and versatility needed to pursue product development.

The absence of SBIR funding would have had a significant impact even on the 13 percent of respondents who said that they probably or definitely would have undertaken the project even without the SBIR award. Half of these respondents reported that the scope of the project would have been narrower,[55] and 62 percent expected the project would have been delayed. More than a fifth of these companies expected a delay of at least 24 months; an additional 49 percent thought the delay would have been at least 12 months.

In addition, 78 percent of all respondents noted that the project would have taken longer to complete in the absence of a Phase II award. Thus, in the absence of the award, there would have been a twofold impact of the firm's efforts to transform an R&D concept into a marketable product: The start of the project would have been delayed and project would have taken longer to complete.

There is therefore little doubt that the SBIR awards had, in the vast majority of cases a significant, often decisive, impact on the company's ability to undertake the research supported by the work.

[55]This last response presumably reflects the narrowing of an R&D concept to mesh with the scope of the topic solicitation announcement.

**BOX 4-2**
**Cybernet—Start-Up Case Study**

SBIR has many different impacts at different firms, but one clear impact has been on the decision to found the firm. Only a relatively small percentage of firms were founded directly "because" of SBIR, but Cybernet is one of them.

Heidi Jacobus, the founder, was working on a doctoral thesis on human-computer interaction at the University of Michigan in 1988. She also worked in the university library on a project developing an indexed reference book on the SBIR program. Soon after, she saw her thesis topic listed as a DARPA topic of interest. She distilled her thesis proposal into an SBIR proposal. Subsequent to the submission of the proposal, she received a telephone call from a DARPA official stating that her proposal was the "best" he had ever read.

The feedback from DARPA was the motivating event that gave Jacobus the courage to found Cybernet. She submitted SBIR proposals to other agencies, receiving awards from NASA and the Army, followed by the award from DARPA that had catalyzed the firm's founding. Cybernet was a bootstrap operation: The firm's offices were housed in her daughter's bedroom. Jacobus had to learn the basics of government contracting and accounting procedures, such as overhead rates, allowable expenditures, and related provisions. She accomplished this by reading manuals obtained at the regional SBA office, purchasing technical assistance from local consultants, and receiving assistance from the regional Small Business Development Center.

Cybernet's core mission is the development and application of robotics technology solutions to human-machine interaction. It applies its work to a diverse set of defense and nondefense industries. The firm has drawn on the rich tradition and ready availability of robotics and related manufacturing expertise in the Ann Arbor region to steadily increase its manufacturing activities, and also sells to the private sector. Cybernet currently derives about 70 percent of revenues from federal contracts and 30 percent from the private sector.

### 4.4.3 Multiple-award Winners and New Firms in the Program

Some critics argue that the DoD SBIR program tends to favor "established" small firms with well-developed ties to decision makers in the funding stream.

In a certain sense, this is entirely natural. Firms that are successful in winning contracts and then in delivering results will tend to seek to duplicate these successful efforts, and awarding officers will become familiar with these firms and confident that they can deliver a quality product in a timely fashion.

In interviews, agency staffers have also noted that SBIR is a critically important mechanism for "proving out" potential vendors: The risk is relatively low, contracts are tightly time bound and relatively small, and positive experiences can be built upon quickly as the timeline between solicitations is short. Thus, some staffers see SBIR very much as a means of validating the capabilities of potential vendors, who when successful can be added to the technical resource pool available for meeting agency needs.[56]

These program characteristics imply that the distribution of awards will be skewed, and that some companies will, over time, win a considerable number of awards—particularly firms with a wide range of technical capabilities who are able to address the requirements for many topics in a given solicitation.

At the same time, however, it is important that the program not be captured by a group of regular winners to the extent that potential new vendors are frozen out, and either fail to win or are so discouraged that they no longer apply. Equally, it is important to ensure that firms do not simply become specialists in winning awards, without ever producing commercial results or in the larger sense addressing agency needs for quality research and product prototypes.

The question of new winners is addressed in Chapters 3 and 6, based on DoD awards data. However, additional information has been developed using the NRC's Phase II Survey.

#### 4.4.3.1 New Firms

Most NRC Phase II Survey respondents had received only a few SBIR awards. Most significantly, just under one-third (29 percent) had no prior Phase I award and were entirely new to the program. This underscores the competitive nature of the program, namely that a significant number of awards have regularly gone to new firms.[57]

Another 42 percent of respondents had had five or fewer prior Phase I awards. Five firms reported 96 or more prior Phase I awards. Excluding these five firms, the remaining 561 firms averaged less than four prior Phase I awards.

The average number of previous Phase II awards for firms in this sample was 2.45. Forty-six percent of respondents had no prior Phase II awards and another 36 percent had five or less. The same five firms accounted for all respondents with more than 35 previous awards. The remainder of the firms averaged 1.7 prior Phase II awards.

Respondents were queried about the number of prior Phase I awards in related technologies to the technology embedded in the Phase II project being surveyed. Firms reported an average of 1.5 related Phase I awards with 46 percent of respondents having no prior related Phase I awards, and an additional 48 percent having five or fewer prior related awards. In terms of related Phase II awards, the average number reported by firms was 0.8: Fifty-six percent has no

[56]Comments by John Williams, Navy SBIR Program Manager, at Navy Opportunity Forum, September 2005.

[57]The substantial percentage of new entrants each year was revealed in the 2000 NRC survey. See National Research Council, *The Small Business Innovation Research Program: An Assessment of the Department of Defense Fast Track Initiative*, op. cit.

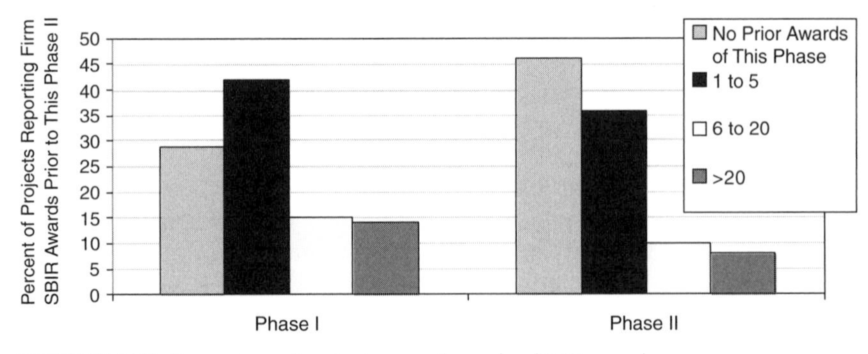


**FIGURE 4-12** Percentage of projects reporting prior SBIR awards.
SOURCE: NRC Phase II Survey.

prior related Phase II award and another 42 percent had 5 or less prior related Phase II awards.

These data make clear that the DoD SBIR program is open to new entrants, with a significant percentage of new entrants for both Phase I and Phase II.

### 4.4.3.2 Understanding Multiple-award Winners

**Commercialization and Multiple-award Winners.** Increased attention to commercialization also has implications for a group of companies that win multiple SBIR awards. As noted above, there are understandable reasons why some firms have become proficient at winning SBIR awards just as some large contractors and some universities become proficient in their domains. One critical question is whether these firms do more than just win awards: Do they provide good value to DoD and do they help the DoD SBIR program meet the congressionally mandated objectives?

This is a harder question to answer than it appears. Firms may be performing work that is important to DoD, but for which there is a limited market both at DoD and commercially. So commercial results in and of themselves are not a sufficient metric, and can indeed be misleading if they become the central focus of analysis.[58]

On the other hand, companies that receive large numbers of awards should be providing work of value to the agency, and one important way to tell whether such work is being done is whether the agency continues to fund the project after Phase II is completed. Again, it is important to consider commercialization metrics only as one indicator of value—an important indicator, but only one of

[58]One important output of research is new ideas that are often picked up and applied by others. This is a major objective of government sponsorship. For more, see Section 4.5 of this chapter on SBIR and the expansion of knowledge.

several. Moreover, case studies clearly show that firms can perform useful work that is not well reflected in commercial results, for example by "answering" a research question of providing a product that has no follow-on sales.

Keeping this complexity of goals and outcomes in mind, commercial results do matter. How well do multiple-winner firms commercialize? Part of the answer lies in Table 4-14 and Figure 4-13. Table 4-14 shows commercialization by number of previous awards. Figure 4-13 focuses on lack of commercialization—the percentage of projects reporting no sales at all, by number of previous awards.

By its three-phase structure, the SBIR program suffers from an implicit linear myth, namely that a single grant for a single project is sufficient to fully develop a technology and drive long-term growth of the company. In practice, a single grant is often not sufficient to commercialize a product. Often multiple related projects, complementary technologies, and varied funding sources are needed, in addition to effective management, to bring a product to market. With regard to additional awards, NRC Phase II Survey data indicate that returns are maximized where firms have received 10–25 previous Phase II awards.

### 4.4.4 Differing Uses of SBIR by Firms

The expansive scope of the NRC study and its use of multiple methods provide an opportunity to place conventional measures of commercialization, such as sales, within the larger context of firm formation, business strategy, and long-term growth.

Earlier assessments have contained typologies of the several different types

**TABLE 4-14** Sales by Number of Previous Awards

| Additional Prior SBIRs | Sales Reported (Percent of Responses) | | | | | | Total Responses |
|---|---|---|---|---|---|---|---|
| | 0 | <$100K | $100K to < $1M | $1M to <$5M | $5M to <$50M | $50M+ | |
| 0 | 46.1 | 13.3 | 22.3 | 13.7 | 4.3 | 0.4 | 256 |
| 1 | 63.2 | 9.2 | 13.2 | 10.5 | 3.9 | 0.0 | 76 |
| 2 | 50.6 | 10.4 | 20.8 | 15.6 | 2.6 | 0.0 | 77 |
| 3–5 | 51.1 | 7.5 | 22.6 | 12.8 | 4.5 | 1.5 | 133 |
| 6–10 | 56.7 | 7.8 | 15.6 | 14.4 | 5.6 | 0.0 | 90 |
| 11–15 | 54.5 | 3.0 | 12.1 | 15.2 | 15.2 | 0.0 | 33 |
| 16–25 | 56.1 | 2.4 | 17.1 | 12.2 | 12.2 | 0.0 | 41 |
| 26–50 | 51.1 | 12.8 | 14.9 | 14.9 | 6.4 | 0.0 | 47 |
| 51–100 | 78.6 | 3.6 | 14.3 | 3.6 | 0.0 | 0.0 | 28 |
| 101+ | 70.6 | 5.9 | 19.6 | 3.9 | 0.0 | 0.0 | 51 |
| Total | 53.7 | 9.4 | 19.1 | 12.6 | 4.8 | 0.4 | 832 |

SOURCE: NRC Phase II Survey.

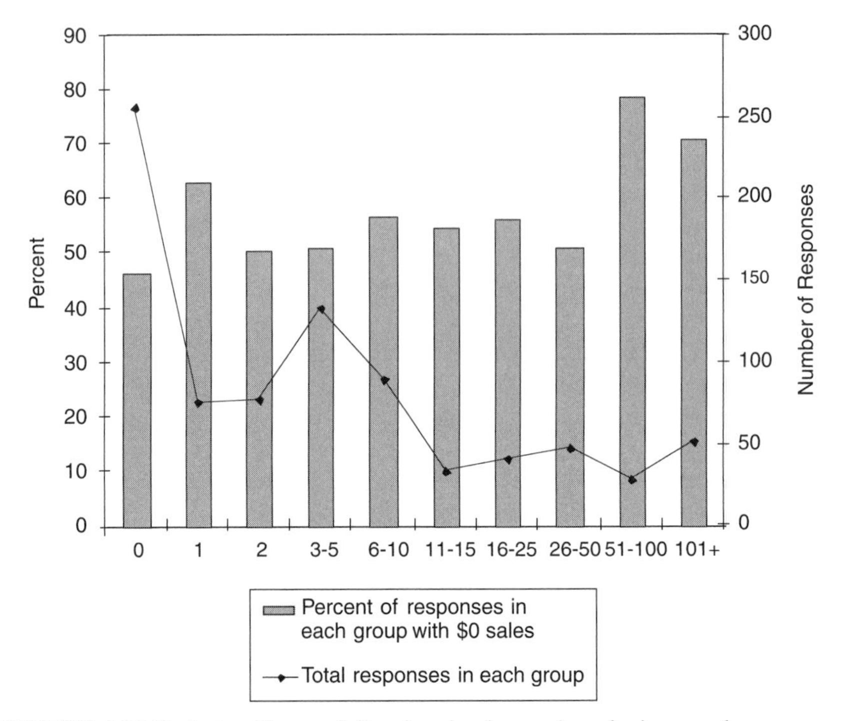


**FIGURE 4-13** Projects with zero dollars in sales, by number of prior awards.
SOURCE: NRC Phase II Survey.

of firms participating in the SBIR program.[59] These typologies are especially useful for extending the analysis beyond the focus on "young, high-technology firms" to the more diverse array of small, high-technology firms that actually participate in the SBIR program. They also help to explain certain outcomes.

Fieldwork conducted during the course of the NRC SBIR study, coupled with review of earlier typologies, produces the following classification of firms that participate in the DoD SBIR program:

- Start-up firms.
- R&D contractors.
- Product-oriented firms.
- Defense-oriented technology-based firms.

[59]See Reid Cramer, "Patterns of Firm Participation in the Small Business Innovation Research Program in Southwestern and Mountain States," in National Research Council, *The Small Business Innovation Research Program: An Assessment of the Department of Defense Fast Track Initiative*, op. cit.

#### 4.4.4.1 Start-up Firms

These are newly formed firms whose early viability is directly connected to receipt of an SBIR award. Some firms were formed specifically to be eligible for SBIR funding; others were existing early-stage firms that would probably have failed without SBIR.

The "start-up firm" category captures the firm in its embryonic form, a transitional stage before the firm evolves into other categories. Firms in this group provide material for exploring SBIR's contribution to new firm formation. In many cases, they also show how SBIR helps to grow the manufacturing base and thus to expand the technological options available to meet DoD mission goals.

For these start-up firms, SBIR Phase I and Phase II awards may provide the initial or critical operating and start-up capital needed to explore leading edge technologies. And, as many interviews—as well as previous NRC studies[60] and academic analyses[61]—indicate, that capital is not readily available from any private sector source. Start-up firms are rarely good candidates for venture capital funding. This is especially true for the relatively narrow, regulatory constrained defense market.

Finally, firms also credit the entrepreneurial behavior of selected SBIR program managers as having catalyzed the firm's formation. Among DoD Services and units, DARPA program managers are most often cited as having played this role. In the case of Cybernet, for example, the interest of a DARPA program officer in the proposal submitted by the firm's founder led the founder to forego an opportunity to complete a Ph.D. and instead venture out to form a firm. Personal encouragements by DARPA program managers to ex-military and ex-industry employees who were employed as consultants to DoD or defense-related firms also are recounted as having led these individuals to form firms based on the prospects of successfully competing for SBIR awards.

One striking aspect of DoD SBIR start-up firms is that many of the founders are former employees of large defense and aerospace prime contractors, who often left because they wanted to run their own business. They describe themselves as refugees from former firms that had grown too large and bureaucratic,

[60]R. Archibald and D. Finifter, "Evaluation of the Department of Defense Small Business Innovation Research Program and the Fast Track Initiative: A Balanced Approach" in National Research Council, *The Small Business Innovation Research Program: An Assessment of the Department of Defense Fast Track Initiative*, op. cit.; D. Audretsch, J. Weigand, and C. Weigand, "Does the Small Business Innovation Research Program Foster Entrepreneurial Behavior" in National Research Council, *The Small Business Innovation Research Program: An Assessment of the Department of Defense Fast Track Initiative*, op. cit., pp. 160-193.

[61]See Lewis M. Branscomb, Kenneth P. Morse, and Michael J. Roberts, *Managing Technical Risk: Understanding Private Sector Decision Making on Early Stage Technology-based Projects*, NIST GCR 00-787, Gaithersburg, MD: National Institute of Standards and Technology, 2000.

**TABLE 4-15** Was Company Founded Because of SBIR Program?

| | Number of Responses | Percent of Responses |
|---|---|---|
| No | 342 | 74.8 |
| Yes | 49 | 10.7 |
| Yes, in part | 66 | 14.4 |
| | 457 | 100.0 |

NOTE: Data reported in Table 4-15 are for firms with at least one DoD award. NRC Firm Survey results reported in Appendix B are for all agencies (DoD, NIH, NSF, DoE, and NASA).
SOURCE: NRC Firm Survey.

or as exiles from small firms that had shelved development of a technology they strongly believed in.[62]

Responses to the NRC Firm Survey indicate that 25 percent of firms were founded entirely or in part as a result of SBIR awards (See Table 4-15). These data are important because they suggest that SBIR funding should be considered one of the more important sources of seed capital for new high-technology companies and entrepreneurs.

This finding was supported by previous analysis. As one scholar reported in the first Academy workshop on the SBIR program, "the picture that emerges is a program that is working effectively and appears to be playing a positive role in stimulating small firm creation." This represents a significant contribution in two respects. As noted in the introduction, from a public perspective this is desirable because there are significant knowledge spillovers associated with R&D, that is, the benefits of R&D do not accrue only to those making the investment.

Second, information problems hamper investors' efforts to identify promising technologies. The large number of companies seeking financing and the uncertainty involved with innovative business proposals pose significant risk assessment challenges for potential investors—challenges that may result in underinvestment in new technologies.[63]

#### 4.4.4.2 R&D Contractors

A second generic type of participant in the DoD SBIR program are R&D contractors that specialize in the performance of R&D, strategically positioning

[62]See for example the remarks of David O'Hara of Parallax Research who left his previous firm, one that he believed was insufficiently focused on the commercial potential of their SBIR awards, in order to found Parallax and pursue commercialization. National Research Council, *The Small Business Innovation Research Program: Challenges and Opportunities*, op. cit., p. 25.

[63]This problem may be especially acute in small firms. See the remarks by Joshua Lerner of the Harvard Business School in National Research Council, *The Small Business Innovation Research Program: Challenges and Opportunities*, op. cit., p. 23.

**BOX 4-3**
**First RF: Addressing New Defense Needs**

FIRST RF Corporation was founded by Farzin Lalezari and Theresa Boone in 2003. Lalezari was born in Iran, and emigrated to the United States in 1971, while a high school student, following the imposition of a death sentence on his father, who was serving as Iran's Minister of Education, by the Khomeini regime. Upon graduation, he joined Ball Aerospace, where he advanced to position of chief scientist and director of research.

At Ball, Lalezari's research led to 25 patents, all assigned to the firm. Lalezari left to form FIRST RF because of disenchantment with the bureaucratization and technological stagnation of large firms, and their overemphasis on short-term profit measures designed to meet the requirements of stock market analysts.

FIRST RF's core technology focus is advanced antennas and RF systems. Lalezari used the SBIR solicitation of topics to focus on a specific problem. In his view, one of the primary benefits of the SBIR program is that it is seen as forcing firms to "think out of the box," while simultaneously providing innovators with access to users.

Lalezari reports writing about 12 SBIR proposals during the firm's first year of operation. The firm received awards on seven of these proposals, a number described as a national record for a start-up company. In late 2003, it submitted a Phase I proposal for an Army-generated topic related to the detection of improvised explosive devices.

By the time its Phase I project was finished, the firm had delivered production prototypes for use by U.S. military forces in Iraq. In 2004, the firm entered a structured competition against 27 other firms, including major defense contractors such as Raytheon and BAE for volume production of IED countermeasure devices. It won the competition, receiving an initial $21.5 million contract from the Army, with delivery scheduled for December 2005.

This Army contract has been followed by several additional contracts with DoD prime contractors.

themselves as a DoD equivalent to private sector industrial R&D labs. R&D contracts constitute a substantial portion of revenues for these firms, and a relatively high percentage of those revenues (especially in the firm's early years) may come from SBIR awards. These awards may in part come from multiple federal agencies. Typically, firm dependence on SBIR declines over time as the firm becomes more of an established supplier of its technical services, gaining revenues from standard procurement contracts, integrating into downstream products, and entering private sector markets.

**SBIR—Main source of federal funding for early-stage technology development.** SBIR provides over 85 percent of federal financial support for early-stage development. SBIR provided over 20 percent of funding for early-stage development from all sources in 1998.

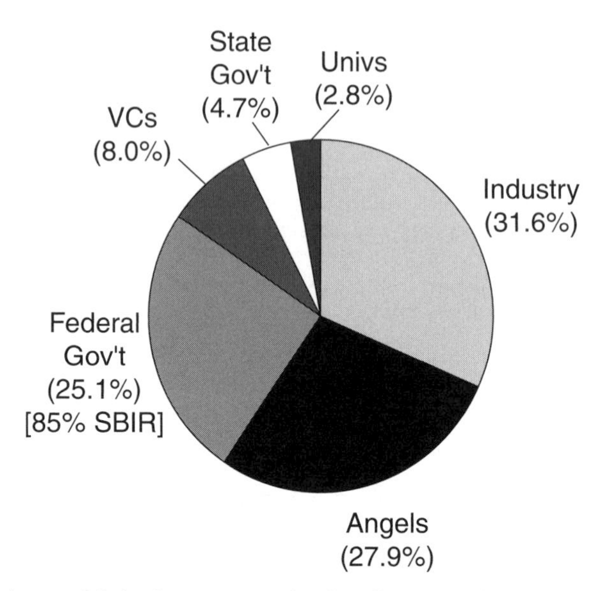


**FIGURE 4-14** Estimate of federal government funding flows to early-stage technology development.
NOTE: Based on total funding for ATP, SBIR, and STTR programs.
SOURCE: Lewis M. Branscomb and Philip E. Auerswald, *Between Invention and Innovation: An Analysis of Funding for Early Stage Technology Development*, NIST GCR 02-841, Gaithersburg, MD: National Institute of Standards and Technology, 2002.

However, even if their work leads to production prototypes, R&D contracting firms often choose not to become extensively involved in downstream manufacturing lest it detract from an emphasis on R&D. They deliberately limit their scope to R&D activities in which they have competitive expertise, plus some selective, (usually) customized manufacturing. Additional revenues may come from licensing proprietary technology or income from spin-off manufacturing subsidiaries.

Such firms operate within self-imposed ceilings on growth. They tend not to seek the external capital usually needed to expand operations because to do so would dilute founder control and equity.[64] This strategy results in a high percentage of revenues from R&D contracts, as the result of a specific commercial strategy.

Even where the venture capital community has expressed interest, not all

[64]E. Roberts, *Entrepreneurs in High Technology: Lessons from MIT and Beyond*, Oxford, UK: Oxford University Press, 1991, p. 328.

**BOX 4-4**
**TRIDENT SYSTEMS: The Challenge of Moving Upstream**

The data strongly suggest that there are different levels and degrees of commercial success, and the Trident Systems case study indicates that stepping up from one level to the next can be challenging.

Trident Systems was established by Nicholas Karangelen in 1985. Initially, the firm operated primarily as a "services" company, consulting to major DoD contractors, such as General Electric and Westinghouse, but lacking its own prime contracts. This work helped build Trident's reputation as a knowledgeable, reliable performer and gave it new insights into DoD requirements.

Trident began submitting SBIR proposals around 1986, submitting four unsuccessful proposals before winning an award. It views the SBIR program as one of the few available contract mechanisms that provide "size appropriate competition"—opportunities for small firms to compete for DoD's R&D and procurement contracts.

In 1988, Trident became a prime contractor on a Navy system development project related to antisubmarine warfare. The stability of the contract, the quality of the work Trident performed under it, and the business relationships developed during performance of the contract launched Trident on the growth trajectory it has experienced since the late 1980s.

Trident has grown primarily by expanding its business around its core competencies in requirements analysis for weapons systems, systems engineering, and, more recently, systems design. The company now serves as prime contractor on many programs and has recently moved further into downstream integration.

**Difficulties in Moving Beyond Phase II.** Trident's success in developing DoD related technologies, especially under the SBIR program, has not led to proportionate successes in landing procurement contracts. In part this may be due to the attitudes of acquisition managers. Trident believes that DoD acquisition offices are reluctant to recognize the value of small firms and their technologies, and are more concerned with maintaining the status quo and avoiding risk. Prime contractors too are seen by Trident as generally unwilling to bring in a promising externally developed (and potentially disruptive) technology when they either have or believe they can develop an internally developed alternative.

SBIR awardees are eager to enter into an agreement lest they be required to relinquish ownership and control. The basic motivation for starting a firm for several of the founders included in this study and the stylized workings of the start-up/venture capital markets diverge at this point. As noted, several of the firm founders in this study in effect are "new world" immigrants from what they see as an "old world" of large, bureaucratic defense and aerospace contractors. As such, they are wary of entering into business relationships that dilute their selection of research topics or require that they meet the profit/sales targets of outside investors, even if the consequence is a smaller firm.

Two hedging, or mid-ground mechanisms, also at times permit founders to maintain autonomy and increase personal incomes. First, they can retain ownership of their initial firm, which retains an emphasis on R&D, while entering into equity agreements with outside investors in the launching of subsidiaries that focus on manufacturing. Second, as a form of life-cycle choice, some founders eschew venture capital and other outside investment during their "work years," and then cash out their founder's profits in the form of equity arrangements as they near retirement.

Besides concern about dilution of ownership prerogatives, another contributing factor in the decision of several firms to refrain from growing via addition of manufacturing and marketing capabilities is that their management perceives the cultures and worldviews of R&D personnel as being distinctively different from (and perhaps superior to) those of personnel engaged in the more downstream activities. The difficulties of integrating the cultures of R&D, manufacturing, and marketing within a single firm were noted by several respondents. One founder observed that in his previous employment settings, he had seen three efforts to reorient a firm from being a service provider to providing both services and equipment; each effort failed. In his view, a firm must decide between one or the other. This philosophy does not, however, rule out the above-noted strategy of spinning off subsidiaries if and when R&D leads to marketable products requiring volume production. Ultimately, firm orientation depends on management preferences—and opportunity—but this does not detract from the versatile and timely contributions successful firms can make to meeting Defense R&D needs.

#### 4.4.4.3 Product-oriented Firm

Product-oriented firms are firms with an existing core technical competency and commercially viable products or services that seek to grow by expanding into, or within, the DoD market.

SBIR topics and funds can assist product-oriented firms in starting new R&D initiatives. They can also help the companies adjust to adverse changes in market conditions. SBIR funds often enable the research small businesses need to make a major course correction when such funding would otherwise be difficult to obtain within the company's existing resources.

For some more established firms, SBIR awards help augment existing product lines and broaden R&D portfolios. SBIR topics both focus a firm's attention on potential new uses for its technologies and expertise, and provide the incentive needed for the firm's management to devote resources to the agency's R&D priorities rather than to other uses.

Product-oriented firms tend to have a mixed portfolio of products and customers, often switching their R&D and commercial activities between defense and nondefense markets as opportunities arise. In some cases, such as ACR (see Box 4-6), SBIR has been critical in helping companies adapt when the high risks

**BOX 4-5**
**PSI Physical Sciences, Inc.**

Physical Sciences, Inc., was established in 1973. The founders left Avco-Everett to start their own firm partly because they sought a smaller research and working environment; Avco-Everett at the time had about 900 employees.

PSI's growth was initially modest, based on contracts with the Air Force and DoE. By the early 1980s, it had approximately $10m in revenues and a staff of 35–50. After a decline in the early 1980s, PSI diversified its federal customer base as well as its range of technological expertise. As it has grown, SBIR awards have contributed a diminishing portion of firm revenues, falling from a peak of about 60 percent in the late 1990s to a projected 35 percent in FY2006.

Since the uses of optical technology have dramatically expanded, the firm's technological and market bases have widened to encompass applied R&D, production operations, and bundling of "hands-on" service delivery with the application of newly developed products, especially in the areas of instrument development, diagnostics, and monitoring. PSI has strategically positioned itself in an R&D market niche defined by multidisciplinary expertise and research infrastructure in specialized high-tech areas too small to attract major investments by large DoD prime contractors, while at the same time too mission-driven to elicit competition from universities.

The firm's successes led to opportunities in new directions, but the founding vision was to maintain owner/employee control of the firm. Hence the firm remains focused on R&D and prototype development rather than manufacturing, which would require additional external capital.

Some of the firm's contracts with DoD involve development of specialized, one of a kind technologies. These can meet critical DoD needs, but may constitute a market with a small sales volume. Other DoD contracts led to the development of technologies, mainly in the area of instruments, that the firm does seek to market to the private sector. For example, PSI's SBIR-funded development of sensor technology to detect methane gas leaks has been sold to gas utilities. In general, sales to the private sector are largely based on technologies developed for DoD under SBIR awards—a classic case of spin-out development.

PSI will engage in limited production of specific instruments for DoD and other federal agencies. When its technological developments lead to commercially viable products, PSI follows a mixed strategy. One approach is to form new firms, with new, independent management, that operate as partially owned spin-outs. Shaping this business decision is the firm's view that the "cultures" and operational needs of contract R&D and manufacturing firms differ sufficiently that it is more efficient to operate them as separate entities rather than attempt to combine them into one larger firm. Conversely, PSI also creates wholly owned subsidiaries, focused on R&D activities, which have become eligible for SBIR competition on their own (as long as PSI remains a small business).

involved in a production-oriented strategy (with attendant high overheads and substantial dependence on specific markets) leads to a crisis for the firm.

For several of the larger (yet still "small business"), private sector firms with established product lines within the case study sample, SBIR funding provides a source of incremental revenue that can be used to determine the feasibility of new technologies without having either to cut back on manufacturing or marketing or seek access to external capital. A pattern evident in the case histories, consistent with the description of the strategic, commercially oriented approach to SBIR topics employed by firms, is for a firm to seek out topics that will permit it to test or demonstrate the applicability of its generic technology to a wider set of applications (and thereby markets). Thus, the SBIR project becomes the test bed to address specific technical problems (e.g., weight, durability, processing speed, luminescence, bandwidth, etc.) of interest to DoD, which if solved could also open up or enhance the competitiveness of the firm's products in different markets (e.g., medical equipment, computer games, first responder equipment, etc.)

#### 4.4.4.4 Defense-oriented Technology-based Firm

These are existing firms whose core competency is the integration of leading edge R&D and product development for DoD. The firms use R&D—including SBIR awards—to advance the performance capabilities of a technology, and then engage in an ongoing search to adapt the core technology to a widened set of applications and users, including initially DoD but also potential customers in the public and private sectors.

SBIR awards can provide an opportunity to explore the broader applicability of core technologies most immediately to DoD's needs, as stated in solicitation topics, as well as to new uses and markets.

The typology described above is useful for understanding some of the different strategies of small high-tech firms participating in SBIR. These can include: (a) seeking to transition from an R&D provider into a supplier of products and processes, subsequently followed by movements into product development, manufacturing, and marketing; (b) failing to make this transition successfully, possibly because of the difficulties of integrating R&D and manufacturing functions (and cultures) within the same organization, and then reverting to its prior R&D specialization; (c) deliberately rejecting the downstream transition strategy by choosing to remain a specialized supplier of R&D services. These points are reflected in the discussion below.

### 4.4.5 How Firms Use SBIR: Commercialization Case Study Results

As noted earlier, different types of firms use SBIR awards in different ways to achieve different objectives. A series of case studies were conducted to flesh out these varied approaches to the program. They explore the workings of the

**BOX 4-6**
**Advanced Ceramics Research (ACR)**

From its inception in 1989, ACR sought to become a product development company, capable of manufacturing products for varied industries. The company knew about, but initially rejected, the SBIR program on the grounds that the 5–7 percent allowable profit was too low.

ACR shifted perspectives on the program, however, and has since participated in SBIR programs at DoD, NASA, DoE, and NSF. SBIR awards accounted for nearly all of ACR's revenues by 1992, but by 2005 only 15–20 percent of its sales were projected to come from STTR/SBIR.

Drawing in part on the advanced research being done at the University of Michigan and its own expertise in both advanced ceramics and manufacturing, ACR developed a general purpose technology for converting AutoCAD drawings first into machine-readable code, and then to direct generation of ceramic, composite, and metal parts.

One market with considerable potential was "flexible carriers for hard-disk drives" for the electronics industry. ACR's aggressive marketing soon helped the company become a major supplier to firms such as SpeedFam Corporation, Komag, Seagate, and IBM. Demand for this product line grew rapidly, enabling the firm to go to a three-shift, seven-day-a-week operation. Demand for ACR's electronic products grew fast during the 1990s, from 5,000 to 60,000 units monthly, and ACR built a new 30,000 square foot plant. The electronics market for ACR's products, however, declined abruptly in 1997, when two of its major customers shifted production to Asia. The loss of its carrier business was a major reversal for the firm. Heavy layoffs resulted, and employment declined to low of about 28 employees in 1998.

The next 2 years are described as a period of reinvention for survival. The firm's R&D division, formerly a money-loser, became its primary source of revenue. The explicit policy was to undertake only that R&D which had discernible profit margins and the opportunity for near-term commercialization.

Since 2000, ACR has received funding from the Office of Naval Research (ONR) to develop a new low-cost, small unmanned aerial vehicle (UAV), initially for whale watching around Hawaii, in support of the Navy's underwater sonic activities. Once developed, the UAV's value as a low-cost, highly flexible, more general purpose battlefield surveillance tool soon became apparent, and ONR provided additional funding to further refine the UAV for use in Iraq.

ACR is now actively engaged in the continued development and marketing of Silver Fox, a small, low-cost UAV, supported by awards under DoD's SBIR (and STTR) program, which have funded collaboration between ACR and researchers at the University of Arizona, University of California-Berkeley, the University of California-Los Angeles, and MIT.

SBIR program, how SBIR awards enter into the business history and strategy of a firm, and the process of technological innovation at the firm level.

#### 4.4.5.1 SBIR and Entrepreneurship

Several themes emerge. The first is the connection between SBIR funding and entrepreneurship in defense-related industries. The NRC Firm Survey data indicate that, for 78 percent of the respondent firms, at least one founder's most recent employment was at a private company. For 28 percent of the firms, at least one founder came directly from a university and, for 7 percent of the firms, at least one founder was most recently employed by the government.[65] The case studies provide additional detail on prior employment. In a majority of the 31 cases, the firm's founder (or founders) was an individual who had worked for a large defense or aerospace firm, and who had left a relatively senior or secure position to develop and commercialize a specific technology.[66]

The founders had several reasons for making this move. In some cases, it reflected a desire to start out on one's own; reflecting what one founder called his "entrepreneurial heritage." In other cases, it reflected a desire to pursue development of a technology that had been sidetracked in the larger R&D portfolio of a prime contractor, as in the case of Trident Systems, above. Some founders revealed that their move was caused by frustration and disappointment that decisions and values within a formerly technologically innovative firm had given way to preoccupation with short-term financial targets.

The high percentage of former defense-related employees among the set of firm founders helps to explain some of the data noted above. It is consistent with the limited involvement of third parties in providing technical assistance, as many SBIR firms were already familiar with DoD's needs and with federal contracting and cost accounting practices. It also reinforces the view that SBIR firms are not necessarily oriented toward rapid financial returns along the lines of the traditional VC model.

Several of the firms explored in the case studies were true "garage" start-ups. Founders located their first activities within garages, basements, or children's bedrooms. Several reported relying on their own savings or funds from relatives to start their businesses. A number recounted a period of serving as consultants to their former employers, customers of those employers, or DoD organizations with whom they had formerly worked, as a transition stage in which they crystallized their technological visions and embryonic business plans while still earning an income.

---

[65]These data are for DoD only. NRC Firm Survey results reported in Appendix B are for all agencies (DoD, NIH, NSF, DoE, and NASA).

[66]Examples here include Applied Signal Technology, Bihrle Applied Research, Custom Manufacturing & Engineering, First RF Corporation, JX Crystals, Physical Sciences, Inc., Scientific Research Corporation and Systems, and Process Engineering Corporation.

These activities served mainly to launch their businesses. The challenge for many was "working capital" to meet payroll for even a few employees, purchase the components necessary to develop and build even rough prototypes (which were seen as a competitive edge toward winning a Phase II award), and to pay for unavoidable overheads. Founders reported mortgaging their homes, drawing down their children's college savings accounts, and foregoing salary for extended periods of time.

At this point in the development process, prospects for venture capital funding were nil. More importantly, their access to bank loans was also limited. Founders reported that a Phase I or even Phase II award usually did not provide sufficient collateral for the firm to qualify for a bank loan. State government programs likewise were seen as better suited to later-stage development, or as dependent on securing funds from a third party, which the state would then seek to leverage. Some firms noted that the payback requirements of state loans were more demanding than those offered by banks.

As a result of all these factors, start-up firms clearly found SBIR Phase I and Phase II awards to be an indispensable source of initial seed capital.

#### 4.4.5.2 Firm Complexity

The case histories also offer a more complex picture of firm behaviors than suggested by the stylized dichotomy of "commercializers" and "mills" that pervades many discussions about the SBIR program, at least with respect to DoD. DoD awards data indicate the presence of a small number of firms with more than 50 Phase I and Phase II awards. Case studies indicate that some of these firms have purposefully positioned themselves to be primarily, but not exclusively, performers of contract R&D. Their business strategy is not necessarily to pursue SBIR awards, but to address the specific, sometimes "one-off," research, testing, and evaluation needs of one or more services and agencies.

As noted above, earlier reductions in DoD 6.2 funding had led services and agencies to turn to the SBIR program to undertake R&D projects that were deleted or delayed as a result of budget cutbacks. Interviews with DoD SBIR managers suggest that although this practice has been reduced by subsequent thrusts to promote dual-use technology and more commercialization, it nevertheless remains part of the portfolio of approaches used by topic authors to achieve mission-oriented R&D objectives.

A group of small, high-tech, defense-oriented firms have positioned themselves to serve these niche R&D markets. Implicitly, the SBIR awards received by such niche firms attest to the needs of DoD Services and agencies for specialized extramural R&D services. The awards also underscore the flexible (but still competition-based) use the R&D managers are making of the SBIR topic generation and solicitation processes to get needed research accomplished in a difficult budgetary environment.

The diversity of these awards highlights the way in which small firms can meet important needs at DoD, not least by addressing problems that provide too small a market to justify the interest of larger defense contractors. These firms may have advantages over academic institutions as well, including specialized equipment, willingness to engage in classified research, and, especially important, the capacity to address demanding time schedules, generated by newly emerging national security needs.

#### 4.4.5.3 SBIR and Competition: A Useful Characteristic for Small Firms

The highly competitive character of DoD's SBIR selection process, described in Chapter 6, is well recognized, and, in many respects, valued by firms. Especially at the Phase I stage, firm executives view the selection process as generating a level playing field. Because they perceive the SBIR award process to be so highly competitive, firms report using a fine mesh filter in deciding whether or not to invest internal resources in preparing an SBIR proposal. During interviews, several firms emphasized that they submitted Phase I proposals only if they could foresee a definable product and market.[67] Unsuccessful firms also noted that they often received valuable feedback from DoD personnel about why their proposal was not funded. The technical commentary contained in this feedback often provided useful guidance for reworking the technology outlined in the initial proposal, akin to "revise and resubmit" commentary found in refereed journal or proposal review panels.

Another recurrent theme to emerge from the case histories is that firms do not see SBIR awards as "free" goods.[68] Preparing a Phase I proposal is not costless to the firm. It involves the direct expenditure of funds for staff time, as well as the allocation of the firm's limited pool of researchers to pursuing SBIR awards rather than alternative contracts.

Firms with DoD awards often have received awards from at least one other agency (NASA being the most frequently cited couplet). Thus, submission of proposals to DoD, or any other agency, not only entail costs in the direct sense of expenditures for staff time and ancillary contract services, but also represents an opportunity cost in submitting proposals to one agency instead of another. Interviewed firms describe careful calculations about how many proposals to write about which topics and to which agencies. Several firms report using in-

[67]This is not to say that these forecasts were accurate. As later described, firm histories highlight how technologies have often evolved into quite different applications than markets and customers had originally conceived.

[68]Sample selection issues affect the generalizability of this account. The case studies recount the experiences of Phase II winners only. Thus, it is possible that the motivations of firms who were unsuccessful in securing Phase I awards or those who had Phase I awards but not Phase II awards differs from this sample. Indeed, one of implicit determining factors between winners and losers of Phase II awards, given a Phase I award, may be the very characteristic of having an a priori commercial target in mind when the Phase I project was initially submitted.

ternal policy and review committees to apportion resources to support staff time to write proposals.

#### 4.4.5.4 Indirect/Nonlinear Commercialization

Another theme to emerge from the case histories is that the initial topic and the resulting commercial use are not always the same. A related subtheme is that this process of matching needs and capabilities often takes several years.

Two examples illustrate the times lags and unexpected outcome that frequently characterize new technologies and small firm growth. One example is Savi Technology's well-known success with radio identification devices, now widely employed in both commercial and defense markets. It was originally intended as a product that would be installed in children's shoes as a way for parents to monitor their location. A second example is Starsys Research Corporation's success as both a performer of R&D and manufacturer of components for launch release systems and satellite capture systems, with diversified funding from the Air Force, MDA, NASA, and the commercial sector. These accomplishments are based on a core technology for nonelectronic thermal control systems originally aimed at the commercial water heater market. Thermacore International's core technology involves the conversion of heat to electricity. Started in 1970 at the time of the shift in national energy and environmental policy from nuclear to solar power, the firm worked on developing heat pipe technologies for solar applications under a series of SBIR and non-SBIR contracts from DoD, NASA, and DoE. As the market for personal computers grew, so did the importance of finding solutions to dissipating the computers' internal heat. The marketplace thus created new uses for the firm's technology, allowing it to quickly expand into a major component supplier to HP, Dell, IBM, and Sun.

#### 4.4.5.5 The Importance of the DoD SBIR Program for the Participating Firms

The single overriding theme that emerges from the case histories is the importance assigned by these firms to the DoD SBIR program. Beyond its contribution to the firm's history, SBIR is seen by the companies as a beneficial national investment strategy in technological innovation, in the birth and growth of small, high-technology firms, and in enhanced national defense capabilities. These assessments are independent of whether or not the firm is currently eligible to submit an SBIR proposal, whether or not it currently has an SBIR award, the absolute and relative dollar amount of SBIR revenues to total firm revenues, and whether or not the firm operates within defense or nondefense markets. Indeed, some of the strongest statements on behalf of the program emanate from firms that are no longer eligible for the program or whose growth and current prosperity is based on DoD awards made a decade ago or more. Looking back, these now-

**TABLE 4-16** SBIR Impacts on Company Growth

| | Number of Responses | Percent of Responses |
|---|---|---|
| Less than 25% | 132 | 29.5 |
| 25% to 50% | 100 | 22.4 |
| 51% to 75% | 78 | 17.4 |
| More than 75% | 137 | 30.6 |
| Total | 447 | 100.0 |

NOTE: Data reported in Table 4-16 are for firms with at least one DoD award. NRC Firm Survey results reported in Appendix B are for all agencies (DoD, NIH, NSF, DoE, and NASA).
SOURCE: NRC Firm Survey.

successful firms credit the DoD SBIR program with providing critical infusions of funds in their formative period.

As reported in the case studies, limited access to capital to scale up production is often a significant hurdle to commercialization. Not unexpectedly, the primary sources of expansion capital reported are retained earnings, bank loans, angel capital, and venture capital, with a small amount of state government augments. Some firms report ongoing discussions with venture capital representatives about new infusions of funds to permit expansion. Other firms report less positive interactions with the venture capital community. Some are too new, too small, or too early in the product development process to have elicited any interest from investors in the venture capital community, hence the importance of SBIR awards as a validation of the technology and, indirectly, of the firm's potential.

### 4.4.6 SBIR and Firm Growth

Estimates about the effect of the SBIR program as a whole (and individual projects in particular) on company growth must involve an element of judgment and one best provided by the firm itself, survey respondents did provide their own estimates of the SBIR program's impact on their development (see Table 4-16).

Almost half of respondents (48 percent) indicated that more than half of the growth experienced by their firm was directly attributable to SBIR. This too is evidence of the powerful influence of SBIR on the development trajectories of firms winning SBIR awards.

## 4.5 SBIR AND THE EXPANSION OF KNOWLEDGE

Quantitative metrics for assessing knowledge outputs from research programs are well known, though far from comprehensive. Patents, peer-reviewed publications, and, to a lesser extent, copyrights and trademarks are all widely used metrics. They are each discussed in detail below. However, it is also important to understand that these metrics do not capture the entire transfer of knowledge involved in programs such as SBIR. Michael Squillante, Vice-President for Research at Radiation Monitoring Devices, Inc., points out that there may be very substantial benefits from the development and diffusion of knowledge that is simply not reflected in any quantitative metric:

> For example, our research led to a reduction in the incidence of stroke following open-heart surgery. Under an NIH SBIR grant we developed a tool for medical researchers who were examining the causes of minor and major post-operative stroke occurring after open-heart surgery.[69]

It is therefore critically important to understand that the quantitative metrics discussed below are an *indicator* of the expansion of knowledge. They reflect that expansion *but do not entirely capture it.* In particular, they say little about the impact of knowledge which generated no patent, no commercial sales, and no impact on the company's bottom line. As can be seen from Squillante's example above, some of these unquantifiable technological developments can remarkably improve outcomes for other actors (like open-heart surgeons).

**BOX 4-7**

"Results! Why, man, I have gotten a lot of results. I know several thousand things that won't work."

Thomas A. Edison

In addition, there is also strong evidence within the literature for the existence of what have been called "indirect effects"—spillover effects that are not captured within the context of a single project or even a single company, but may nonetheless make an important contribution to the field. Even if a project fails to reach the market, and is eventually shuttered, knowledge gained can be important in several ways: it can help other companies (and DoD itself) avoid technological dead ends (as exemplified by Edison's quote above); it can create knowledge that is then used in subsequent projects, inside or outside the company; it can expand human capital, by helping the PI to learn more; and it can support the transition from a solely scientific orientation to one with more commercial understanding. None of these important effects can be easily captured in a quantitative analysis using the currently available tools.

Many commercial uses of new knowledge are discussed above in the sections on commercialization. But new knowledge may also be made a public good via open publications and presentations, even when such dissemination could

[69]Michael Squillante, testimony presented to the NRC research team, June 11, 2004.

**TABLE 4-17** Patents, Copyrights, Trademarks and/or Scientific Publications

| Number Applied For/Submitted | | Number Received/Published |
|---|---|---|
| 836 | Patents | 398 |
| 71 | Copyrights | 62 |
| 211 | Trademarks | 176 |
| 1,028 | Scientific Publications | 990 |

SOURCE: NRC Phase II Survey.

limit the firm's ability to later claim intellectual property. Limits on the scope of patents and multiple forms of "leakage' of proprietary knowledge may also lead to spillover benefits, as others employ the new knowledge without paying the inventing firm for its use.[70]

As shown in Table 4-17, the 816 firms responding to this question on the NRC Phase II survey applied for 836 patents, and had, at the time of the survey, received 398. These firms reported having published 990 related scientific publications, with an additional 38 under review. In addition, 593 Phase I awards that did not directly result in a Phase II generated 108 patents, with eight applications in review, and 157 scientific publications, with an additional four under review.[71]

Of the 564 DoD respondents that answered this question, 42 percent indicated zero patents, and 28 percent reported receiving one or two patents. The average number of patents per firm was 3.5. A few firms invested heavily in patenting SBIR innovations, however. Slightly less than ten percent of the firms reported ten or more patents, and four firms reported receiving 50 patents or more.[72] The activities of the 25 firms included in the case study portion of the study suggest the probability of sizeable increases in coming years in the number of patents received by SBIR firms.

The absence of comparison groups as well as case study findings point to the need to exercise great care in interpreting these survey data. Lacking a comparative yardstick that provides patent or publication data on other, comparable firms, it is not possible to say whether the reported numbers for SBIR awardees are high or low.[73]

Patents are seen as indispensable forms of intellectual property rights protec-

---

[70]For an informed discussion of this phenomenon, see A. Jaffe, *Economic Analysis of Research Spillovers: Implications for the Advanced Technology Program*, NIST GCR 97-708, Gaithersburg, MD: National Institute of Standards and Technology, 1996.

[71]NRC Phase I Survey.

[72]These data indicate the tendency of firms which have received DoD SBIR awards, to patent, not the number of patents exclusively tied to DoD only awards. Some number of these reported patents are likely attributable to SBIR/STTR awards from other agencies.

[73]Examples include number of patents per million dollars of sales, or number of patents per number of SBIR awards, to gauge the effectiveness of the SBIR program across agencies.

**TABLE 4-18** Projects Reporting Patent Applications and Awards

| | Projects with Patent Applications | | Projects with Patents Awards | |
|---|---|---|---|---|
| | Number | Percent | Number | Percent |
| Yes | 281 | 34.4 | 205 | 25.1 |
| No | 535 | 65.6 | 611 | 74.9 |
| Total | 816 | | 816 | |

SOURCE: NRC Phase II Survey.

tion for most of the case study firms. Many SBIR awardees generate products and services that are inputs into the larger weapons systems being managed by large defense contractors. Without strong patent protection, firms are wary about entering into subcontracting relationships with prime contractors. They fear that the larger firm will imitate, reverse engineer, or preemptively patent functionally equivalent technologies, thus eliminating the SBIR firm's major or only market.[74]

However, some case study firms are resource constrained in pursuing an active patent protection strategy. Other firms see patents as a limited, relatively ineffective or inefficient form of intellectual property protection given their technology domain or the size of the prospective market. Thus, some firms do not pursue patents, relying instead on trade secrets and know-how to protect their intellectual property. Choice of this strategy is influenced in part by the assessment that patents provide little economic benefit if the dominant customer of the firm's product is the U.S. government, which under SBIR is entitled to royalty-free use of resulting technology. Also, this strategy avoids the costs of patenting and the associated public disclosure of related proprietary technological knowledge.

### 4.5.1 Patents

As noted above, SBIR awards are generally a significant patent stream. Table 4-18 reports findings on patents and publications, as well as other forms of intellectual property rights protection, received by NRC Phase II Survey respondents as a result of Phase II awards. The NRC Phase II Survey data indicate that about one-fourth of respondents received patents related to the relevant

[74]The *Night Vision Corp. v. The United States of America* patent case filed before the United States Court of Federal Claims represents a variation on these concerns. In this case, Night Vision sued the United States claiming that the Air Force shared with another firm, Insight, prototypes that it had developed, and that Insight had then disassembled the prototypes, violating its SBIR data rights. See U.S. Court of Federal Claims, No. 03-1214C. 25 May 2005. See also U.S. Court of Appeals for the Federal Circuit, 06-5048, November 22, 2006.

SBIR-funded project, and about 34 percent of projects generated at least one patent application.

These data are similar, though slightly lower, than those reported at NIH. They indicate that at a minimum, one-third of projects generated knowledge that was judged by the firm to be sufficiently unique and commercially important to be worth the significant expense of patent filing. And a quarter of all projects reported that government examiners agreed and awarded at least one patent.

This is a very significant finding. It can be viewed as addressing the tip of the knowledge iceberg. Only a small percentage of the knowledge generated during a research project meets the relatively stringent tests indicated above. Most research outcomes are not sufficiently unique to qualify for patent protection. And even unique knowledge must pass formidable internal hurdles before patent protection is sought, as the process is expensive and time-consuming.

Most companies file only one project-related patent. However, a few file many, as shown by Figure 4-15.

Because it is sometimes assumed that smaller firms have more limited access to the funds and expertise necessary to file patents, we examined the relationship between patent filing and firm size (see Figure 4-16).

By developing a "patenting ratio"—the ratio of firms with at least one patent awarded to all responses, by size of firm—the data seems to show that projects at firms with more than 75 employees are less likely to generate patents than those with less than 75 employees. The former generate 0.23 patents per project, the latter more than twice as many, 0.56, although this may be in part because larger firms are more likely to engage in manufacturing and other activities in addition to R&D. It may also reflect differences among sectors—for example, between information technology and less patent prone sectors.

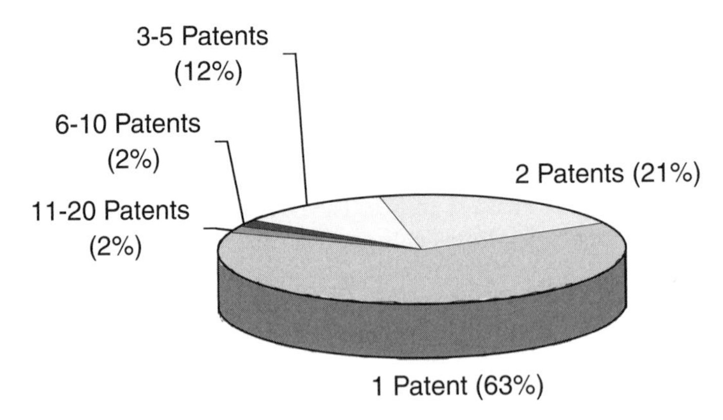


**FIGURE 4-15** Distribution of patenting activities—Responding projects.
SOURCE: NRC Phase II Survey.

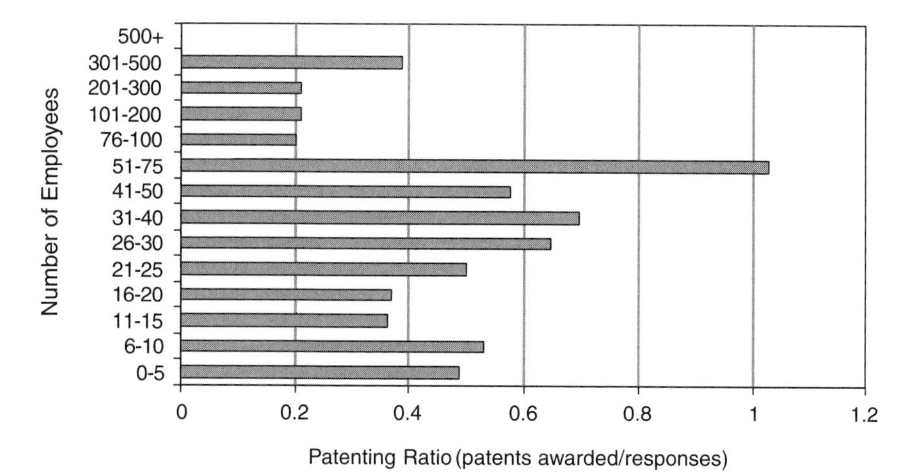


**FIGURE 4-16** Patents and firm size at time of survey.
SOURCE: NRC Phase II Survey.

This is somewhat counterintuitive, as larger firms tend to have more resources, and patenting is expensive. Case studies do suggest that for these larger firms with better access to financial resources, SBIR projects tend to be deployed on research that is not in the company's core strategic plan, and is therefore less likely to be seen as having immediate commercial benefit (necessary to attract the internal resources for patenting).

Analysis of the scientific importance of the patents listed has not been possible, as the patents themselves were not disclosed in the course of the survey.

#### 4.5.1.1 Knowledge Generation vs. Commercialization?

It has also been suggested that there might be a disconnect between research-oriented firms and commercially oriented firms, and that SBIR programs should be adjusted to focus more effectively on the latter. This is—partly—the purpose of including the CAI in SBIR proposal reviews.

The NRC Phase II Survey can help to test the hypothesis that commercially oriented companies were focused on areas different from research-oriented firms. If this hypothesis was correct, we would anticipate a distinction between firms that report IP-related activities (filing patents, copyrights, and trademarks, and seeking publications) and those undertaking marketing activities (preparation of marking plans, hiring marketing staff, etc.).

In fact, Figure 4-17 shows that no such distinction is observed. 40 percent of respondents report both IP- and marketing-related activities, and a further 22 percent report neither. About 28 percent appear to fit the "commercially oriented"

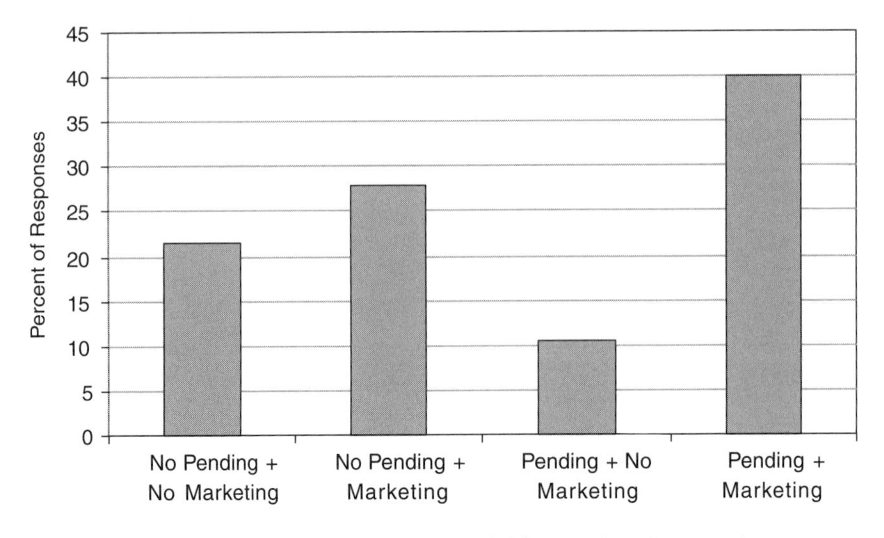


**FIGURE 4-17** Patenting and marketing activities tend to be complements not substitutes.
SOURCE: NRC Phase II Survey.

model with no IP activities, and just over 10 percent fit the "research-oriented" model with no marketing activities.

It thus appears that the SBIR program has been quite successful in encouraging firms engaging in early-stage research to focus on the commercial applications that might be drawn from that work. Two-thirds of the respondents in this sample report some specific marketing activities related to the project.

### 4.5.2 Scientific Publications

Publications fill two important roles in the study of SBIR programs. First, they provide an indication of the quality of the research being conducted with program funds. More than half of the DoD-funded projects were of sufficient value to generate at least one peer-reviewed publication. Second, scientific publications are themselves the primary mechanism thorough which knowledge is transmitted within the scientific community.

The existence of papers based on SBIR projects is therefore direct evidence that the results of these projects are being disseminated widely, which in turn means that the congressional mandate to support scientific outcomes is being met.

Unlike NIH, where scientific publication is at the core of the enterprise for both the agency and the investigators (who are overwhelmingly drawn from academic environments), publication is not always viewed as an unmixed blessing in the DoD environment. Even where founders had advanced degrees, their professional careers in the military and industry had moved them away from emphasizing peer-reviewed publications as a mode for establishing priority or disseminating research findings. More common in the security industry are presentations at professional meetings, or briefings with sponsors and users. Knowledge in general seems to be viewed more pragmatically and commercially, being released publicly in some cases but not in others.

Bearing these points in mind, considerable scientific publication still comes from DoD-funded SBIR research. Out of the 816 projects responding to the relevant question in the NRC Phase II Survey, 348 (42.6 percent) reported at least one scientific peer-reviewed publication related to the project, and some reported many more (see Figure 4-18).

This compares with 53.5 percent at NIH. About 15 percent of DoD projects with publications had published a single paper, but one company had published 114 papers on the basis of its SBIR, two others had published at least 100, and two more had published between 50 and 100 papers.

These data fit well with case studies and interviews, which suggested that SBIR companies are proud of the quality of their research, and justifiably so given their success in publication. Publications are featured prominently on many company Web sites, and companies like SAM—among many others—made a point during interviews that their work was of the highest technical quality as measured in the single measure that counts most in the scientific community, peer-reviewed publications.

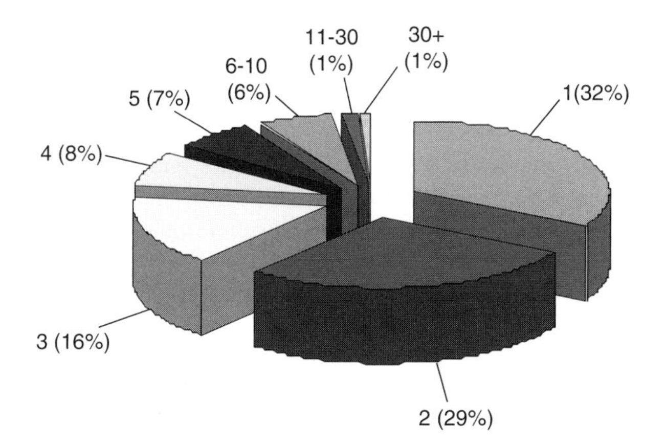


**FIGURE 4-18** Project-related scientific peer-reviewed publications.
NOTE: This figure provides a breakout of the number of publications among those that reported at least one scientific peer-reviewed publication.
SOURCE: NRC Phase II Survey.

### 4.5.3 SBIR and the Universities

There is anecdotal evidence from case study interviews and papers presented at the 2004 Technology Transfer Society Conference in Albany that university views about SBIR have begun to change.[75] Traditionally, universities have been wary of SBIR, partly because it might "distract" faculty from the pure pursuit of knowledge, and partly because it may lure faculty away from academic work altogether into commercial research. More fundamentally, SBIR has been seen as diverting resources from the other activities of research agencies—notably NSF and NIH—whose funds would likely find their way to universities or university researchers.

This view is changing. As universities themselves have become much more concerned about the commercial side of research—witness the dramatic increase in university patenting and licensing activities reported by NSF and the widespread introduction of IP-related clauses into faculty contracts—they have begin to see SBIR as a positive source of funding for research. Universities are increasingly looking toward mechanism of technology transfer as they increase their ties to their surrounding locales. As a part of this effort, many universities now make a conscious effort to inform faculty about SBIR, even helping faculty put together SBIR applications.

A quarter of projects indicated that there had been involvement by universities' faculty, graduate students, or the university itself in developed technologies. For the response to a follow-up question on the variety of relationships this encompasses, see Table 4-19.[76]

The wide range of roles played by university staff and students indicate once more the multiple ways in which SBIR projects increase the knowledge base of the nation. Involvement in these projects provides different opportunities for university staff than those available within the academy.

The results of this change in perspective were certainly indicated in the course of case studies, where a number of SBIR recipient firms indicated the importance of their ties to universities.

These stories suggest that the flow of information and funding between small businesses and universities working within the SBIR framework is neither simple nor unidirectional. The constant flow of feedback, testing, and insights between university researchers and staff at SBIR awardees helped to move those companies forward toward product deployment into new research areas.

One further impact of SBIR has been to facilitate transitions of both tech-

---

[75] See <*http:///www.t2society.org*>.

[76] See questions 30 and 31 in the NRC Phase II Survey, presented in Appendix B. Of the 837 respondents to these questions, 25 percent responded in the positive (and 75 percent in the negative) to question 30 as to whether there was any involvement by university faculty, graduate students, and/or university developed technologies in executing this award. Responses to Question 31, shown in Table 4-19 address any relationships between the respondent's firm on the Phase II project being surveyed the same 837 respondents who answered Question 30, not just those who answered "yes."

**TABLE 4-19** University Involvement in SBIR Projects

| | |
|---|---|
| 1.3% | The Principal Investigator (PI) for this Phase II project was at the time of the project a faculty member. |
| 1.3% | The Principal Investigator (PI) for this Phase II project was at the time of the project an adjunct faculty member. |
| 13.6% | Faculty member(s) or adjunct faculty member(s) work on this Phase II project in a role other than PI, e.g., consultant. |
| 11.4% | Graduate students worked on this Phase II project. |
| 9.2% | University/college facilities and/or equipment were used on this Phase II project. |
| 2.2% | The technology for this project was licensed from a university or college. |
| 3.9% | The technology for this project was originally developed at a university or college by one of the percipients in this Phase II project. |
| 12.5% | A university or college was a subcontractor on this Phase II project. |

NOTE: Survey respondents could check more than one category.
SOURCE: NRC Phase II Survey (n = 837).

nologies and researchers from university labs to the commercial environment. Data from the NRC Firm Survey (using data for all agencies) strongly support this hypothesis, with 66 percent of SBIR companies including at least one academic as founder, and 28 percent having more than one academic as a founder (see Figure 4-19). The same survey found that about one-third of founders were most recently employed in an academic environment before founding the new company.

These data and evidence from case studies strongly indicate that SBIR has indeed encouraged some academic scientists to work in a more commercial environment.

What is not clear from this research is the extent to which universities themselves see SBIR as a mechanism for technology transfer, commercialization, and additional funding for university researchers. These questions are beyond the scope of the current study but merit additional research.

### 4.5.4 Inventions and Indirect Knowledge

This analysis has understandably focused on the data available from the NRC Phase II Survey about the IP-related activities of firms. However, it must be stressed that these are only the *formal* IP-related activities. Every project generates a very wide range of less formal, less easily captured knowledge effects, which are nonetheless important despite being very difficult or even impossible to quantify. The case of Thermacore, described in Box 4-9, provides some insights into this kind of program effect.

**BOX 4-8**
**Brimrose Corporation**

Brimrose Corporation was founded by Dr. Ronald G. Rosemeier while still a Ph.D. student in Johns Hopkins University material science program. After graduation, working as a post-doctoral student at the University of Maryland, he started writing SBIR proposals.

After three years of submitting unsuccessful proposals, he was awarded four Phase I's, whose total value approximated $200,000. He started hiring his first employees and began applying for loans at banks. Because banks were not willing to give him loans backed by the SBIR awards, he amassed charges of $100,000 on credit cards. Six months later he wrote the follow-on Phase II proposals, receiving awards on three of them for a total of $1.5 million.

With these funds, he was able to hire additional employees, and expand operations. At that time 10 percent of the firm's revenues were from commercial sales (selling X-ray imaging at tradeshows) and 90 percent from the SBIR awards. As the firm started commercializing new products, this percentage shifted to 80 percent commercial revenue and 20 percent SBIR revenue. Most of the R&D team and few of the support staff were hired under SBIR related activities. Brimrose began operations with 6 employees; by 2005 its employment level had reached approximately 60 employees. Overall, throughout its history, Brimrose has received 65 Phase I and 28 Phase II SBIR awards.

The firm's major lines of business are industrial process control spectroscopy in the pharmaceutical and petrochemical industries, nondestructive testing and evaluation and novel opto-electronics devices. Its business model is to specialize in applied R&D. A few of the SBIR programs have directly resulted in commercial products but most have led to product improvements.

The firm's commercialization strategy emerged from and has been greatly enhanced through its participation in SBIRs. The Phase I and Phase II SBIR funding allowed them to determine the feasibility of new technology and develop it to the point of prototype development without allocation of significant internal resources. Following prototype development, Brimrose used internal funds from previous commercial sales to bring the technology to the point of commercial availability. Thus, the SBIR funds lowered the company's financial burden by decreasing the risks associated with new technology development.

## 4.6 UNDERSTANDING OUTCOMES: EMPIRICAL FINDINGS

Overlapping methods were used to address questions relating to commercialization of DoD's SBIR research. First, a Web-based survey was conducted of firms that had received at least one Phase II award between the years 1992–2001.[77] Second, case histories were prepared of a sample of firms that had received Phase II awards. Thirty-one firms representing a cross-section of DoD sponsoring agencies and states were selected for study. Third, an additional Web-

[77]Appendix B contains a full account of the survey methodology.

**BOX 4-9**
**Thermacore: Creating Knowledge and Capacity**

Many of the benefits to participants and their clients are hard to quantify. Thermacore, for example, believes that its experiences under the SBIR program provided it with a "brain pool" of "know-how" related to manufacturing reliability. These tacit skills have contributed to the firm's ongoing competitive position even as patents on its initial core technologies have expired.

Thermacore was founded by Yale Eastman, an RCA employee, in 1970. The firm started as a "garage" start-up, focused on RCA-abandoned heat pipe technologies for solar applications. Throughout most of the 1970s, the company remained small, with no more than ten employees, working on industry and government R&D contracts. In the 1980s, it began to grow via non-SBIR and SBIR R&D contracts primarily from NASA, DoD, and DoE.

**Thermacore and SBIR.** Thermacore began active pursuit of SBIR awards in the early 1980s. While it remained eligible, Thermacore received 82 SBIR awards from several government agencies, including DoD, NASA, and DoE. This substantial and repeated support, provided by several agencies over a number of years, highlights the way in which SBIR supports technologies that are complex and require a number of incremental technological advances to transition an R&D concept into a viable commercial product.

Thermacore describes its growth as a case of the marketplace catching up to its technology. In the early 1990s, Thermacore was approached by Intel to discuss the possibility of mass producing its heat pipe technology for use in the rapidly growing market for personal computers. With financial support from a venture capital firm, Thermacore took the risk of setting up a production line before receiving orders. Subsequently, it received large orders from several major computer manufacturers, such as HP, Dell, IBM, and Sun.

Reflecting its transition from an R&D to a production-oriented firm, contract R&D projects and OEM work now account for only 6 percent of revenues; 94 percent comes from sale of commercial products.

After its sale in 2001, following the retirement of its founder, Thermacore became ineligible for SBIR awards because it was a wholly owned subsidiary of a larger company. It does however continue to do some SBIR-funded research as a subcontractor to small firms conducting Phase I and II research.

based survey addressed Phase I awards. Additional data were collected from the DoD CAI index database.

Collectively, these methods strongly suggest that DoD companies funded by SBIR do commercialize their results, and that the rate of commercialization appears to be increasing. Thus, the charge of low commercialization by multiple-award winners no longer appears to be supported by the data.[78] If it even was a

[78]The 1992 GAO study that identified this "trend" is often cited by the GAO investigators, multiple caveats are not. For example, the limited time frame of the study for several of the years studied, e.g., 1987 awards reviewed in 1990. See U.S. Government Accounting Office, *Federal Research: Small Business Innovation Research Shows Success but Can Be Strengthened*, GAO/RCED-92-37, op. cit.

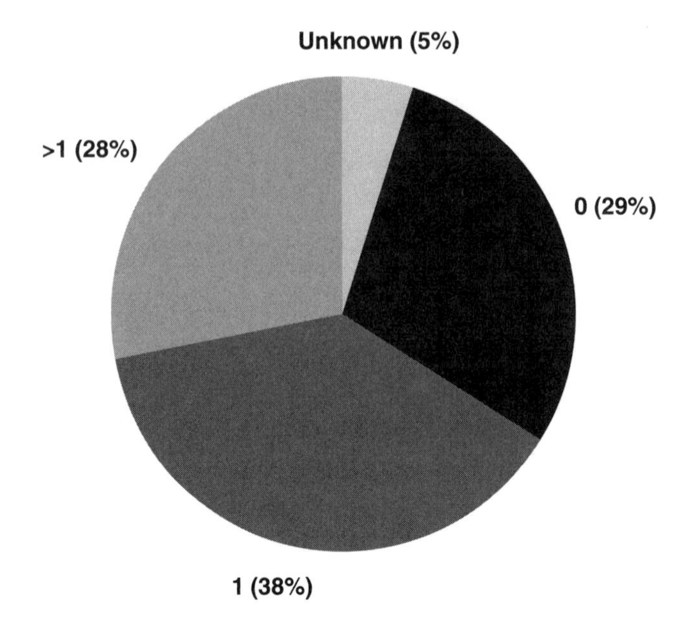


**FIGURE 4-19** Number of academics as founders.
SOURCE: NRC Firm Survey.

problem in the past, it is now to a considerable extent resolved, with some very large winners graduating from the program in various ways (e.g., Foster-Miller's purchase by a foreign firm, which made it ineligible) and also clear evidence that some multiple-award winners are now reducing their dependence on SBIR, and to generate more substantial commercial results (e.g., Radiation Monitoring, PSI).

The findings from the three approaches are generally consistent with one another and complementary. The case studies provide explanations for findings generated via the two sets of surveys. The findings from the separate approaches at times though tend to emphasize different aspects of SBIR's program impacts mainly because they frame the question of commercialization differently. The survey findings, in keeping with congressional and Executive Branch usage, define commercialization in terms of economic outcome measures, such as sales, investment, employment, and patents. The case histories contain data on these variables, but also narratives about firm formation, business strategies, intellectual property rights strategies, and processes of technological innovation and diffusion of innovations. They also contain limited data on the gestation processes connecting SBIR awards and commercial outcomes.

The review of outcomes described in this chapter provides an overview of an effective program, meeting a wide variety of program goals. The DoD awards are generating new knowledge evidenced in publications and successful patenting activity, new products and processes to maintain equipment, improve supply management, and develop new products to better support and defend those charged with combat missions. The program also provides valuable linkage between university professors and students and the commercial and defense market place. By growing and nurturing the defense industrial base, the program is also encouraging high-tech entrepreneurship, thereby increasing innovation, encouraging competition, and offering greater choices. The SBIR program is helping DoD to meet the new and often sudden challenges of a turbulent world.

# 5

# Phase III Challenges and Opportunities

## 5.1 CHARACTERISTICS OF PHASE III

Phase III is defined in the authorizing legislation as commercialization of SBIR technologies beyond Phase II. It differs from Phase II in that the set-aside SBIR funding may not be used for Phase III; funding for this phase must come from elsewhere in agency budgets, or from nongovernmental sources.

At DoD, Phase III is especially important because it encompasses two of the primary objectives of the program: commercialization, and—more importantly to DoD—the transition of technologies from SBIR projects into DoD acquisition

**BOX 5-1**
**Definition of Commercialization**

"Commercialization is *the process* of developing marketable products or services and producing and delivering products or services for sale (whether by the originating party or by others) to government or commercial markets. A 'Phase III' is work that derives from, extends, or logically concludes effort(s) performed under prior SBIR funding agreements (Phase I & II). Phase III contracts are not SBIR funds and may be for products, production, services, additional R/R&D, or any combination that is funded by the government, defense or nondefense commercial vendors, or individuals."

---

SOURCE: U.S. Small Business Administration, *SBIR Final Policy Directive*, September 2002.

programs. Small businesses are critically important to technology development at DoD. According to Michael Caccuitto, DoD SBIR program administrator, after assessing 255 industrial capabilities, DoD concluded that 36 percent of the companies with relevant products have less than 100 employees.[1]

"Phase III" is not funded by any line item. It is a phrase that describes post-Phase II commercialization or agency acquisition of SBIR-sponsored technology. There is no formal program or budget for "Phase III."

In the early years of the SBIR program, Phase III was not a very high priority. SBIR topics were defined and awards were made largely in line with the interests and activities of the wider R&D programs—for example, the Army Research Labs. During the 1990s, following the renewal of the program, growing pressure from Congress, and changes in priorities of the leadership in the Pentagon, gradually shifted the SBIR program's emphasis toward serving the warfighter more directly, and specifically to the issue of Phase III.

### 5.1.1 Congress

Over the past fifteen years, Congress has repeatedly directed SBIR programs generally, and DoD in particular, to emphasize commercialization and to promote the use of SBIR-sponsored technologies in acquisition programs.

Congress has considered the Phase III component of SBIR at the time of each reauthorization. In 1992 the SBIR Reauthorization[2] increased the emphasis on commercialization. In 1999, Sec 818 of Defense Authorization Act required "favorable consideration [for SBIR projects] in acquisition planning process."[3] More recently, the 2005 Defense Authorization Act, House Armed Services Committee (HASC) report "directs USD (AT&L) to encourage DoD acquisition managers and prime contractors to make significantly more SBIR Phase III contract awards . . . and to report on DoD Phase III contracts during last three years."[4]

The 2002 SBIR law reauthorization directed the SBA to strengthen SBIR guidelines by mandating Phase III commercialization "whenever possible."

House Report 108-491 accompanying the National Defense Authorization Act—FY2005, directed the Under Secretary of Defense for Acquisition, Technology & Logistics to encourage acquisition program managers and prime contractors to make significantly more SBIR Phase III contract awards and to report to the congressional defense committees on actions taken by March 31, 2005.

---

[1]Presentation by Michael Caccuitto, DoD SBIR/STTR Program Administrator, at National Research Council Symposium on *SBIR: The Phase III Challenge*, June 14, 2005.

[2]PL 102-564.

[3]Sec 812 of the 2000 Act, House Report 106-244, and Senate Report 106-50 all emphasized increased use of Phase III contracts by acquisition programs.

[4]Presentation by John Williams, Navy SBIR Program Manager, October 15, 2005. Available at *<http://www.onr.navy.mil/about/conferences/rd_partner/2005/docs/past/2004/2004_williams_navy_tap.pdf#search=%22Navy%20primes%20initiative%22>*.

These efforts highlight the longstanding interest in Congress in the success of Phase III at DoD, and the consistent congressional encouragement to the Defense Department's management to take the steps needed to support this phase of the program.

## 5.2 PHASE III OUTCOMES

The 2005 symposium on the Challenge of Phase III Commercialization at the National Academy of Science, arranged as part of this study, was the first gathering of programs officers, small businesses, prime contractors, and researchers focused specifically on Phase III issues. Many of the comments at the meeting highlighted successes but also the difficulties that different actors had with Phase III transitions.

As with many aspects of the program, data on Phase III activities are very limited. DoD analysis is focused almost exclusively on reporting via the DD350 form—a form completed by contracting officers for all RDT&E contracts at DoD. The form has a check box to indicate that the project in question is a Phase III, or results from a Phase II (see definition above). Data from the DD350 suffer from serious deficiencies. For example, contracting officers are often unaware that a contract is a Phase III. In other cases, there may be insufficient emphasis on careful reporting. As a result, DD350 data tends to undercount the real number of SBIR-related RDT&E contracts.

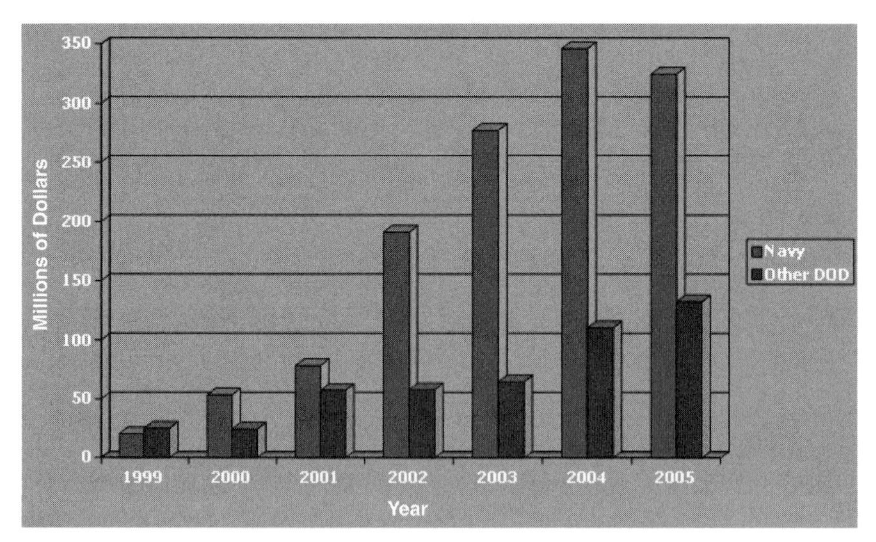


**FIGURE 5-1** Phase III awards total in millions of dollars, 1999–2005.
SOURCE: John Williams, Navy SBIR Program Manager, April 7, 2005.

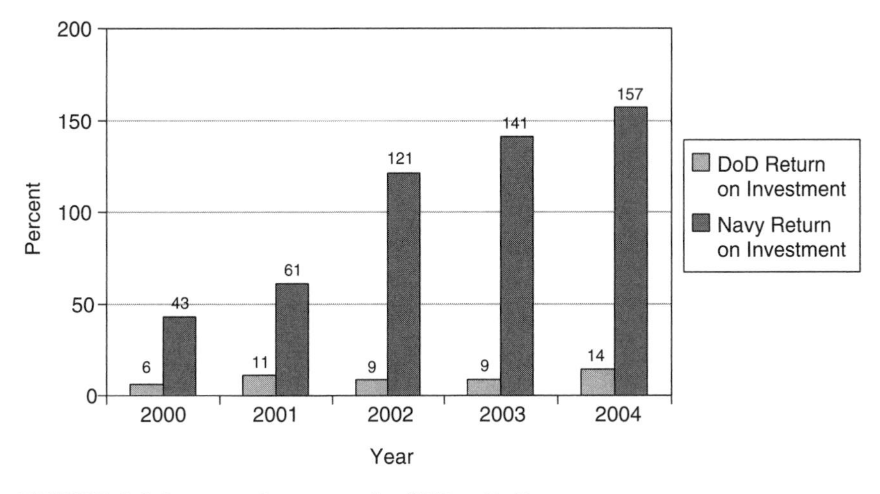


**FIGURE 5-2** Return on Investment for SBIR at DoD.
NOTE. The chart only includes Phase III dollars from DD350 for that fiscal year. Phase III funds that go to the firm indirectly via prime contractors or funding that is not marked as Phase III on DD350 are not included. (Return on Investment = Phase III dollars divided by Phase I-II dollars.)
SOURCE: John Williams, Navy SBIR Program Manager, April 7, 2005.

Nonetheless, the DD350 data do show that the amount of Phase III contracts generated have been climbing steadily in recent years, particularly at the Navy.

According to these data, the Navy accounted for about 70 percent of all DoD Phase III contracts in FY2005 (with PEO SUBS accounting for about 86 percent of Navy's total). Navy's Phase III awards started to grow very rapidly in FY2002, and continued to grow until a slight decline in FY2005. However, it is worth noting that these substantial results are based on a relatively low number of actual Phase III awards. These data are also reflected in Navy efforts to calculate the return on investment for SBIR funding by dividing Phase III awards by the total of Phase I and Phase II funding (see Figure 5-2).

For DoD as a whole, Michael Caccuitto, DoD SBIR program administrator, also noted that the amount of commercialization generated from SBIR projects leads the total amount spent on SBIR, with about a 4-year lag.

For DoD, with its focus on getting technology into production for use at DoD, the distribution of commercialization is also important. The DoD data in Figure 5-4 indicate the distribution of Phase III sales by sector and show that while there is a strong focus on DoD and the prime contractors is unsurprising, more commercial activity occurs with the private sector outside DoD. Only 44 percent of Phase III contracts can be attributed to DoD and DoD primes. Forty-seven percent comes from the private sector.

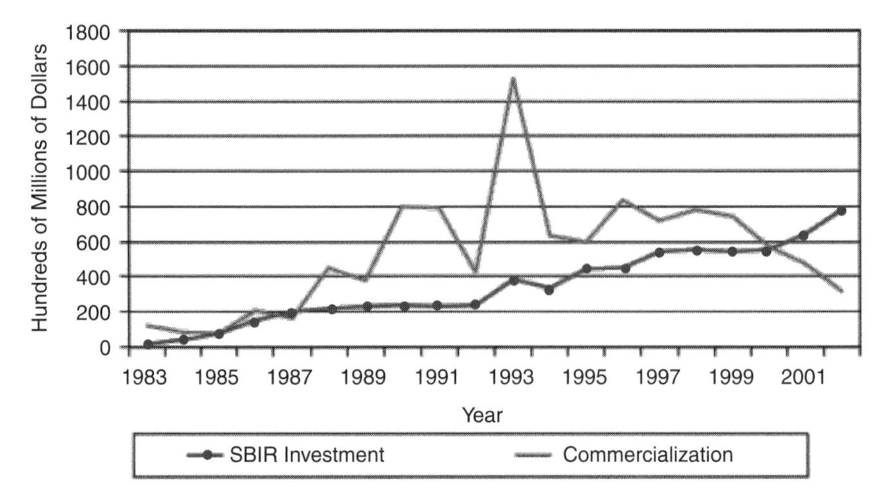


**FIGURE 5-3** Reported commercializations vs. SBIR budget.
SOURCE: Michael Caccuitto, DoD SBIR/STTR Program Administrator, Presentation to SBTC SBIR in Rapid Transition Conference, September 27, 2006, Washington, DC.

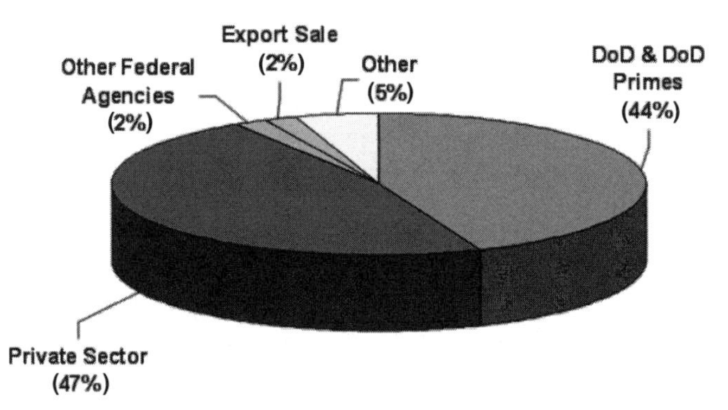


**FIGURE 5-4** Distribution of Phase III sales.
SOURCE: Michael Caccuitto, DoD SBIR/STTR Program Administrator, Presentation at NRC Conference *SBIR: The Phase III Challenge*, June 14, 2005, Washington, DC.

## 5.3 PHASE III OPPORTUNITIES AND NEEDS

From an agency perspective, SBIR offers important and unique opportunities, which should be reflected in a strong and growing Phase III program. In particular,

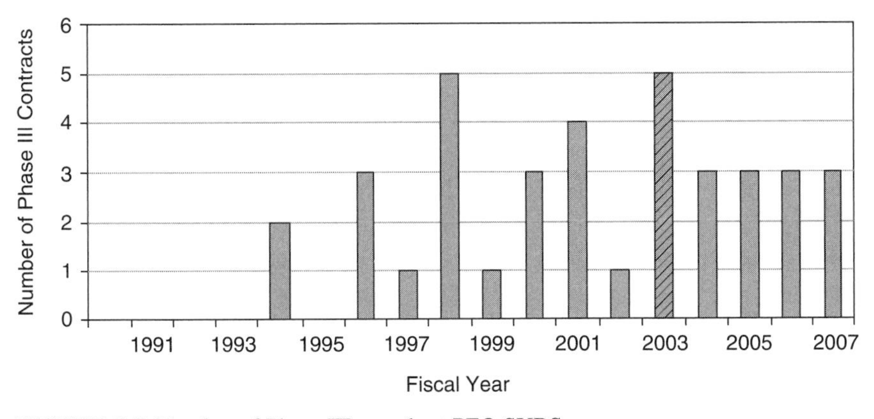


**FIGURE 5-5** Number of Phase III awards at PEO SUBS.
SOURCE: Richard McNamara, PEO SUBS, Presentation to SBTC SBIR in Rapid Transition Conference, Washington, DC, September 27, 2006.

- **Flexibility.** SBIR offers an unusual degree of execution year flexibility, unlike most RTD&E accounts which have to be described in detail in the President's budgetary message.[5]
- **Shorter Planning Cycle.** SBIR allows a much shorter planning horizon. Most R&D programs at DoD had to be planned years ahead of the budget cycle. And in some cases, agencies have taken advantage of that flexibility. The Navy issues a "quick response IED topic" in 2004, and had made 38 Phase I awards within 5 months of topic development. These have developed into 18 Phase II awards, and results from those were to be available in 2006–2007. The first prototypes were expected in Iraq in fall 2006.
- **Faster Development Time.** Products and services from SBIR can often be developed within a relatively short time frame. In the view of John Williams, Navy SBIR Program Manager, the notion that it takes 5–10 years to commercialize most technologies—and that Phase III results could take 5–10 years—is a myth. He argues that, for the Navy, if some Phase III funding (sales or further development funding) is not achieved by two years after the end of Phase II effort, the probability of a Phase III success is very low.[6]

The substantial increase in Phase III activity at Navy in recent years suggests that there may be room for similar increases elsewhere—both in other services and

[5]Presentation by Michael McGrath, Navy, at National Research Council Symposium on *SBIR: The Phase III Challenge*, June 14, 2005.
[6]John Williams, Navy SBIR Program Manager, April 7, 2005.

within Navy at some of the other commands. More than 86 percent of NAVSEA's Phase III awards are accounted for by SUBS.

Put another way, the substantial recent successes at Navy (and at SUBS in particular) would suggest that there are opportunities available elsewhere in the Department for similar levels of success if the senior management and the relevant Program Executive Officers are encouraged to identify and exploit those opportunities.

## 5.4 PHASE III CONCERNS

In theory, there would be a smooth flow of technology and funding from Phase II to Phase III and then into systems eventually adopted by the agencies for use by warfighters. In reality, this process is much more complex, requiring multiple champions at different phases in addition to effective management and product development by the SBIR firm. The process can, and does, work. There are important success stories. Nonetheless, there are substantial barriers that impede Phase II projects from successfully transitioning into Phase III.

To begin with, acquisitions officers have traditionally viewed SBIR more as a tax on their other research projects than an opportunity. This may be an inevitable result of the flow of funding generated by SBIR where the set-aside funding for SBIR draws more from the applied part of the technology development spectrum. Table 5-1 shows that 84 percent of SBIR funding comes from the acquisitions-dominated elements of the Navy development cycle (6.4–6.7) rather than from earlier in the technology development cycle.

Perhaps as a result of these attitudes, small businesses are still not as fully engaged in the work of defense acquisitions as they might be. Allocation of R&D funds from above remains quite centralized. The top 10 DoD contractors received 62 percent of DoD RDT&E funding in 2003, and the top 100 (including

**TABLE 5-1** Sources of Navy SBIR Funding

| Percent | | BA (Stage in Technology Development) | Title (Description of Level of Technology Development) |
|---|---|---|---|
| 16 | 3 | 6.1 | Basic Research |
| | 6 | 6.2 | Applied Research |
| | 7 | 6.3 | Advanced Tech. Development |
| 84 | 21 | 6.4 | Adv. Component Dev. Prototypes |
| | 51 | 6.5 | System Dev. And Demonstration |
| | 2 | 6.6 | RDT&E Management Support |
| | 10 | 6.7 | Operational System Development |

SOURCE: Navy FY 2003 SBIR Assessment.

one acquired small business) received 88.9 percent of 2003 RDT&E funding—up from 85.5 percent in 2001, according to the Small Business Technology Council (SBTC).[7] Conversely, according to a 2005 Small Business Administration report,[8] small businesses generated 60 to 80 percent of net new jobs annually over the last decade, employ 39 percent of high tech workers, produce 13 to 14 times more patents per employee than large patenting firms. Small businesses also account for a significant percentage of nonfederal expenditures of R&D.[9] And, according to DoD, an assessment of 255 industrial capabilities determined that 36 percent of the companies with relevant products have less than 100 employees.[10]

At the same time, DoD is strongly committed—on paper—to the integration of SBIR into acquisitions: the Interim Defense Acquisition Guidebook directly addresses use of SBIR technology in Sec. C2.9.1.5:

> The Program Manager shall develop an acquisition strategy that plans for the use of technologies developed under the SBIR program, and gives favorable consideration for funding of successful SBIR technologies. At milestone and appropriate program reviews for ACAT I programs, the PM shall address the program's plans for funding the further development and insertion into the program of SBIR-developed technologies.[11]

Comments made by all the stakeholders at the NRC Phase III Symposium, and in discussions with case study companies and DoD officers, all underline the problems and difficulties faced by companies in making the Phase III transition. Some of the more notable issues and concerns are discussed below.

### 5.4.1 The TRL Gap

The *Interim Defense Acquisition Guidebook*[12] includes the DoD Technology Readiness Level (TRL) table. This tool, derived from NASA practice, is the accepted means of classifying the maturity of technologies. As the TRL table in Box 5-2 shows, technologies must be at TRL 8 for effective transition into an acquisition program system. TRL 3-5 is common for DoD SBIR technologies at

---

[7]Small Business Technology Coalition, *Fighting an Unconventional Enemy*, January 20, 2005.

[8]SBA Office of Advocacy (2005) data drawn from U.S. Bureau of the Census; Advocacy-funded research by Joel Popkin and Co. (Research Summary #211); Federal Procurement Data System; Advocacy-funded research by CHI Research, Inc. (Research Summary #225); Bureau of Labor Statistics, Current Population Survey; U.S. Department of Commerce, International Trade Administration.

[9]Several of these issues are discussed in Robert-Allen Baker, "Incentives and Technology Transition: Improving Commercialization of SBIR Technologies in Major Defense Acquisition Programs," SBTC White Paper, Washington, DC, September 21, 2005.

[10]Michael Caccuitto, Department of Defense SBIR Program Manager, April 2005.

[11]USD (AT&L), *Interim Defense Acquisition Guidebook*, October, 2002, p. 46. The guidebook has now become an online decision support system, at <*http://akss.dau.mil/dag/DoD5000.asp?view=document*>. It was last updated December 16, 2004.

[12]USD (AT&L), *Interim Defense Acquisition Guidebook*, October, 2002, p. 41.

**BOX 5-2**
**TRL Definitions**

**Technology Readiness Level Description table**

1. **Basic principles observed and reported.**
   Lowest level of technology readiness. Scientific research begins to be translated into applied research and development. Examples might include paper studies of a technology's basic properties.
2. **Technology concept and/or application formulated.**
   Invention begins. Once basic principles are observed, practical applications can be invented. Applications are speculative and there may be no proof or detailed analysis to support the assumptions. Examples are limited to analytic studies.
3. **Analytical and experimental critical function and/or characteristic proof of concept.**
   Active research and development is initiated. This includes analytical studies and laboratory studies to physically validate analytical predictions of separate elements of the technology. Examples include components that are not yet integrated or representative.
   Basic technological components are integrated to establish that they will work together. This is relatively "low fidelity" compared to the eventual system. Examples include integration of "ad hoc" hardware in the laboratory.
4. **Component and/or breadboard validation in laboratory environment.**
   Basic technological components are integrated to establish that the pieces will work together. This is relatively "low fidelity" compared to the eventual system. Examples include integration of "ad hoc" hardware in a laboratory.
5. **Component and/or breadboard validation in relevant environment.**
   Fidelity of breadboard technology increases significantly. The basic techno-

the end of the Phase II SBIR process.[13] This underscores the higher-risk nature of DoD SBIR programs, especially from an acquisitions perspective.

The "gap" between TRL 3-5 and TRL 6 can be characterized as the "TRL Gap," and it is a critical element in the difficulties experienced in transitioning Phase II technologies through Phase III into the mainstream acquisition process.

Essentially, SBIR can fund technology development to the point of TRL 3-5, and the acquisitions programs, through their own RDT&E programs and funding, can "pull" technologies from the pool generated through SBIR (and of course outside SBIR) into acquisitions. But the TRL gap must still be bridged, and there are major difficulties in doing so.

[13] A conclusion confirmed in case studies and in discussions with program executive officers (PEOs) responsible both for SBIR programs and for Phase III and eventually for acquisitions.

logical components are integrated with reasonably realistic supporting elements so it can be tested in a simulated environment. Examples include "high fidelity" laboratory integration of components.

6. **System/subsystem model or prototype demonstration in a relevant environment.**
Representative model or prototype system, which is well beyond that of TRL 5, is tested in a relevant environment. Represents a major step up in a technology's demonstrated readiness. Examples include testing a prototype in a high-fidelity laboratory environment or in simulated operational environment.

7. **System prototype demonstration in an operational environment.**
Prototype near, or at, planned operational system. Represents a major step up from TRL 6, requiring demonstration of an actual system prototype in an operational environment such as an aircraft, vehicle, or space. Examples include testing the prototype in a test bed aircraft.

8. **Actual system completed and qualified through test and demonstration.**
Technology has been proven to work in its final form and under expected conditions. In almost all cases, this TRL represents the end of true system development. Examples include developmental test and evaluation of the system in its intended weapon system to determine if it meets design specifications.

9. **Actual system proven through successful mission operations.**
Actual application of the technology in its final form and under mission conditions, such as those encountered in operational test and evaluation. Examples include using the system under operational mission conditions.

SOURCE: Defense Acquisitions Handbook, 10.5.2. Technology Maturity and Technology Readiness Assessments.

### 5.4.2 Risk and Risk Management

Bridging the TRL Gap is to a considerable extent a question of risk and risk management. Just as once upon a time, "no-one ever got fired for buying IBM," so in the world of defense contracting, "no-one ever got fired for contracting with a prime contractor." This caution is embedded directly in the DoD acquisitions manual:

> *If technology is not mature, the DoD Component shall use alternate technology that is mature . . .* [our italics and emphasis][14]

Bridging the TRL Gap is expensive. Costs rise as a technology matures, and the testing and evaluation (T&E) needed to move a technology from TRL 3-5 to TRL 6 can be very costly. Moreover, bridging the TRP Gap requires that a DoD program executive assume risk that would not be associated with a technology with a higher TRL.

[14]USD (AT&L), *Interim Defense Acquisition Guidebook*, October, 2002, pp. 9-10.

Who pays for technology risk mitigation? Both the *Department of Defense Instruction 5000.2 (May, 2003)* and the *Interim Defense Acquisition Guidebook* address this issue somewhat inconclusively. One formulation, found in various contexts in both baseline documents, suggests the impropriety of making industry pay:

> The PM shall structure the acquisition strategy to promote sufficient program stability to encourage industry to invest, plan, and bear risks. However, the PM shall not use a strategy that causes the contractor to use independent research and development funds or profit dollars to subsidize defense research and development contracts . . .[15]

So, as noted by many speakers at the NRC Phase III Symposium, the Phase III transition remains fraught with difficulties. This view is summarized by Anthony Mulligan, CEO of Advanced Ceramics Research (ACR), a successful SBIR company that uses ceramics technology for several DoD systems:

> ACR is just one example of the small businesses that are succeeding in developing technologies and capabilities that can provide significant cost savings to a wide and diverse array of military weapons systems. The difficulty is how these new technologies developed by small businesses can be transitioned into military program offices and picked up by the prime contractors. There are currently very few mechanisms, if any, to help ensure that this technology transition happens quickly. Military program offices and large program offices do not have efficient methods to fold new technologies into programs once the program has been road-mapped and already started.[16]

### 5.4.3 Small Business Perspectives

From a small business perspective, the lack of a defined and funded Phase III program makes Phase III transition a difficult and confusing matter. As noted by Anthony Mulligan, CEO of ACR, there is "no effective bridge between the acquisition community and those who are developing innovative technologies." A number of different concerns emerged at the NRC Phase III Symposium:

- **Timing.** Small businesses are often blocked by the very slow pace of acquisition partly because they do not have the resources to survive long stretches without revenue.
- **Complexity.** The acquisition process is both complex and unique, and small firms face a steep learning curve.
- **Phase III Funding Beyond DoD.** Few small firms have the staff or resources to do the market analysis necessary to attract funding from venture

[15]USD (AT&L), *Interim Defense Acquisition Guidebook*, October, 2002, p. 46.

[16]HASC Subcommittee on Tactical Air and Land Forces, hearing on small business technologies, June 29, 2005.

**BOX 5-3**
**Prime/SBIR Success Stories**

Representatives from the primes at the NRC Phase III Symposium also noted that there had been some important success stories in working with SBIR projects, including:

- Virtual Cockpit Development Program (Boeing).
- Advanced Adaptive Autopilot project, part of the Joint Direct Attack Munitions Program (JDAM) (Boeing).
- Cruise Missile Autonomous Routing System (CMARS) for the Tomahawk Mission Planning System. (Boeing).
- Mark 54 Torpedo Array Nose Assembly (Raytheon).
- Exo-Atmospheric Kill Vehicle (Raytheon).

capitalists, which are not, in any case, attracted to government contracting opportunities for a variety of reasons such as the expected limited market size, long lead times, and the level of regulatory "red tape" encountered in the procurement process. Similarly, firm founders are often reluctant to dilute their equity position to accommodate the needs of venture investors.

- **Phase III Funding Size.** Phase III may not be large: Mr. Mark Redding, CEO of Impact Technologies, noted that his company had successfully won more than 30 Phase III awards—but that these had averaged only $50,000 each (often not enough to get the technology to TRL 6).
- **Planning.** A number of agency staff and prime contractors noted that companies needed to be concerned with commercialization and Phase III activities right from the start—even during Phase 0, before the first Phase I was awarded. If an acquisitions program was unaware of a promising technology until after the Phase II had been completed, the relatively slow pace of acquisition meant a very substantial delay before the technology could be integrated into the program, even in the best of circumstances.
- **Roadmaps.** Much technical planning in DoD acquisition is driven by roadmaps developed by program officers and prime contractors. Failure to integrate SBIR and small businesses generally into these roadmaps means that they are likely to be excluded from acquisition programs, regardless of their technological success.
- **Contract Downsizing.** Even once a substantial Phase III has been awarded, there are no guarantees that the budget will be maintained at the contracted level.[17]

[17]For example, Orbitec's $57 million NASA Phase III contract was reduced by more than 80 percent after its first year.

- **Budget Squeeze.** In general, small businesses may lack the influence to maintain budget levels when agencies change priorities—and this can be devastating for companies with few other resources.

### 5.4.4 Prime Contractor Perspectives

There is considerable evidence that prime contractor interest in—and engagement with—the SBIR program has been growing rapidly in recent years (see Chapter 6). This is reflected in growing contractual linkages. Raytheon, for example, estimates that the value of technology leveraged through SBIR jumped from $3.8 million in 2004 to $11.6 million in 2005, and looks set to grow as rapidly in 2006. Raytheon is involved with 36 Phase I projects, 17 Phase II projects, four Phase III projects (with three more in the works), and has been a subcontractor on other projects.[18]

A number of points regarding the relationship of prime contractors to SBIR projects were expressed at the NRC Phase III Symposium. These included:

- **Increased interest in SBIR.** Strikingly, representatives of prime contractors testified both that there was already a substantial amount of prime involvement with the SBIR program, and also that recently several primes had made significant efforts to increase their levels of involvement. For example, Boeing had recently decided to increase its emphasis on SBIR—its lead SBIR liaison was now working on the program 100 percent time (up from 20 percent). At Raytheon, some divisions (e.g., Integrated Defense Systems) had been working formally with SBIR for some years. Half to two-thirds of a typical program for Raytheon Missile Systems was outsourced to subcontractors, and more than half of the companies involved meet the SBA's small business definition.
- **Agreement on the TRL Gap problem.** Speakers from Lockheed observed that the key to the transition from TRL 4-5 to TRL 6-8 was the presence of available funding on hand. This reflected the comments of many speakers that smoothing the funding path across the route from TRL 4-5 to TRL 6-9 was the single most important step to improved take-up of SBIR projects into acquisition programs.[19]
- **Lack of efficient links to small firms.** Many speakers noted the important example of the Navy Opportunity Forum as a means of making connections between the agency and program officers, SBIR program officers, primes, small businesses, and other funders such as venture capitalists. However, the Forum was seen as unusual, largely because other DoD agencies do not make the funds available for similar activities.

---

[18]Lani Loell, SBIR Program Manager, Raytheon Integrated Defense Systems, presentation to SBTC SBIR in Transition Conference, September 27, 2006, Washington, DC.

[19]Presentation by Mario Ramirez, Lockheed Martin, at National Research Council Symposium on *SBIR: The Phase III Challenge*, June 14, 2005.

- **Inadequate SBIR database for awards and solicitations.** Prime contractors need better capabilities for matching up their technology needs with the capacities of small firms.
- **Lack of evidence and cases.** Cases that demonstrate involving small business can lead to a real positive return on investment for the primes might help generate more such partnerships.
- **Real concerns about risk.** The risk of working with a small business extends far beyond issues of technology. As Mr. Fisher of General Atomics noted

**BOX 5-4**
**SBIR and Boeing—Expanding Prime Activities**

Boeing is one prime contractor that has taken a substantial and increasing interest in SBIR. According to Rich Hendel, SBIR program manager at Boeing, two primary divisions at Boeing—Phantom Works and Integrated Defense Systems (IDS)—are currently working with 32 small businesses on SBIR contracts, including five Phase I, 26 Phase II, and one Phase III. Over the past ten years, Boeing has worked with more than 100 companies on more than 200 projects.

**Topic development.** DoD sometimes requests that Boeing provide ideas for potential SBIR topics. Boeing solicits ideas from researchers and programs within the company, while working to make sure that potential topics are aligned with Boeing strategic roadmaps.

**SBIR advisory council.** In October 2005, Boeing formed an SBIR advisory council, with members from 7 IDS and Phantom Works groups. The council meets monthly to map out SBIR strategy.

**Phase III tactics.** Boeing sees success at Phase III resulting from a number of key components, including:

- Finding champions early within the government agency who will see technology development through to insertion and implementation.
- Finding champions within Boeing to push development and implementation.
- Availability of funding for Phase III at both customer (agency) and prime.
- Early establishment of the project team—small business, prime, customer.
- Resolution of proprietary IP issues.

**Key Phase III challenges:**

- Difficult to find the right technology and the right provider among the sea of possible partners and technologies.
- Technologies at low TRL levels (TRL 3-4 after Phase II).
- Provider track records, especially within Boeing.
- Funding for Phase III.

SOURCE: Presentations by Mr. Rich Hendel, Boeing SBIR Program Manager at National Research Council Symposium on *SBIR: The Phase III Challenge*, June 14, 2005, and SBTC SBIR in Rapid Transition Conference, September 2006.

recently, "Even the best technology cannot overcome small business' financial instability concerns, particularly on fixed price contracts."[20]

- **Cycle time mismatches.** SBIR projects can be completed "too soon" for entry into acquisition programs, leaving a timing gap that could stretch into years.
- **Difficulties in integrating SBIR projects into the planning process for acquisitions.** The relatively high level of technical risk involved in many SBIR projects means that it is not clear *ex ante* that SBIR-funded technologies would be sufficiently successful for eventual inclusion in the acquisitions program.
- **Agreement that VC funding is not likely at DoD.** Speakers from the prime contractors also noted that venture capitalists were unwilling to step into the Phase III funding gap partly because government contracts might not be large enough to ensure commercial viability, and partly because the longer time horizons and significant uncertainty involved in government contacting did not fit with the relatively short time horizon and private market focus of venture capital firms.
- **Misalignment between agencies, primes, and small business.** Several speakers, including Senate staff, noted that communication was not always good between the agencies, the primes, and the small business research community. Primes often had difficulty identifying the technology assets of small businesses. Small businesses often had weak linkages into the primes. Boeing noted that it was eager to partner with small businesses and had a significant track record in doing so within the SBIR program, but small businesses rarely came to Boeing seeking partnerships.

### 5.4.5 Program Officer Perspectives

A considerable part of the NRC Phase III Symposium focused on the role of program acquisition officers, and their difficulties in participating fully in Phase III. A number of speakers noted that program officers were critical to effective transition because they controlled acquisition funds needed to eventually move SBIR technologies into weapons systems.

Several speakers observed that acquisition program managers did not traditionally see SBIR as part of their mainstream activities. The CEO of Trident, Inc., Mr. Karangelen, observed more specifically that 89.9 percent of all federal R&D was currently being performed by the 100 largest contracting companies. Less than 4 percent went to small businesses. Only about 0.4 percent of all R&D generated by the government went to small technology businesses, even though one-third of all U.S. scientists and engineers were employed there.

---

[20]Fisher, General Atomics, presentation at the SBIR in Transition conference, Washington, DC, September 27, 2006.

Many speakers noted that few program managers and Program Executive Officers had historically taken an interest in SBIR. There appear to be real barriers to overcome in this area. Most notably, program officers were trained to reduce risk to the minimum, and SBIR-based projects offered a number of added technical and reputational risks, compared to working with prime contractors.

Risk aversion is entirely understandable at DoD, where lives are often ultimately at stake. It is therefore important to understand that from a program officer's perspective, introducing an SBIR project into acquisition programs can carry with it numerous risks. These include:

- **Technical risks**, including the possibility that the technology will not prove sufficiently reliable for use in weapons systems.
- **Company risks**, in that SBIR companies are by definition smaller and have fewer resources to draw on than prime contractors. In addition, many SBIR companies have only a very limited track record, which limits program manager confidence in their ability to deliver.
- **Funding risks.** The $850,000 maximum for Phase II may not be sufficient to cross the TRL Gap, or to fund a prototype appropriate for subsequent take up by Program Executive Offices.
- **DoD-specific risks**, as SBIR companies are often unfamiliar with the very high level of testing and engineering necessary to meet DoD acquisition requirements.
- **Timing risks.** DoD planning, programming, and budgets work in a two-year cycle, and it is difficult for Program Executive Officers to determine whether a small firm will be able to actually produce to meet perceived needs, even if the research is successful.
- **Effort/resource risks.** Carol Van Wyk (Navy) noted that program managers had a negative view of SBIR partly because they saw it as involving substantial effort, especially in terms of guiding small businesses through the DoD acquisition process, while larger companies were already well versed in these matters.[21]

In short, it appears that resolving the Phase III transition challenge at DoD will require a substantial effort, similar to that undertaken by the Navy, with regard to the role of acquisition program managers. There are few reasons to believe that such a shift would occur in the absence of renewed focus and incentives at both the DoD and service/agency level.

[21]Carol Van Wyk, Navy, presentation to PMA-209, May 25, 2005.

## 5.5 PHASE III INITIATIVES

DoD has long been aware that change will be needed if the full potential of the SBIR program is to be unleashed. This potential has been recognized and thus been the object of numerous initiatives over the years, at both the DoD and the agency level.

Given the wide range of challenges that intersect at the Phase III transition, many DoD initiatives bear on the problem, even if they have not been exclusively and explicitly focused on Phase III itself. Thus the Technology Transition Program is focused on technology transfer from DoD S&T programs into acquisitions—the specific policy challenge of Phase III—but it is not limited to SBIR firms, and is not focused specifically on small business.

Considerable further analysis concerning which of these programs is having a substantial impact on SBIR Phase III transition is needed; the data to support such an analysis do not currently appear to exist, but such data should be generated.

These initiatives include:

- **The Technology Transition Initiative (TTI)**, **Defense Acquisition Challenge Program (DACP)** and **Quick Reaction Fund (QRF)** ($64 million in combined funding for FY2005).[22,23]
- DoD improvements:[24]
    - **Company Commercialization Report (1993)**, (Standard Report Form (1997)). The CCR was the first effort at any SBIR agency to develop a way of systematically tracking post-SBIR Phase II outcomes. It was enhanced by development of the standard reporting form in 1997.
    - **Fast Track (1995).** The Fast Track program was an effort to reward firms for finding third-party funding, which both leveraged the original Phase II investment and also added additional validation of the quality of the research.
    - **Solicitation Pre-release (1996).** Among the most important initia-

---

[22]GAO notes a limited number of successful outcomes: U.S. Government Accountability Office, *Defense Technology Development: Management Process Can Be Strengthened for New Technology Transition Programs*, GAO-05-480, Washington, DC, U.S. Government Accountability Office, June 2005, pp. 3-4.

[23]GAO says that the purpose of TTI is to "Facilitate the rapid transition of new technologies from DoD science and technology programs into acquisition programs;" DACP is to "Identify and introduce innovative and cost-saving technology or products from within DoD's science and technology community as well as externally into existing DoD acquisition programs;" and QRF is to "Identify and rapidly field-test promising new technologies within DoD's budget execution years." U.S. Government Accountability Office, *Defense Technology Development: Management Process Can Be Strengthened for New Technology Transition Programs*, GAO-05-480, op. cit., Table 1, p. 6.

[24]Unless otherwise footnoted, these topics were mentioned by Michael Caccuitto, DoD SBIR/STTR Program Administrator, in his presentation at the National Research Council Symposium on *SBIR: The Phase III Challenge*, June 14, 2005. Descriptions provided by NRC.

tives at DoD, the pre-release provides a period during which interested small businesses can directly contact topic authors, to determine both the feasibility of a possible proposal, and also to define in more detail the precise nature of the customer's needs. Although there are no data to measure the results of this initiative, interviews suggest overwhelmingly that small businesses see this is a tremendous time- and resource-saver, and agency staff see pre-release as an important way of ensuring high quality proposals.

◦ **DoD SBIR/STTR Help Desk (1996).** The Help Desk is another important initiative. Now entirely manned by civilian contactors, the Help Desk provides a one-stop shop where small businesses can address many of the nontechnical, but nonetheless daunting, aspects of DoD proposals and contracting. This initiative effectively transfers many "hand-holding" functions from topic authors and SBIR management to professional support staff.

◦ **Uniform DoD-wide Topic Review Process (1997).** The uniform review process has strengths and weaknesses. While undoubtedly eliminating a number of poor quality topics, the added layer of review inevitably means further delays in the final publication of a topic, which can reduce its saliency.

◦ **DoD SBIR Web Site (1997).** Under more or less constant redevelopment since 1997, creation of a very high quality Web-based information delivery system has been a key to bridging the information gap between DoD and small business—and especially small businesses that have never done business with DoD before. It is telling that since 1997, approximately one-third of the companies winning SBIR awards are first-time recipients of SBIR contracts. High-quality/low-cost information is the first step toward encouraging new firms to enter the SBIR competition.

◦ **DoD SBIR/STTR Desk Reference (1999).** The Desk Reference is a comprehensive manual and help documentation, available over the Web. It too provides low-cost high-quality information for applicants.

◦ **Endorsement and/or Sponsorship of Acquisition Program Topics (1999).** Starting in 1999, these acquisition community offices currently sponsor or endorse more than half of all DoD topics. Across DoD, 60 percent or more of topics are now either sponsored or endorsed by program managers or Program Executive Officers.[25] At Navy, the acquisition-driven model of topic development has been expanded further. Dr. McGrath, Navy Deputy Assistant Secretary for RDT&E, noted that 84 percent of Navy topics came from the acquisition community, and that Program Executive Officers in the Navy systems commands participated in selecting proposals and managing

[25]Presentation by Charles J. Holland, Department of Defense, at National Research Council Symposium on *SBIR: The Phase III Challenge*, June 14, 2005.

**BOX 5-5**
**The Navy Primes Initiative**

The Navy Primes Initiative builds partnerships with Navy contractors to enhance new technology insertion in key programs by leveraging SBIR/STTR resources, in accord with best business practices of our partners.

**Actions**

- Established POC's at major primes offices.
- Held multiple site visits with primes and helped them to identify strong potential partners.
- Improved Search Database.
- Identified opportunities to cost-share demonstrations and integrations with SBIR.
- Prime and Acquisition Resources.
  - Two trial Primes Initiatives programs launched: *Lockheed MS2 Ship Systems, Raytheon IDS.*
  - One trial *Partnering Workshop* launched with DD(X) focus: *PEO Ships-PMS 500, Northrop Grumman Ship Systems, Raytheon.*
- Transitions Newsletter published which profiles Prime/Navy/SBIR accomplishments.

SOURCE: John Williams, Navy SBIR Program Manager.

them through Phase I and Phase II.[26] The impact of this change on topic take-up on Phase III success is captured in Figure 5-6.

○ **Commercialization Achievement Index (2000).** The CAI for the first time provided a quantitative analysis of commercialization outcomes from prior SBIR awards—even awards at other agencies. There is some evidence (from interviews) that the Index is being used as part of the proposal assessment process.

○ **Phase II Enhancement (2000).** The new Phase II enhancement program offers companies which can show matching funds additional SBIR funding, as an effort to partly bridge the TRL/Phase III gap. Its utilization appears to be growing, and it may be effectively replacing Fast Track as the option of choice for SBIR companies.

○ **Direct Program Executive Office (PEO) sponsorship pilot.** A 2005 Army pilot program to allocate 10 topics to PEO's has had the side-effect of driving SBIR toward applied research, the normal horizon of PEO's. This constituted a shift away from the traditional Army Research Office focus on more basic research.

[26]Presentation by Michael McGrath, Navy, at National Research Council Symposium on *SBIR: The Phase III Challenge*, June 14, 2005.

**BOX 5-6**
**DoD Commercialization Pilot Program (CPP)**

CPP is a new program to accelerate the transition of technologies, products, and services developed under SBIR to Phase III, including the acquisition process. The program was authorized under the National Defense Authorization Act for Fiscal Year 2006, section 252.

The program asks the services to find ways to accelerate the transition of SBIR-funded technologies to Phase III, partly by improving communications between the stakeholders. It allows agencies to spend up to 1 percent of SBIR program funds on these pilot activities.

The agencies have responded in a range of ways (see above for the Navy program):

**Air Force**

- Hiring "Transition Agents" for each product center with the responsibility to act as a bridge between the laboratory and product centers.
- Redistributing topic ownership more to product centers.
- Establishing a link between laboratory and acquisition.
- Ensuring selected Phase II topics meet needs of a program of record.
- Tracking and documenting successful transitions.
- Ensuring SBIR projects are included in program roadmaps.

**Army**

- Assessing and identifying SBIR projects and companies with high transition potential that meet high priority requirements.
- Providing market research and business plan development.
- Matching SBIR companies to customers and facilitate collaboration.
- Preparing detailed technology transition plans and agreements.
- Providing additional funding for select SBIR projects.
- Applying metrics and measure results.

**DARPA**

- Providing Management/Technical and manufacturing mentoring to Virginia SBIR Phase II contractors.
- Providing Regulatory/Management and Manufacturing mentoring to DARPA-selected SBIR Phase II contractors outside of Virginia.
- Providing accounting/business plan assistance/business management mentoring to new Phase I winners located in California.

Key elements of all the plans include an effort to develop better metrics and tracking capabilities, and improved information flows between stakeholders.

---

SOURCE: Michael Caccuitto, DoD SBIR/STTR Program Administrator and Carol Van Wyk, DoD CPP Coordinator, presentation to SBTC SBIR in Rapid Transition Conference, Washington, DC, September 27, 2006.

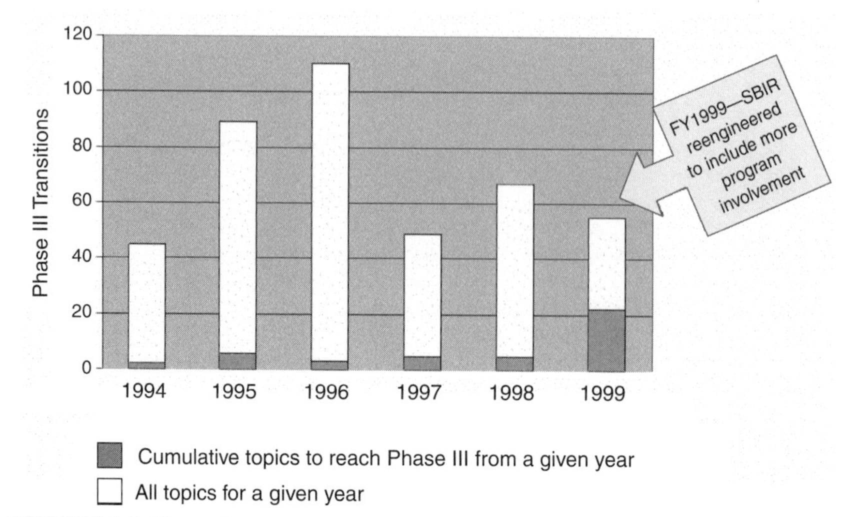


**FIGURE 5-6** Share of topics attracting Phase III funding.
SOURCE: Adapted from presentation by Carol Van Wyk, Navy, at Navy Opportunity Forum, 2005.

Beyond initiatives undertaken across DoD as whole, the DoD services and agencies have introduced their own initiatives to address the Phase III problem. These include:

- **The Navy "Primes Initiative" (2002).** This is an active outreach effort to connect primes to the SBIR program in more formal way. Primes have become increasingly interested in more access to the SBIR program.
    - **Extra-large awards** (beyond $750,000) have sometimes used at the Air Force, partly as a way of "exciting the program officers."[27]
    - **The Transition Assistance Program** (TAP) in the Navy provides mentoring and a management assistance program for supporting commercialization, i.e., transition through the Phase III maturation process. The Navy Opportunity Forum—part of the TAP—is another major initiative that brings together SBIR firms, primes, and PEOs/PMs, offering important networking opportunities. (See Chapter 6 for details.)
- **Training and education.** The Air Force has implemented a training and education program for primes and program offices.[28]
- **New funding initiatives** such as OnPoint, the Army's venture capital initiative. OnPoint makes equity investments in small entrepreneurial companies,

[27]Presentation by Mark D. Stephen, Air Force, at National Research Council Symposium on *SBIR: The Phase III Challenge*, June 14, 2005.
[28]Ibid.

**BOX 5-7**
**The NAVSUB Program at Navy**

Under the leadership of Richard McNamara, NAVSUB has developed a cohesive program aimed at providing strong incentives for program managers to use SBIR to help solve their technical programs, while developing processes that make it easier to award Phase III contracts. NAVSUB uses the following measures to promote SBIR projects:

- **Acquisition involvement.** PEO SUB is the most successful Phase III program at DoD. It advertises SBIR opportunities through a program of "active advocacy." PMs compete to write topics to solve their problems.
- **Topic vetting.** Program Executive Officers keeps track of all topics. PM's compete in rigorous process of topic selection. SBIR contracts are seen as a reward, not a burden.
- **Treating SBIR as a program**, including follow-up and monitoring of small businesses to help keep them alive until a customer appears. This encourages program managers to demonstrate commitment by paying half the cost of a Phase II option.
- **Providing acquisition coverage**, which links all SBIR awards to the agency's acquisition program.
- **Awarding Phase III contracts** within the $75 million ceiling that avoids triggering complex Pentagon acquisition rules.
- **Brokering connections** between SBIR and the primes.
- **Recycling** unexploited P1 awards, a rich source for problem solutions.

See Figure 5-5 and details in discussion of NAVSUB SBIR operations in Chapter 6.

including those that would otherwise not be doing business with the Army. It is focused on mobile power and energy for the soldier.[29]

- **Roadmaps.** Initiatives focused on developing joint technology maps and coordinated planning processes, including:
  - The Navy Advanced Technology Review Board process for evaluating across programs.[30]
  - The Joint Strike Fighter Technology Advisory Board, which reviews program priorities and includes a program office, the contractor team, and the S&T organizations of every service partner.[31]

[29]Available at <*http://www.onpoint.us/*>.

[30]David Bailey, Advanced Technology Review Board, Process Overview Brief for ONR Partnership Conference, August 5, 2004.

[31]Available at <*http://www.jsf.mil/*>.

- **Modifying the Company Commercialization Report** to identify manufacturing innovations and collect Phase III contract numbers.[32]
- **The Commercialization Pilot Program (CPP)—See Box 5-6**

Other DoD initiatives could also have a positive impact on SBIR companies, which might be able to find Phase III funding via these new programs. These initiatives include:

- **The Technology Transition Initiative (TTI).** Although this initiative is focused on technologies at TRL 6 or 7, it could provide a bridge mechanism for some projects across at least part of the TRL Gap.
- **The Defense Acquisitions Challenge (DAC).** This is an agency-wide initiative focused on identifying and supporting technologies that could quickly improve affordability, manufacturability, performance, or capabilities, with proposals that "challenge" existing technologies or methods.[33]

## 5.6 BEST PRACTICES

The structure of the DoD SBIR program can make it hard for companies to transition effectively into Phase III (as described earlier in this chapter). However, among the many initiatives discussed above, a number have emerged that could be considered best practice within DoD. These include:

- **"Returning the tax."** Over time, a growing number of SBIR topics have been "set aside" for the needs of program officers. By effectively returning the SBIR funding to acquisitions programs—with the continuing proviso that research be allocated via the SBIR mechanism—program officers can see a more direct connection between the SBIR program and their own needs.
- **Acquisitions involvement in topic development and selections.** The direct impact of this change can be seen in a chart from Navy showing the jump in the number of topics that eventually attract Phase III funding (See Figure 5-6).
- **Closer acquisitions involvement in the "downselect" process.** To the extent that acquisitions officers participate in the decisions on which Phase I projects should be funded at Phase II, it is more likely that Phase II recipients will get Phase III contracts from DoD.
- **Linking information flows between small businesses, primes, DoD acquisition offices, and SBIR programs.** One theme of the NRC Phase III Symposium was the difficulty of sharing information between stakeholders. A variety of efforts have been made to remedy this situation, with some success. These efforts include:
    - Pre-release of topics.

[32]Presentation by Michael Caccuitto, DoD SBIR/STTR Program Administrator, at National Research Council Symposium on *SBIR: The Phase III Challenge*, June 14, 2005.

[33]See <*https://cto.acqcenter.com*> for more information on these initiatives.

- ○ Electronic communication systems.
- ○ The Navy Opportunity Forum and other SBIR gatherings.
- ○ The SBIR database.
- ○ The Primes Initiative.

- **Commercialization training.** The Navy Transition Assistance Program (TAP) is a 10-month program and the most ambitious of all SBIR training programs. TAP is designed to focus small businesses on transition, to mitigate risk, and to improve return on Navy's' investment. While outcomes data are not yet fully available, participation rates have increased substantially each year. (See Chapter 6 for details.)
- **"Focused call" approach to solicitations.** Navy has developed what it calls a "focused call" approach to solicitations. This involves defining five to six related topics (e.g., Sensors, Algorithms, Materials, Manufacturing), through which are allocated a total of about 20 Phase I awards and 10 Phase II awards, of varying sizes. According to Carol Van Wyk, then-SBIR Program Manager at NAVAIR, a focused call might provide two small Phase II awards (~$300,000), two medium sized Phase II awards (~$500,000), four standard Phase II awards (~$750,000), and two large Phase II awards (~$1.5 million). The focused call approach encourages strategic planning, is seen as cost-effective, and reduces staff workload.
- **Contracting improvements.** NAVAIR pioneered IDIQ (Indefinite delivery/indefinite quantity) contracts within the DoD SBIR program. These mesh well with SBIR, allowing increased program manager flexibility and speed to delivery, outside the normal competitive bidding process. The Universal Phase III Contract outlined in Figure 5-7 could be another significant step forward.

## 5.7 RECOMMENDATIONS

Recommendations for improving Phase III activities and outcomes can be grouped into a few major categories. Within these, the initiatives and best practices above provide a menu from which programs and agencies might wish to develop pilot programs or more for their own use.

- **Developing and gathering metrics on Phase III.** The growing focus on Phase III contracting dollars as the key metric of program performance appears to be gathering momentum across DoD. However, implementation of this metric is uneven, and better data are needed. Data from Phase III contracts funded via the primes continues to be absent. The 2005 GAO Report has already recommended that "DoD develop data and measures that can be used to assess short- and long-term impacts of the programs and take other actions to further strengthen selection, management and oversight . . ."[34]

[34]U.S. Government Accountability Office, *Defense Technology Development: Management Process Can Be Strengthened for New Technology Transition Programs*, GAO-05-480, Washington, DC: U.S. Government Accountability Office, June 2005, pp. 3-4.

## Transition Enabler: Phase III Universal Contract

- Flexible tasking options
  - √ Exploratory application study
  - √ Further R&D
  - √ System Integration Analysis
  - √ Customized prototype for specific platform needs
  - √ Test & evaluation
  - √ Production buy
  - √ Support
  - √ Training
- 5 years duration
- $25 million in total funding

- Advantages
  - √ Streamlines negotiations
    - Creates central Phase III contract
    - Pool for PMAs
    - Easier for PMAs to use SBIR
  - √ Capitalizes on positive SBIR attributes:
    - Rapid response in fast moving technology market
    - Innovative solutions to benefit the warfighter
    - Cost effective
    - Flexibility in competitive environment
  - √ Permits one agency to leverage another

**FIGURE 5-7** Phase III Universal contract.
SOURCE: Carol Van Wyk, NavAir.

- **Incentives.** Better incentives may be needed in several areas. Acquisitions officers need career-oriented incentives that will reduce risk and enhance the benefits of participating actively in the SBIR program. However, incentives might also be needed to encourage the primes to participate more fully in the program. This could include financial incentives of several kinds for meeting SBIR inclusion goals on major contracts, or even—as suggested by SBTC—requirements that large contracts meet certain SBIR goals, similar to current targets for woman- and minority-owned businesses. Some programs similar to this are already in place. For example, PEO SUB offers a small business subcontracting incentive in its *Virginia*-class program through a formal plan incorporated as clauses in the *Virginia*-class construction contract which allow for the payment of a Small Business Subcontracting Incentive Fee (SBSIF) for increasing the level of small business subcontracting participation.

### 5.7.1 Improving Program Officer Use of SBIR

A number of symposium speakers noted that acquisition officers were the key to moving SBIR to Phase III. They controlled the funding, and their involvement was critical. And active championing by Program Executive Officers seems to be a critical ingredient in Phase III success. A clear cultural shift was observed at Navy once Program Executive Officers became active champions of SBIR involvement in acquisitions.

- **Senior acquisitions involvement.** For DoD, there appears to be no substitute for the systematic support and involvement of acquisitions officers, especially at the most senior level. PEO SUB has driven major changes in the SBIR culture within that component of Navy, with transformative results. Only senior managers can insist that all program managers integrate SBIR fully into their acquisition programs—and give them strong incentives to do so.
- **Follow-on funding.** Some suggest that further improvement in this area include the development of a fund for the provision of matching funds for Phase III (which would reduce the risk level for program managers, and would follow NASA practice).
- **Tools for better integration with acquisitions.** There are very real barriers to the smooth flow of information within DoD and among key SBIR stakeholders. Companies are naturally reluctant to share important technical information, and small businesses in particular are well aware of the potential dangers of sharing key intellectual property with companies that could easily turn out to be competitors (i.e. the primes). This leaves SBIR companies in many cases with Hobson's choice: share their IP and hope for the best, or stay private and be frozen out of partnerships with the primes that dominate DoD spending. Finding ways to address these issues—possibly through better protection of small business IP, or at least better training for acquisitions officers about SBIR IP protections, is an important area for further exploration.
- **Educating Program Executive Officers** that the SBIR-supported technologies can be big time- and money-savers, and that small companies can produce to scale and on time. Richard Carroll, then-CEO of Digital System Resources, noted that SBIR training had been part of the general Program Executive Officer training curriculum for one year, but was later deleted.[35] In the Navy, SBIR management has tried to provide a consistent message to Program Executive Officers and program managers: "SBIR provides money and opportunity to fill R&D gaps in the program. Apply that money and innovation to your most urgent needs."[36] In essence, SBIR's unique advantages can be used to solve specific kinds of problems for acquisition officers.
- **Incentives.** The evidence above suggests that DoD needs to find ways to reduce the risk to program mangers of utilizing SBIR Phase II technology. Various options might be considered, not least on a pilot basis.

### 5.7.2 Roadmaps and Technology Planning

Because the integration of subprojects (such as those funded by SBIR) into larger weapons systems is such a complex and long-cycle process, speakers from the primes stressed that coordination is key:

[35]National Research Council, Symposium on *SBIR: The Phase III Challenge*, Washington, DC, June 14, 2005.

[36]Presentation by John Williams, Navy SBIR Program Manager, at National Research Council Symposium on *SBIR: The Phase III Challenge*, June 14, 2005.

- **Roadmaps are a key to successful coordination** of small business activities with the primes: "To make successful transitions to Phase III, SBIR technologies must be integrated into an overall roadmap."[37] For example, Lockheed Martin uses a variety of roadmaps, including both technical capability roadmaps and corporate technology roadmaps. Roadmaps allow program officers to generate effective "pull," via the leads to the prime and to smaller subcontractors.
- **Start early.** The long development cycle of major weapons systems means that for SBIR projects, panning activities must start very early in the technology development cycle—if possible during Phase 0—the stage at which topics are developed.

### 5.7.3 Outreach and Matchmaking

Suggestions focused on the need for more events like the Navy Opportunity Forum, on better communications channels, and on improved databases that shared technology results more effectively across agencies:

- **Improved information flow.** New electronic tools are needed to help share technologies and opportunities between and among stakeholders. Current databases are not sufficient.
- **Very-early-stage outreach.** As stakeholders have noted the importance of very early planning, new mechanisms may be needed to bring small business into the planning process at an earlier stage than is currently the case.
- **More funding for outreach.** The Navy's TAP program appears to constitute a best practice model. It would therefore be appropriate to provide the funding necessary to support similar activities at other agencies and components.

### 5.7.4 Integrating the Primes and SBIR

Elements of an improved relationship between SBIR programs, acquisitions offices, and the primes are already in place, though these elements are scattered across DoD. Some of the reforms that might improve relationships among these parties are:

- Extensive outreach by SBIR program managers to primes' Technical Management and Strategic Sourcing staff.
- Education for managers of the prime contractors about the competitive advantages of participating in the SBIR program, and about congressional interest in the success of Phase III.
- Improved mechanisms for participation of the primes in topic development.

[37] Presentation by Mario Ramirez, Lockhead Martin, at National Research Council Symposium on *SBIR: The Phase III Challenge*, June 14, 2005.

- Improve primes' subcontract reporting to include a separate breakout for SBIR firms, similar to those currently provided for woman-owned and minority-owned firms.
- Improved reporting could be matched by expanded requirements, for example that all new contracts over a specified size should include SBIR subcontracting goals and incentives.
- Making the SBIR subcontracting plan part of the evaluation criteria for major contracts.
- Make sure primes are paid for "Technology Insertion," and that it is a major element of their contract.

### 5.7.5 Funding for Program Management

- **Add management funding.** The success of the Navy Phase III effort is at least partly predicated on the extensive and expensive outreach and commercialization support activities it has implemented. While funding is currently provided by the Navy out of its administrative budget, similar funding has not been available at other components and agencies—and hence similar programs have not developed. Additional funding for these purposes should be provided.

In contrast, the entire Air Force SBIR program is managed by four staff members at Wright-Patterson AFB. While the program has experienced 70 percent growth since 2000, there has been no additional funding for transition assistance or program administration. As a result, the Air Force has no funds to document or track success—which is an important component in helping acquisition program managers see the value of the program.[38]

### 5.7.6 Training

- **Acquisitions officers.** Improved understanding of SBIR among acquisitions officers is probably a necessary condition for overall increase in Phase III success rates. Several possible options here include:
  - Requiring SBIR training through the Defense Acquisition University.
  - Requiring Phase III reporting by acquisition offices.
  - Requiring acquisition programs to include SBIR projects and the planned transition path in milestone reviews.

### 5.7.7 Reduce Time from Topic Selection to Award

- Ensure acquisition offices are aware of and leverage Phase II to Phase III gap-funding programs.

[38]Presentation by Mark D. Stephen, Air Force SBIR Program Manager, at National Research Council Symposium on *SBIR: The Phase III Challenge*, June 14, 2005.

### 5.7.8 A Flexible Approach to Other Possible Agency Initiatives and Strategies

A number of suggestions seem to be best addressed through the design and rollout of carefully designed pilot programs. This would require in some cases waivers from SBA for activities not otherwise permitted under the SBA Guidelines. SBA should be encouraged to take a highly flexible view of all agency proposals for pilot programs. Some possible options that could be explored in this way include:

- **Small Phase III awards.** These could be a key to bridging the "valley of death" between technology development and commercialization. Providing even small Phase III awards—perhaps enough money to fly a demonstration payload—for a technology not ready for a full Phase III might be explored.
- **Unbundling larger Phase III awards.** Organizing larger contracts into smaller components would open Phase III opportunities. For example, the unbundling of a large contract for a complex life sciences module being competed by Lockheed and McDonnell Douglas in 1995 led to Orbitec's major $57 million Phase III award.
- **Redefining T&E within SBIR.** DoD and SBA could adopt a wider view of RTD&E, so that SBIR projects could qualify for limited T&E funding. That in turn would help fund improvements in readiness level.
- **Spring loading Phase III**, by putting place in milestones that could help to trigger initial Phase III funding. This could occur in the context of larger, staged, Phase II awards in which additional stages fund more Demonstration and T&E when non-SBIR funds or resources are leveraged (beyond current Phase II-plus).

This chapter's focus is on a key phase of the program, the transition from a successful Phase II to Phase III. It describes the multiple challenges participating firms, program managers, and senior management face in maximizing the returns on the SBIR program at DoD. Also described, however, are a wide range of measures developed over a decade to meet the transition challenge and address congressional concerns about the need for greater commercialization. This active experimentation, and the flexibility that permits it, are hallmarks of the SBIR program at DoD.

The recommendations made here are intended to contribute to enhanced output from the program that is increasingly seen as an asset by Program Executive Officers and others in the Defense acquisition process. The growing interest of the prime contractors, the new incentives provided by Congress, and the growing recognition of SBIR companies as a valuable source of innovation are all positive trends that these recommendations are intended to enhance.

# 6

# Program Management

## 6.1 INTRODUCTION

Management of the DoD SBIR program is characterized by two central elements: (1) the tremendous diversity of objectives, management structures and approaches among the different services and agencies that fund SBIR programs at DoD; and (2) the consistent pursuit of improvements to the program to enable it to better meet its objectives.

The review that follows is focused on describing the mainstream of DoD practice, and where relevant, divergences from it among the agencies and components. It concentrates on describing current practices and recent reforms.

These reforms also impact the way in which assessment must be made. The significant lags between award date and commercialization means that comprehensive outcomes are only now available for awards made in the mid to late 1990s. However, management practices have changed—often significantly—since the time of those awards.[1] Hence it is methodologically not possible to build a one-to-one relationship between outcomes and management practices.

[1]For example, a number of major internal changes followed the 1995 Process Action Team (PAT) review. These led to a reduction in the lag between receipt of proposals and award announcement from 6.5 months to 4 months for Phase I, and from 11.5 months to 6 months for Phase II. The Fast Track Program was also established, which both accelerates the decision-making process and increases the level of funding for Phase II projects which obtain matching funds from third-party investors. DoD also required all SBIR Phase II proposals to define a specific strategy for moving their technology rapidly into commercial use.

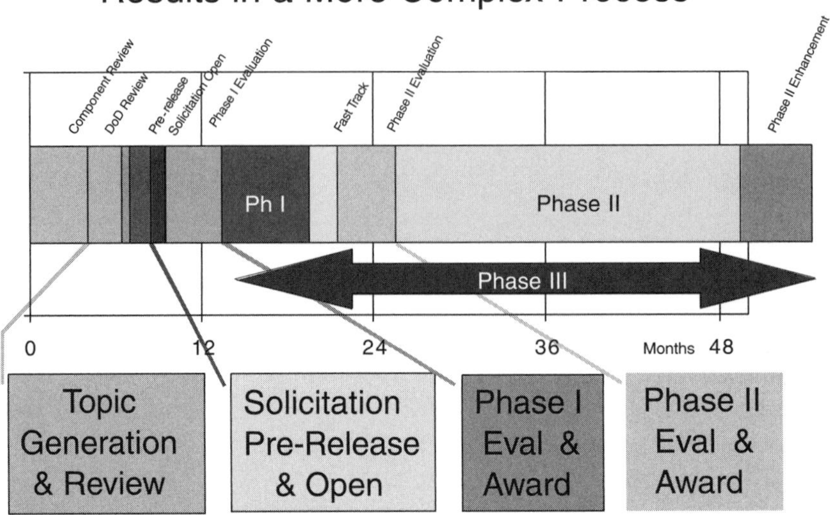


**FIGURE 6-1** SBIR timeline at DoD.
SOURCE: Michael Caccuitto, DoD SBIR/STTR Program Administrator and Carol Van Wyk, DoD CPP Coordinator. Presentation to SBTC SBIR in Rapid Transition Conference, Washington, DC, September 27, 2006.

The latter must therefore be assessed primarily through interviews, focused on current practice, with awardees, agency staff, and other stakeholders.[2]

Finally, it is worth noting that DoD processes are quite complex—unsurprising, given the high volume of proposals and awards, and the wide variety of Service and Agency objectives. However, it is possible to provide an overview of core activities, as seen in Figure 6-1. Each phase of the SBIR program will be reviewed in turn. Figure 6-1 shows the significant pre-solicitation activities focused around topic development, some of the funding initiatives in place (Fast Track and Phase II Enhancements), and the potential role of Phase III which, as we shall see, should be part of very early activities within the SBIR framework.

[2]The continuing, at times incremental nature of these changes set against the longer term, often circuitous processes of firm growth and commercialization of SBIR awards complicates efforts to relate program management techniques to performance outcomes. Thus, results measured for awards that occurred ten years ago may not adequately describe how well a service or agency is managing in its SBIR program today.

## 6.2 TOPIC GENERATION AND PROCEDURES

Management of the DoD SBIR program has been largely decentralized to individual services and agencies. The exception is that the Office of the Deputy Director of Research and Engineering (DDR&E) uses the topic review process to exert centralized control over the definition of the SBIR topics included in official solicitations.

Informal DoD topic review under the lead of the DDR&E began in 1996, following the recommendations of the 1995 PAT review. A formalized process for topic review began in 1997. It was designed to promote the closer alignment of service and agency R&D with overall DoD R&D priorities, to avoid duplication, and to maintain the desired degree of specialization in the R&D activities of the respective services and agencies.

Ultimate decision authority on the inclusion of topics in a solicitation lies with the Integrated Review Team, which contains representatives from each of the awarding components. Topics are reviewed initially at DDR&E and then returned to the agencies for correction of minor flaws, for revision and resubmission, or as discards.

This review process is not necessarily popular with topic authors or program managers, as it limits their authority. Some senior managers have stated that they believe the DDR&E offices are not close enough to the programs to make these kinds of decisions effectively. The process also reduces responsiveness to

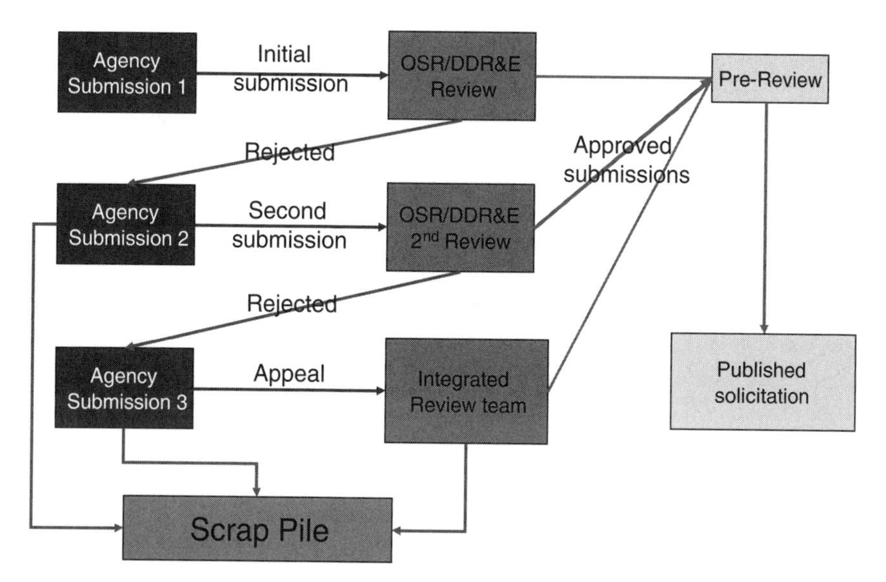


**FIGURE 6-2** Topic review process.
SOURCE: Developed from interviews with DoD staff.

**BOX 6-1**
**Acquisition Liaisons**

To further foster coordination between its R&D and acquisitions programs, DoD mandates that each major acquisition program must designate as SBIR liaison an individual who is:

- knowledgeable about the technology needs of the acquisition program and
- responsible for technology infusion into the program.

These liaisons interface with the SBIR program managers within DoD and with the SBIR contractor community. Their role is to integrate SBIR technologies into their acquisition programs.

Contact information for the liaisons is listed on a DoD SBIR Web Site so that both DoD laboratory personnel and SBIR contractors have—in theory at least—an efficient means of communicating with their end customers in acquisition programs at all stages of the SBIR process. The liaisons may author topics or cause them to be authored.

However, speakers at the NRC Phase III Conference observed that agencies sometimes worked around the mandate by assigning numerous liaison roles to a single individual as a pro forma matter, making the function effectively useless.[a]

[a]See National Research Council, *SBIR and the Phase III Challenge of Commercialization*, Charles W. Wessner, ed., Washington, DC: The National Academies Press, 2007.

changed circumstances (such as 9/11) as it increased the time required for approval of possible topics. However, the review process has, according to DoD staff, improved clarity in topic descriptions and forced the technical monitors (TPOCs) to address how the possible applicants might transition into Phase III.[3] DoD has tried to find ways of mitigating the negative effects of review—for example, by providing means to reduce delays for "hot" topics, such as the Navy quick response topics awarded in the aftermath of 9/11.

Topics originate in service laboratories or in program acquisition offices. The laboratories are focused on developing technologies to meet the ongoing research needs of their organization. Some awarding organizations within DoD do not have their own laboratories. In practice, these organizations frequently turn to the "in-house" expertise of the service laboratories both to transform mission requirements into R&D topics, and to suggest topics relevant to the organizations' requirements.

Topic authors frequently serve in a dual capacity. After their topic has been accepted and an award made, they become the technical monitors for the contract that results. Thus, even though these technical monitors are often insulated

[3]Presentation by Carol Van Wyk, presentation to Navy Opportunity Forum, Reston, VA, 2004.

from the day-to-day needs of the front-line commands, they are charged with the responsibility of developing technologies that are relevant to DoD's overall performance and to the specific mission needs of the funding components.

Further changes in topic generating procedures are under way. Missile Defense Agency (MDA) representatives, for example, described a reengineering of topic generation procedures intended to shift the focus of projects away from those based on program elements to what was termed a more MDA-centric approach, aimed at generating topics that enhance the agency's technical performance capabilities and fill agency-wide gaps in existing systems.

Primes are also invited to suggest SBIR topics through informal discussions with laboratory personnel or SBIR program managers at scientific meetings, technology conferences, and trade shows, as well as at DoD's own outreach workshops. Other channels for input include prerelease discussions with topic authors, and ongoing contacts between firms and technical monitors for current SBIR awards. Case study interviews with firms indicated that these informal[4] channels are a recognized and generally accepted facet of the SBIR program. Firms do express some ambivalence about the proprietary of these informal channels, and about the frequency with which they affect the selection of topics, and thus the distribution of awards.

There is no formal process within DoD through which firms can suggest topics, so SBIR program managers have no information about how often firm-suggested topics are adopted. The iterative review and revision process for DoD topics is also such that the ultimate topic released may differ substantially from that originally proposed. And, of course, many proposed topics do not make it through the review process.[5]

Overall, the vast majority of topics are agency-driven. DoD has established seven criteria[6], which are used in the review of potential SBIR topics:

- Topics will solicit R&D and not procurement.
- Topics must involve technical risk; i.e., technical feasibility is not yet established.
- Topics will fall within one or more of the DoD key technology areas.

[4]The Army section of the Solicitation 06-2 contains the following "Small Businesses are encouraged to suggest ideas that may be included in future Army SBIR solicitations. These suggestions should be directed to the SBIR points-of-contact at the respective Army research and development organizations listed in these instructions."

[5]It is important to note there are no hard data on the extent of firm influence on topics, and there are unlikely to be any in the future. When a firm wins an award on a topic that it is perceived to have initiated, other firms suggest that selected SBIR competitions are "wired." However, if a firm fails to win a topic it suggested, that information rarely becomes public, so information asymmetries result in an unbalanced perspective.

[6]See National Research Council, *The Small Business Innovation Research Program: An Assessment of the Department of Defense Fast Track Initiative*, Charles W. Wessner, ed., Washington, DC: National Academy Press, 2000.

- Topic will allow the performing company "significant flexibility," stressing innovation and creativity.
- Topics will include examples of possible phase III dual-use applications.
- Topics will not duplicate each other.
- Topics will be clearly written.

The services and agencies initially submit draft topics to OSD 6 months before the scheduled closing date of a solicitation. OSD and DDR&E conduct a one month detailed review using the review criteria. The topics are then accepted or returned to the originating service or agency for revision. The originators may revise and resubmit for a second DDR&E review. Topics that fail the second review are returned to the services and agencies, who may appeal the rejection to an integrated review team.

All ultimately approved topics are returned to the originators for final certification, followed by publication of the pre-release.

In addition, each Service and agency has its own review process that precedes the DoD-wide OSD review. The duration of these processes depends on numerous factors, including the size of the agency (larger agencies have more topics to review). Thus, the Army topic review is a centralized online process that takes 4 months, while the less centralized, but also online, Air Force SBIR Topic Submission Module may be active six months ahead of the OSD review process.

## 6.3 PRE-RELEASE

An overriding consideration at DoD is that its procedures for managing the SBIR solicitation process comply with Federal Acquisition Regulations (FAR). At the same time, DoD wants to provide as much information as allowable to firms interested in submitting proposals. FAR prohibits contact between an agency issuing a solicitation and prospective bidders once a solicitation has been issued, other than written questions to the contracting officer. The contracting officer in turn must make the question and answer available to all prospective bidders. This balancing act is managed through the pre-release process.

Pre-release is an important DoD SBIR initiative, which has won considerable praise from small businesses. The pre-release posts the entire projected solicitation on the Internet about two months before the solicitation is to open. Each topic includes the name and contact information of the topic author/monitor. Interested firms may discuss with the topic author/monitor the problem that the government wants to solve, as well as their intended approach.

From the firm's perspective, a short private discussion with the topic author can often help avoid the cost of preparing a proposal if the firm's capabilities do not match those required to compete successfully. Alternatively, the discussion can help refine the firm's understanding of the technical requirements embedded in the solicitation and lead it to prepare a more focused and competitive proposal.

Discussions can also open new possibilities for the firm, as it learns that its capabilities are better suited for other topics, other acquisition programs, or the unmet needs of prime contractors. Firms new to SBIR often get procedural information, and are steered to the appropriate DoD Web sites for further information. In all, the pre-release period has generally been viewed positively both by DoD and by participating firms.

Pre-release concludes when the solicitation is formally opened, about 45 days before the closing date. Almost all released topics coincide with those in the pre-release, although occasionally a few topics from the pre-release are not included in the solicitation. At this stage, all mention of topic authors is removed from the topics. The formal solicitation is posted on the Internet, at <*http://www.dodsbir.net/*>. As mentioned above, firms can ask questions, but the answers from the contracting officer are posted for view by all potential proposing firms.

## 6.4 SELECTION PROCEDURES

### 6.4.1 Phase I Contract Selection

Since the SBIR program's inception at DoD, all SBIR awards have been contracts awarded on a competitive basis. The solicitation identifies the evaluation criteria for both Phase I and Phase II. Contracts generally require a deliverable, and for most Phase II SBIR contracts in recent years, the deliverable is a prototype. Having a prototype is often the first step in demonstrating commercial potential. Only firms that are completing a Phase I project can be considered for a Phase II award.

Beginning in 2003, DoD put a single contracting officer in the Defense Contracting Office, Washington, DC, in charge of the solicitation process. This civilian position is designed to provide prompt proper attention to logistical problems that might adversely affect the timely submission of proposal—such as the January 2003 overloading[7] of the electronic site that caused some firms to miss the original submission deadline. The lead contracting officer can now make decisions to extend the proposal deadline or otherwise modify requirements.

Once the Phase I proposal deadline date has passed, each DoD component takes charge of the proposals submitted in response to its topics. The selection

[7]DoD SBIR Solicitation 2003-1 was the first to require all proposals to be submitted electronically. It cautioned firms; "As the close date draws near, heavy traffic on the web server may cause delays. Plan ahead and leave ample time to prepare and submit your proposal." However most firms waited until the closing 24 hours, resulting in network overload and a subsequent decision to reopen the solicitation. Subsequently, DoD substantially increased server capacity, and no subsequent solicitation has encountered similar problems of overload and delay.

process varies considerably among services and agencies, from a centralized decision process in the Army to decentralized processes in Navy and Air Force. SBIR technical monitors are involved in the evaluation teams and recommendation of proposals to varying degrees. Contracting officers make the final selections and awards.

The actual selection process is quite diverse. DoD has about 30 separate awarding elements (e.g., the Navy has 8 to 10). There are published criteria for evaluation in each solicitation, and written evaluations are prepared by each evaluator for each proposal, but there is no published standard procedure as to how an element picks who will evaluate, though generally there are three reviewers per proposal, nor what happens after the initial evaluation. Proposals are evaluated on their merit, not necessarily in comparison to other proposals before the agency decides how many awards to make. Thus there may be one or more tiers of technical management within an element making recommendations before the proposals reach the contracting officer. Ultimately, some topics will see only one award while others may see multiple awards. These decisions may be based solely on the quality of the proposal, or may include the diversity of technical approaches and the importance of the topic, as well as available funding.

A contracting officer is designated as the Source Selection Authority (SSA),[8] with responsibilities defined in Federal Acquisition Regulations (FAR).[9] Adherence to these regulations is necessary to avoid protests about selection procedures being filed with the General Accounting Office (GAO). Proposal evaluations are legally based solely on the factors specified in the solicitation. These include:

a. The soundness, technical merit, and innovation of the proposed approach and its incremental progress toward topic or subtopic solution.

b. The qualifications of the proposed principal/key investigators, supporting staff, and consultants. Qualifications include not only the ability to perform the research and development but also the ability to commercialize the results.

c. The potential for commercial (government or private-sector) application and the benefits expected to accrue from this commercialization.

Where technical evaluations are essentially equal in merit, cost to the government is considered as a tiebreaker. The solicitation also states that, "Final decisions will be made by the DoD component based upon these criteria and

[8]Unless the agency head appoints another individual for a particular acquisition or group of acquisitions which it does rarely for SBIR.

[9]For example, in accordance with FAR 15.303(b)(1), the SSA shall "establish an evaluation team, tailored for the particular acquisition, that includes appropriate contracting, legal, logistics, technical, and other expertise to ensure a comprehensive evaluation of offers."

consideration of other factors including possible duplication of other work, and program balance."[10]

In general, firms speak positively of the fairness of the award selection process. Some interviewees privately note that some firms have extensive contact with DoD officials, and are thus better able to have their specific technologies "built into" the topic selection process and thus have an inside track in selected competitions. Importantly, they also report that this advantage does not automatically lead to an award.

### 6.4.2 Phase II Selection Procedures

A Phase II proposal can be submitted only by a Phase I awardee, and only in response to a request from the agency. The latter condition is unique to DoD's SBIR program. Phase II application is *not* initiated by a solicitation, or by the awardee. Although the formal evaluation criteria remain the same, the commercialization factor is more important for Phase II selection.

DoD components use different processes to determine which firms to invite for the Phase II competition. These vary from a decision made by the technical monitor for Phase IIs at DARPA to a centralized process like that used at MDA.[11] The latter provides a template for Phase II decisions, where recommendation is based on several criteria.[12]

- The Phase II prototype/demonstration (what is being offered at the end of Phase II?).
- Phase II benefits/capabilities (why it is important?).
- Phase II program benefit (why it is important to an MDA program?).
- Phase II partnership (who are the partners and what are their commitments? Funding? Facilities? This also can include Phase III partners).
- Potential Phase II cost.

These criteria address the basic business case for a Phase II invitation. Providing answers requires communication between the program office, the Phase I SBIR awardee, and the Phase I technical monitor.

Selection processes may be centralized, with a fixed date for submission of all of the Phase II proposals for that year (as in the Army), or decentralized to component commands or laboratories, as in the Navy and Air Force, where deci-

[10]This quote from section 4.1 of DoD SBIR Solicitation 2005.3 had been identical in every solicitation since 1983. It was changed slightly in 2006 such that the other factors are now specified in section 4.2.

[11]An MDA program begins the process for a Phase II invitation by making a recommendation (all MDA topics are sponsored by MDA programs).

[12]Criteria provided by Mike Zammit, MDA SBIR Program Manager, in an interview on September 22, 2005.

sions are made as the individual proposals are received and evaluated. Scoring procedures vary among components, with some using primarily qualitative assessment ratings and others a more quantitative approach to scoring.[13]

### 6.4.2.1 Commercialization Review

Under the 1992 Reauthorization, DoD established a Company Commercialization Report (CCR) as a part of any SBIR proposal for all firms which had 15 or more Phase II awards over the previous five years. DoD extended the CCR requirement to all firms in 1997, and made submission electronic in 1999. By 2000, the DoD CCR required firm information in addition to the sales and funding information on all prior Phase II awards. Firm information includes identification information as well as annual revenue, number of employees, issuance of an IPO—all indicators that can be used to gauge firm development. The CCR also requires firms to state the percentage of revenue derived from SBIR, which measures dependency on SBIR.

The CCR permits firms to provide additional information—such as the noncommercial impact (mission impact, cost savings, reliability improvements, etc.) of its SBIR projects. These factors, coupled with specific results (sales including customer, additional funding by source, identification of incorporation into a DoD system) from prior Phase II awards, along with the numerical score of the Company Achievement Index (CAI), are used to evaluate the past performance of a firm in commercializing their prior SBIR.

The CAI compares how well a firm has commercialized its Phase II compared to other firms with like number of contracts awarded in the same timeframe. Although external discussion often focuses solely on the numerical CAI, the CCR actually provides the evaluator with a comprehensive picture in which the CAI is a component. Even when the CAI is extremely low, theoretically denying one half of the commercialization score, the denial may be overridden based on the more complete picture.

In addition to the required Company Commercialization Report, each Phase II proposal must contain a two-page commercialization strategy, addressing the following questions:

- What is the first product that this technology will go into?
- Who will be your customers, and what is your estimate of the market size?

[13]To give an example of how this works at a component, at MDA the TPOC recommends a Phase II invitation. The recommendation goes to the MDA SBIR Working Group, and on approval then goes to the MDA SBIR Steering Group (which decides based on the same criteria plus funding availability). The steering group recommendation then passes to the MDA Selection Official, who has the final authority.

• How much money will you need to bring the technology to market, and how will you raise that money?
• Does your company contain marketing expertise and, if not, how do you intend to bring that expertise into the company?
• Who are your competitors, and what is your price and/or quality advantage over them?

The commercialization strategy must also include a schedule showing estimated commercial returns (i.e., amount of additional investment, sales revenue, etc.) one year after the start of Phase II, at the completion of Phase II, and after the completion of Phase III.

Finally, proposed cost-sharing by a third party has been an accepted tie-breaker between equivalent proposals since the program's inception. In the early 1990s, MDA (then known as SDIO/BMDO) began emphasizing co-investment as evidence of commercialization potential. Matching funds became a formal requirement for some parts of DoD SBIR with the implementation of Fast Track in 1996. The ratios used and the source requirements for third-party funds vary among components.

### 6.4.3 Composition of Selection Panels

Selection panels are comprised of DoD personnel. Two or three technical experts at the laboratory level review each proposal. Proposals are judged competitively on the basis of scientific, technical, and commercial merit in accordance with the selection criteria listed above.

Responsibility for each topic has been clearly established prior to the Phase I solicitation, so reviewers can access their proposals electronically immediately after the solicitation closes. This significantly shortens the decision cycle. If a proposal is beyond the expertise of the designated reviewers, the person with overall topic responsibility will obtain additional reviewers.

### 6.4.4 Fairness Review

Firms whose proposals were rejected can request a debriefing, which indicates how the proposal was scored on each specific evaluation criterion. The criteria discussed at debriefings must include only those that can fairly and properly be used for determining source selection. If practicable, the contracting officer and at least one engineer or scientist knowledgeable in the applicable field of technology conducts the debriefing, offering feedback on the weaknesses and strengths of the proposal, and how it might have been improved. The debriefing aims to ensure that the applicant fully understands why the proposal was not selected.

As recounted by firms and SBIR program officers, submitting an SBIR proposal to DoD is a learning process. Firm interviews indicate that many were initially unsuccessful, and that all had "losers" as well as "winners." There are many workshops available where firms can learn how to submit good SBIR proposals, but experience—including debriefings—is often the best teacher.[14]

### 6.4.5 Program Manager Role

Currently, the role of the SBIR program manager at the awarding agencies and components is largely administrative. It entails monitoring award decisions, reporting, and the expenditure of contract funds. Notably, program managers do *not* currently make award decisions at any of the components.

This role has changed over time. For example, prior to 1993, the Army program manager did decide who to fund, largely based on which R&D organizations first submitted sound recommendations for funding. Through the late 1990s, the MDA SBIR program manager had considerable influence over final decisions on awards.

In many cases, the maximum award given by a component is smaller than that allowed in SBA guidelines. Successful Phase II outcomes, which demonstrate the value of additional funds, are also often the basis for the addition of non-SBIR program funding. Because of the way DoD records awards, this makes it appear that DoD is awarding contracts much larger than SBA SBIR guidelines. But selection procedures and authority for additional funds lie with the acquisition program or the R&D organization, not the SBIR program manager.

### 6.4.6 Resubmission Procedures and Outcomes

If a Phase I proposal is not selected for award, firms may submit a very similar proposal for a topic in a subsequent solicitation, or submit a proposal in response to the solicitation of a different agency. A firm may also submit a very similar proposal to two or more DoD components or other agencies in the same solicitation if each component had an appropriate topic. In these cases, the firm must note that the other proposals are being submitted. If any proposal is awarded, the firm must inform the other agencies.

Resubmission of rejected Phase II proposals is more difficult. In most of DoD, aside from the Army, a rejected Phase II cannot be resubmitted for the same Phase I topic. The Army allows resubmission of a rejected Phase II proposal or submission of a Phase II proposal on a Phase I proposal from a prior solicitation year. Navy also encourages the Navy staff to find relevant Phase Is that did not go to Phase II, both from the Navy and other agencies and services, to meet new, related needs in a more timely fashion.

[14]When an agency makes an award, if only one proposal was submitted for a topic, that must be reported to SBA. DoD does not award single proposals. Not every topic results in an award, and all awards result from competition.

## 6.5 POST-AWARD TRAINING AND ASSISTANCE

DoD provides considerable information about sources of assistance for potential participants and awardees on its Internet sites. Program managers participate in workshops at national and regional SBIR conferences, and at various outreach activities to provide training to firms interested in participating or improving their performance in the SBIR program.

At the Navy, initial participation is required for all Phase II recipients, although not all choose to complete the entire program. The Navy Transition Assistance Program (TAP), formerly the Commercialization Assistance Program (CAP), was recently reoriented. The name change emphasizes the mission orientation of this program.

TAP is a 10-month program offered exclusively to SBIR and STTR Phase II award recipients. The program aims to (1) facilitate DoD use of Navy-funded SBIR technology; and (2) assist SBIR-funded firms to speed up the rate of technology transition through development of relationships with prime contractors, and by supporting preliminary strategic planning for Phase III. TAP also underwrites the Navy's Opportunity Forum, an annual event attended by prime contractors, other private sector companies, and representatives from various DoD agencies as well as SBIR awardees.

## 6.6 OUTREACH: PROGRAM INFORMATION SOURCES

The DoD Web site[15] provides extensive information that supports the preparation of proposals and negotiation of the contracts. Detailed information available via the web site provides in-depth information on the DoD program.

The DoD SBIR Help Desk, 1-866 SBIR HELP, is available to answer general and administrative questions. During pre-release, technical monitors answer technical questions about topics and agency needs.

DoD sponsors or participates each year in National SBIR Conferences. In addition, when state or regional activities sponsor SBIR events, one or more DoD SBIR program managers (dependent on the size of the event) generally participate. Such events provide information on the program including classes on specific aspects and usually provide opportunities for firms to have one-on-one meetings with a DoD program manager to address individual questions.

The schedule below was taken from the DoD Web site in December 2006. National conferences are published a year in advance, whereas other events are not usually known more than a quarter in advance.

---

[15]Available at <*http://www.acq.osd.mil/osbp/sbir/*>.

### National SBIR Conferences

- 2006 Fall National SBIR/STTR Conference, Milwaukee, WI, November 6–9, 2006.
- Beyond Phase II: Ready for Transition Conference, Crystal City, VA, August 20–23, 2007.
- 2007 Spring National SBIR Conference, Research Triangle Park, NC, April 30–May 3, 2007.
- 2007 Fall SBIR Conference, Richardson, TX, October 29–November 1, 2007.

### Other Events Where DoD SBIR Will Be Present

- Innovative Transitions 2006 Virginia's 12th Annual SBIR Conference, Herndon, VA, December 4–5, 2006.

## 6.7 FUNDING GAPS AND FUNDING INITIATIVES

Funding gaps between Phase I and Phase II proposals continue to present a financial problem for many SBIR awardees, especially for start-up and other smaller firms. The standard adjustment for firms addressing this gap is to reduce work on the project and to redeploy personnel to other funded projects. Larger firms with multiple SBIR awards or considerable prior experience with the program appear to treat the gaps as routine, if annoying, business liquidity problems.

For firms that do not have other sources of funding, funding gaps can require managers to shut down projects, lay off staff, and, go without salary for several months. An especially irksome aspect of the funding gap reported by some firms is that delays in funding do not always lead to adjustments by DoD technical monitors in the scheduling of Phase II deliverables.

Over the years, DoD has implemented a number of initiatives to help address these funding gap issues. Some of these are discussed below.

### 6.7.1 Reducing the Time to Contract

DoD has now formally introduced the objective of reducing the Phase II funding gap from an average of 11.5 months to 6 months.

DoD has limited influence over the actual pace of work under the Phase I award and how quickly firms prepare their Phase II submission following completion of this initial work. As with Fast Track, DoD can encourage early submission. However Phase I research can itself result in a change of direction for Phase II—so an early Phase II proposal may sometimes be inappropriate.

When DoD uses a centralized selection process, all Phase II proposals for that component are due the same day. But since the Phase I contracts are awarded by different contracting officers, some Phase I contracts are awarded before others; thus this part of the gap may vary. If the process is not centralized, Phase II evaluation may begin as soon as a proposal is received, which eliminates part of the potential gap.

The Phase II selection process itself is not the primary source of the Phase II funding gap. Most of the Phase II funding gap occurs *after* the Phase II award selection. While Phase I awards are small enough for the contracting officer to apply simplified contracting procedures, Phase II awards are too large for such procedures, and require a complex process consistent with FAR regulations.

Since 1996, DoD has substantially reduced the Phase I–II gap by speeding the evaluation process and conducting most of the post selection procedures in parallel rather than sequentially. One of the most time consuming activities is the audit of the firm's accounting procedures to ensure compliance with the FAR. After the audit relating to a firm's first Phase II is completed, no subsequent pre-award audits are required. However, since DoD attracts so many new entrants each year, many awardees do require an audit. The time involved includes scheduling an extremely busy Defense Contract Audit Agency, conducting the audit, the firm changing procedures if required and reinspections if needed. Firms new to the SBIR program are informed of the requirement prior to Phase I, and are provided with information on what is required in accounting; they are encouraged to begin the process early. All components have reduced the gap to six months or less, and have established procedures to provided gap funding.

### 6.7.2 SBIR Fast Track

As early as 1992, DoD's Ballistic Missile Defense Organization (BMDO) had begun to reward applications whose technologies demonstrated commercial private sector interest in the form of investment from outside sources. This BMDO "co-investment" initiative was effectively an informal "Fast Track" program.

In October 1995, DoD launched a broader Fast Track initiative to attract new firms and encourage commercialization of SBIR-funded technologies throughout the department. The initiative aims to improve commercialization through preferential evaluation and by providing tools for closing the Phase I–Phase II funding gap. The program expedites review of, and gives continuous funding during the traditional funding gap to Phase II proposals that demonstrated third-party financing for their technology. Under Fast Track, third-party financing means investment from another company or government agency; or investment in the firm from venture capital or some other private source. Internal funds did not qualify as matching funds.

The matching rates depend on whether the proposing firm has won previous SBIR Phase II awards.

Projects that obtain such outside investments and thereby qualify for the Fast Track will (subject to qualifications described in the solicitation):

- Receive interim funding of $30,000 to $50,000 between Phases I and II;
- Be evaluated for Phase II award under a separate, expedited process (the expedited decision-making process is acceptable to the agency because outside funding provides an additional form of validation for the commercial promise of the technology); and
- Be selected for Phase II award provided they meet or exceed a threshold of "technically sufficient" and have substantially met their Phase I technical goals.

Fast Track focuses SBIR funding on those projects that appear most likely to be developed into viable new products that DoD and others will buy and that will thereby make a contribution to U.S. military or economic capabilities. More broadly, the Fast Track program seeks to shorten government decision cycles in order to interact more effectively with small firms focused on rapidly evolving technologies.

Outside investors may include such entities as another company, a venture capital firm, an individual investor, or a non-SBIR, non-STTR government program; they do not include the owners of the small business, their family members, and/or affiliates of the small business.

Small companies report that they have found Fast Track to be an effective tool for encouraging investors to provide additional funds, by offering the opportunity for a match of between $1 and $4 in DoD SBIR funds for every $1 of third-party investment. Investors are essentially acquiring additional nondiluting capital with their investment.

Based on commissioned case studies, surveys, and empirical research, the National Academy's 2000 Fast Track report found that the Fast Track initiative was meeting its goals of encouraging commercialization and attracting new firms to the program,[16] as well as increasing the overall effectiveness of the DoD SBIR Program. The Academy recommended that Fast Track be continued and expanded where appropriate.

In recent years, the data suggest that firms and program managers are increasingly preferring to use Phase II Enhancement rather than Fast Track. Using the award year of the original Phase II as a baseline,[17] the data indicate that for Phase II awards made in 1997, 7 percent were Fast Track and 2 percent were subsequent winners of Phase II Enhancement. For 2002, 4 percent were Fast

[16]It is important to note the limitations to this research. The first limitation concerns the relatively short time that the Fast Track program has been in place. This necessarily limited the Committee's ability to assess the impact of the program. Secondly, although the case studies and surveys constituted what was clearly the largest independent assessment of the SBIR program at the Department of Defense, the study was nonetheless constrained by the limitations of the case-study approach and the size of the survey sample.

[17]Phase II Enhancements for a 2002 Phase II are actually awarded in 2004.

Track and 18 percent were Phase II Enhancement. For 2003 Fast Track awards fell to 2 percent.[18]

### 6.7.3 Phase II+ Programs

Phase II+ or Phase II SBIR Enhancement programs began in 1999 in the Army and the Navy. Army provided a dollar-for-dollar match up to $100,000 against third-party investment funds, in a project aimed at extending Phase II R&D efforts beyond the current Phase II contract to meet the needs of the investor, and to accelerate the Phase II project into the Phase III commercialization stage. The Navy program provided a 1:4 match against third-party funding of up to $250,000. Other services and agencies soon followed suit.[19]

The services and agencies vary widely in their implementation of enhancement programs, and these programs have also changed over time. The Army now defines "third-party investor" to mean Army (or other DoD) acquisition programs as well as the private sector. The Air Force selects a limited number of Phase II awardees for the Enhancement Program, which addresses technology barriers that were discovered during the Phase II work. These selected enhancements extend the existing Phase II contract award for up to one year, and provide a 1:1 match against up to $500,000 of non-SBIR funds.

The Navy essentially breaks its overall Phase II funding into a smaller than maximum Phase II contract plus an option. The latter is expected to be fully costed and well defined in a Phase II proposal, describing a test and evaluation plan or further R&D. Navy Phase II options typically fund an additional six months of research.

The Navy has now introduced a new Phase II Enhancement Plan to encourage transition of Navy SBIR-funded technology to the fleet. Since the law (PL102-564) permits Phase III awards during Phase II work, the Navy will provide a 1:4 match of Phase II to Phase III funds that the company obtains from an acquisition program. Up to $250,000 in additional SBIR funds can be provided against $1,000,000 in acquisition program funding, as long as the Phase III is awarded and funded during the Phase II.[20]

MDA also has a Phase II Enhancement policy. While not guaranteed, MDA may consider a limited number of Phase II enhancements on a case-by-case basis. Both the MDA and Navy programs are focused exclusively on supporting the

---

[18]DoD awards database.

[19]DoD's FY2006 solicitation states: To further encourage the transition of SBIR research into DoD acquisition programs as well as the private sector, each DoD component has developed its own Phase II Enhancement policy. Under this policy, the component will provide a Phase II company with additional Phase II SBIR funding if the company can match the additional SBIR funds with non-SBIR funds from DoD acquisition programs or the private sector. Generally, enhancements will extend an existing Phase II contract for up to one year and will match up to $250,000 of non-SBIR funds.

[20]DoD Small Business Resource Center, available at <*http://www.dodsbir.net/ft-ph2/*>.

transition of technologies into the services, not into private sector commercialization. The Air Force program has similar requirements.

### 6.7.4 DoD Programs for Closing the Phase I-Phase II Gap

DoD services and agencies vary in the upper level of support they provide using Phase I awards. This impacts whether they see a need to have separate gap funding initiatives in addition to Fast Track. DoD Phase I awards are typically $60,000 to $100,000 in size, and generally last for a period of six to nine months. Table 6-1 contains a summary of the provisions of each component's Phase I and Phase II awards.

#### 6.7.4.1 Navy

The Navy only accepts Phase I proposals with a base effort not exceeding $70,000 to be completed over six months. Options for contract extensions not exceeding $30,000 and three months are available to help address the transition into the Phase II effort. Phase I options are only funded after receipt of a Fast Track proposal or after the decision to fund the Phase II has been made. The Navy has thus effectively divided the permitted Phase I funding into two components; the second component is used as bridge funding between Phase I and Phase II as necessary.

#### 6.7.4.2 Air Force

The Air Force Phase I proposal covers a nine month effort, and can cost no more than $100,000 in total. Submission of the Phase II proposal at six months along with an interim Phase I report provides an additional funded period of three months while the Phase II proposal is being evaluated.

#### 6.7.4.3 Army

The Army has implemented a Phase I Option that can be exercised to provide gap funding while a Phase II contract is being negotiated. The Phase I maximum at Army is $70,000 over six months. The Phase I Option—which must be proposed as part of the Phase I proposal—covers activities over a period of up to four months with a maximum cost of $50,000. Only projects that receive an Army Phase II award are eligible to exercise the Phase I Option. Phase II funding is then reduced to keep the total cost for SBIR Phase I and Phase II at a maximum of $850,000.

**TABLE 6-1** Summary of Agency Specific Proposal Submission Requirements

| Component | Upload Technical Proposal | Online Form Cost Proposal Preparation | Phase I | | Phase II | |
|---|---|---|---|---|---|---|
| | | | Cost | Duration | Cost | Duration |
| Army | Required | Required | Base: NTE $70,000<br>Option: NTE $50,000 | 6 months<br>4 months | Year1 + Year2 NTE<br>$730,000 | 24 months |
| Navy | Required | | Base: NTE $70,000<br>Option: NTE $30,000 | 6 months<br>3 months | No limit, but in general<br>Base: $600,000<br>Option: $150,000 | 24 months<br>6 months |
| Air Force | Required | Required | NTE $100,000 | 9 months | No limit, but in general $750,000 | 24 months |
| DARPA | Required | | NTE $99,000 | 8 months | Base + Option: No Limit, but in general $750,000 | 24 months |
| MDA | Required | Required | NTE $70,000 | 6 months<br>(min) | Base: NTE $750,000<br>Option: NTE $250,000 | 24 months<br>(min) |
| DTRA | Required | | NTE $100,000 | 6 months | NTE $750,000 | 24 months |
| SOCOM | Required | | NTE $100,000 | 6 months | NTE $750,000 | 24 months |
| NIMA | Required | | NTE $100,000 | 9 months | Base: NTE $250,000<br>Option: NTE $250,000 | 12 months<br>12 months |
| OSD/Army<br>OSD/Navy<br>OSD/AF<br>OSD/SOCOM<br>OSD/DHP | Required<br>Required<br>Required<br>Required<br>Required | Required | NTE $100,000 | 6 months | NTE $750,000 | 24 months |
| CBD/Army<br>CBD/Navy<br>CBD/AF<br>CBD/SOCOM | Required<br>Required<br>Required<br>Required | Required<br>Required<br>Required<br>Required | Follows the Army, Navy, Air Force, or SOCOM cost and duration requirements shown above | | | |

SOURCE: Department of Defense Web site, accessed May 2007.

#### 6.7.4.4 DARPA

Phase I proposals cannot exceed $99,000, and cover a six month effort. Phase I contracts can only be extended if the DARPA TPOC decides to "gap" fund the effort to keep a company working while a Phase II proposal is being generated. The amount of gap funding depends on the funding available to the TPOC.

#### 6.7.4.5 MDA

MDA accepts Phase I proposals not exceeding $100,000, covering six months' work. Fast Track applications must be received by MDA 120 days prior to the Phase I award start date. Phase II proposals must be submitted within 180 days of the Phase I award start date. Phase I interim funding is not guaranteed. If awarded, it is usually limited to a maximum of $30,000. However, this funding is in addition to the $100,000 maximum awarded for Phase I.

#### 6.7.4.6 USSOCOM

The maximum amount of SBIR funding for a USSOCOM Phase I award is $100,000 and the maximum time frame for a Phase I proposal is 6 months.

## 6.8 DOD SBIR PROGRAM INITIATIVES

Chartered by the Principal Deputy Under Secretary of Defense for Acquisitions, the first solicitation of FY1996 marked the start of new initiatives resulting from the Process Action Team (PAT). These initiatives attempted to reduce the time between the start of proposal evaluation and eventual funding, and to address the need for improved communications between DoD and potential or current applicants.

### 6.8.1 Enhanced Applicant Information and Communications

One important initiative to improve information flows between DoD and applicants, the establishment of pre-release consultations, has been discussed above. Companies were also given access to better information and answers to DoD SBIR questions via DoD Web sites. A copy of a successful SBIR proposal was posted electronically, as were model Phase I and Phase II contracts.

Program outreach activities were enhanced by initiating pre-release on the Internet and in the Commerce Business Daily, where proposed solicitation topics are made available about 45 days prior to the formal release of the solicitation. OSD—in coordination with the component programs and OSADBU—also advertises the SBIR program at conferences likely to reach minority- and woman-owned small technology companies.

In 1997, the DoD SBIR Home Page (*<http://www.acq.osd.mil/osbp/sbir/>*) began offering electronic access to SBIR proposals and contracts, abstracts of ongoing SBIR projects, solicitations for SBIR and STTR programs, the latest updates on both programs, hyperlinks to sources of business assistance and financing, and other useful information. The early posting of Phase I abstracts shortly after award notification allowed potential investors to identify potential Phase I projects in which to invest. The 1997 solicitation also established the Commercialization Achievement Index (CAI) format for commercialization review.

DoD also established a 1-800 SBIR hot line[21] to answer general questions about the DoD SBIR program. This hot line was expanded in 1996 to provide assistance/information relevant to proposal preparation strategy, contract negotiation, government accounting requirements, and financing strategies.

### 6.8.2 Electronic Submission

In FY1999, the Navy required, and the Ballistic Missile Defense Office (BMDO)[22] encouraged, electronic submission of proposal cover sheets and abstracts. The Navy also directed that future Phase I and Phase II summary reports be submitted electronically. By the second solicitation of that FY1999, DoD had established a submission site (*<http://www.dodsbir.net/submission>*), which required all proposers to register and provide commercialization information on their prior Phase II awards electronically.

In 2000, the first entirely electronic submission of proposals occurred in DoD. CBD required, and USSOCOM allowed, complete proposals to be submitted electronically. The 2000 solicitation also stressed that DoD was using commercialization of technology (in military and/or private sector markets) as a critical measure of performance.

The last paper version of a DoD solicitation was printed and distributed during October 2002. All DoD SBIR solicitations have been available electronically since 1997. After 2002, the only source for the DoD solicitation was the submissions Web site.

In the first full use of electronic submissions in January 2003, DoD received substantially more proposals than were expected. The large number of submissions in the last three hours before the deadline[23] led to computer problems that resulted in several companies submitting late proposals. DoD reopened the

---

[21]By the first solicitation of FY1997, the hotline had been renamed the SBIR/STTR Help Desk and both a fax number and an email address were provided in addition to the phone number to provide alternate means for obtaining answers to SBIR questions.

[22]BMDO was the follow on organization to SDIO and the predecessor of MDA.

[23]Prior to this submission, most components required a mailed hard copy in addition to the electronic submission. Since the hard copy had to arrive by the closing date, most small businesses had to complete their proposal on line one or two days before the deadline to allow for mail delivery time. This first solicitation of 2003 was the first time no hard copy was required, resulting in many last minute submissions.

solicitation briefly to allow these companies to compete. DoD has not suffered similar problems since.

## 6.9 REPORTING REQUIREMENTS

Phase I final reports are generally required within 210 days of the award. Most are filed earlier, since Phase I funding is generally complete in 180 days, and the report is needed for a Phase II evaluation. As of 2004, all Phase I and Phase II reports must be submitted electronically on the DoD submission site.

Reports fill the contractual requirement for a deliverable. Their use varies widely based on the initiative of the technical monitor and the specific technology being investigated. However, discussions with agency staff suggest that more use could be made of these reports, especially if better tools were available for allowing interested parties to search them.

## 6.10 EVALUATION AND ASSESSMENT

The 1996–1997 Study of Commercialization of DoD SBIR was the only formal study conducted.[24] Each solicitation cycle, firms must submit their Company Commercialization Report as a part of their proposals.

In addition to use in evaluation, the DoD SADBU aggregates some of the information in the CCR and uses it to brief the DoD Principal Deputy for Acquisition and Technology on progress in the SBIR program. Using the information in the CCR, components identify successful projects and contact the firms to develop information for success stories and outreach brochures. Several components conduct annual awards ceremonies to recognize outstanding SBIR projects.

Starting in 1992, GAO has conducted a number of external reviews of the program, or aspects of the program. These include:

- **GAO/RCED-92-37.** *SBIR Shows Success but Can Be Strengthened.* **This is the first** baseline study of the program. It surveyed 100 percent of all Phase II awards from 1984–1987. It was conducted in 1990–1991.
- **GAO/RCED-95-59.** *Interim Report on SBIR.* Based on agency interviews conducted in 1994 and 1995, this report examined the quality of research and the duplication of projects.
- **GAO/RCED-98-132.** *Observations on the SBIR.* This report compared BRTRC's 1996 DoD survey (100 percent of Phase II awards from 1984–92) with the original GAO 1991 survey. It included an agency SBIR award database and interviews.
- **GAO/RCED-99-114.** *Evaluation of SBIR Can Be Strengthened.*

[24]BRTRC, Commercialization of DoD Small Business Innovation Research (SBIR) Summary Report, October 8, 1997, DoD Contract number DAAL01-94-C-0050, Mod P00010.

This assessment focused on use of commercialization records in proposal evaluation.

- **GAO-07-38.** *Small Business Innovation Research: Agencies Need to Strengthen Efforts to Improve the Completeness, Consistency, and Accuracy of Awards Data.*

In response to a congressional mandate for a review of SBIR at the five leading agencies, DoD has commissioned the NRC to undertake the current study. This review follows the previous NRC report on the Fast Track program at DoD which compared Fast Track firms with the regular SBIR program at DoD. During the NRC study's gestation, DoD program managers also commissioned a smaller, more focused study by RAND that was just recently completed.[25]

- NRC Fast Track.
- Navy Output Report (private).
- PART.
- Program report (50 slides).
- NavAir S&T report.

## 6.11 ADMINISTRATIVE FUNDING

The decentralized organization of SBIR at Defense makes it difficult to precisely determine how much administrative funding is spent on SBIR, or where that funding comes from. The DoD SBIR office is currently engaged in an effort to gather this information, but does not believe that precise accounting is likely, given the wide variety of inputs into the selection and management process, almost all of which is not directly charged to any SBIR budget line.[26]

Prior to the establishment of SBIR, each agency was presumed to be adequately staffed and funded to administer its R&D budget, and SBIR constituted only a change of direction, not an increase in R&D spending, so no additional administrative funding was anticipated. The SBIR legislation prohibits federal agencies from using any of the SBIR set-aside to administer the program. DoD thus incurs costs to administer the SBIR program—and interviews with staff suggest that it is more expensive to operate a program with hundreds of small contracts than with a single large contract—but receives no offsetting line item appropriation.

Each service and agency has had to absorb the costs of managing its SBIR program out of existing budgets. Within the components, this decentralization continues. For example, the Navy SBIR program office controls the budget for

[25]Bruce Held, Thomas Edison, Shari Lawrence Pfleeger, Philip Anton, and John Clancy, *Evaluation and Recommendations for Improvement of the Department of Defense Small Business Innovation Research (SBIR) Program*, Arlington, VA: RAND National Defense Research Institute, 2006.

[26]Interview with Michael Caccuitto, DoD SBIR/STTR Program Administrator, November 27, 2006.

its office; each major Navy component (such as NAVSEA or NAVAIR) controls its own SBIR program budget, and so on, down to the laboratory level.

At the service or agency level, there is an SBIR program manager and perhaps a program office, which includes contract staff support, as well as a budget that covers travel expenses. Within the larger components there are SBIR managers (and offices in some cases) at lower level commands and development agencies. Some positions are full time; other SBIR managers have additional duties as well.

At the project level, there are large numbers of technical monitors (TPOCs), who work part time on one or more SBIR projects. Their salary and travel are not specifically associated with SBIR in the components. Similarly, no separate budgets exist to support the contracting officers and legal support necessary for the operation of the SBIR program.

At the DoD level, the DoD SADBU controls the budget for that office. Similarly the DDR&E controls its budget. Neither of these SBIR-associated offices allocates or controls the SBIR administrative budget of any component.

Even if line item amounts were available for contract, legal, audit and finance support, these budgets would likely not include salaries, travel, and other expenses for the hundreds of technical monitors throughout DoD, who may spend five to fifty days a year writing topics, reviewing SBIR proposals, or monitoring SBIR awards as the Contracting Officer's Technical Representative (COTR). Imputations of the costs incurred by these activities are possible, but have not been done. Thus, no estimate of the cost to DoD of managing its SBIR program currently exists.

Yet while precise budgeting is not possible under the current organization and financial architecture, it is clear that some agencies provide substantially more administrative funding than others.[27] The Navy's SBIR program in particular has been funded at a level of approximately $20 million.[28] This has allowed the Navy SBIR program to innovate in important ways—via the TAP program, for example, and also through enhanced evaluation and assessment efforts. This level of agency commitment is not matched at other agencies, where significantly less administrative funding is available.

Both for purposes of evaluation and management, it is important to better understand the program's operations and the impact of various procedures of program innovation. To do so, more management and evaluation resources are required, as the Navy has demonstrated. Given the substantial size of the current SBIR program at Defense, additional management funds would seem to be warranted and are likely to be cost effective.

---

[27]It should, however, be noted that close comparisons are not self-evident, because each DoD agency funds its administrative work differently, especially in the SBIR program where so many other functions (TPOCs, administrators , topic reviewers, proposal reviewers) work on SBIR and other projects without being attached to any SBIR line item.

[28]John Williams, Navy SBIR Program Manager, Private communication.

# Appendixes

# Appendix A

# DoD Data Book

## AWARDS

**TABLE App-A-1** DoD Phase I and Phase II Awards, per Year

| Year | Number of Phase I Awards | Number of Phase II Awards | Total |
|---|---|---|---|
| 1992 | 1,065 | 433 | 1,498 |
| 1993 | 1,303 | 591 | 1,894 |
| 1994 | 1,370 | 406 | 1,776 |
| 1995 | 1,262 | 575 | 1,837 |
| 1996 | 1,372 | 611 | 1,983 |
| 1997 | 1,526 | 638 | 2,164 |
| 1998 | 1,286 | 672 | 1,958 |
| 1999 | 1,393 | 568 | 1,961 |
| 2000 | 1,220 | 626 | 1,846 |
| 2001 | 1,310 | 702 | 2,012 |
| 2002 | 2,162 | 661 | 2,823 |
| 2003 | 2,113 | 1,078 | 3,191 |
| 2004 | 2,075 | 1,173 | 3,248 |
| 2005 | 2,344 | 998 | 3,342 |
| Total | 21,801 | 9,732 | 31,533 |

SOURCE: Department of Defense.

**TABLE App-A-2** DoD Phase I Awards by Component, per Year

| | Number of Phase I Awards | | | | | | | | | | | | | |
|---|---|---|---|---|---|---|---|---|---|---|---|---|---|---|
| Year | AF | ARMY | BMDO | CBD | DARPA | DSWA | DTRA | MDA | NAVY | NGA | NIMA | OSD | SOCOM | Total |
| 1992 | 230 | 224 | 0 | 0 | 169 | 24 | 0 | 209 | 209 | 0 | 0 | 0 | 0 | 1,065 |
| 1993 | 470 | 288 | 0 | 0 | 70 | 19 | 0 | 147 | 306 | 0 | 0 | 0 | 3 | 1,303 |
| 1994 | 390 | 175 | 0 | 0 | 328 | 9 | 0 | 161 | 290 | 0 | 0 | 0 | 17 | 1,370 |
| 1995 | 427 | 252 | 0 | 0 | 31 | 11 | 0 | 116 | 373 | 0 | 0 | 42 | 10 | 1,262 |
| 1996 | 310 | 247 | 0 | 0 | 237 | 15 | 0 | 140 | 361 | 0 | 0 | 56 | 6 | 1,372 |
| 1997 | 431 | 291 | 0 | 0 | 146 | 19 | 0 | 189 | 380 | 0 | 0 | 59 | 11 | 1,526 |
| 1998 | 471 | 200 | 167 | 20 | 44 | 0 | 9 | 11 | 229 | 0 | 3 | 115 | 17 | 1,286 |
| 1999 | 416 | 197 | 170 | 20 | 119 | 0 | 26 | 0 | 406 | 0 | 2 | 29 | 8 | 1,393 |
| 2000 | 367 | 198 | 230 | 26 | 82 | 0 | 20 | 0 | 216 | 0 | 6 | 56 | 19 | 1,220 |
| 2001 | 373 | 281 | 161 | 18 | 82 | 0 | 24 | 0 | 241 | 0 | 2 | 108 | 20 | 1,310 |
| 2002 | 427 | 282 | 2 | 34 | 120 | 0 | 23 | 520 | 578 | 0 | 4 | 128 | 44 | 2,162 |
| 2003 | 449 | 352 | 0 | 25 | 90 | 0 | 7 | 454 | 550 | 0 | 2 | 156 | 28 | 2,113 |
| 2004 | 527 | 356 | 0 | 27 | 155 | 0 | 0 | 315 | 585 | 2 | 0 | 83 | 25 | 2,075 |
| 2005 | 608 | 705 | 0 | 21 | 74 | 0 | 40 | 240 | 466 | 2 | 0 | 163 | 25 | 2,344 |
| Total | 5,896 | 4,048 | 730 | 191 | 1,747 | 97 | 149 | 2,502 | 5,190 | 4 | 19 | 995 | 233 | 21,801 |

SOURCE: Department of Defense.

**TABLE App-A-3** DoD Phase II Awards by Component, per Year

| | Number of Phase II Awards | | | | | | | | | | | | | |
|---|---|---|---|---|---|---|---|---|---|---|---|---|---|---|
| Year | AF | ARMY | BMDO | CBD | DARPA | DSWA | DTRA | MDA | NAVY | NGA | NIMA | OSD | SOCOM | Total |
| 1992 | 110 | 172 | 0 | 0 | 41 | 0 | 12 | 46 | 52 | 0 | 0 | 0 | 0 | 433 |
| 1993 | 199 | 123 | 0 | 0 | 83 | 0 | 6 | 60 | 120 | 0 | 0 | 0 | 0 | 591 |
| 1994 | 152 | 78 | 0 | 0 | 43 | 0 | 3 | 23 | 107 | 0 | 0 | 0 | 0 | 406 |
| 1995 | 191 | 131 | 0 | 0 | 84 | 0 | 6 | 34 | 127 | 0 | 0 | 0 | 2 | 575 |
| 1996 | 217 | 100 | 0 | 0 | 78 | 0 | 3 | 36 | 156 | 0 | 0 | 16 | 5 | 611 |
| 1997 | 221 | 113 | 0 | 0 | 57 | 0 | 6 | 51 | 164 | 0 | 0 | 23 | 3 | 638 |
| 1998 | 243 | 111 | 0 | 0 | 71 | 0 | 7 | 69 | 136 | 0 | 0 | 33 | 2 | 672 |
| 1999 | 212 | 105 | 0 | 6 | 44 | 0 | 7 | 43 | 107 | 0 | 0 | 37 | 7 | 568 |
| 2000 | 195 | 112 | 0 | 8 | 45 | 0 | 5 | 56 | 186 | 1 | 0 | 13 | 5 | 626 |
| 2001 | 221 | 180 | 0 | 7 | 53 | 0 | 6 | 59 | 136 | 3 | 0 | 29 | 8 | 702 |
| 2002 | 233 | 124 | 0 | 11 | 47 | 0 | 6 | 28 | 170 | 1 | 0 | 36 | 5 | 661 |
| 2003 | 234 | 323 | 0 | 10 | 66 | 0 | 3 | 184 | 193 | 1 | 0 | 50 | 14 | 1,078 |
| 2004 | 317 | 273 | 0 | 11 | 75 | 0 | 3 | 211 | 212 | 2 | 0 | 60 | 9 | 1,173 |
| 2005 | 339 | 123 | 0 | 7 | 80 | 0 | 4 | 102 | 290 | 1 | 0 | 38 | 14 | 998 |
| Total | 3,084 | 2,068 | 0 | 60 | 867 | 0 | 77 | 1,002 | 2,156 | 9 | 0 | 335 | 74 | 9,732 |

SOURCE: Department of Defense.

**TABLE App-A-4** DoD Total SBIR Awards by Component, per Year

| Year | Number of SBIR Awards | | | | | | | | | | | | | |
|---|---|---|---|---|---|---|---|---|---|---|---|---|---|---|
| | AF | ARMY | BMDO | CBD | DARPA | DSWA | DTRA | MDA | NAVY | NGA | NIMA | OSD | SOCOM | Total |
| 1992 | 340 | 396 | 0 | 0 | 210 | 24 | 12 | 255 | 261 | 0 | 0 | 0 | 0 | 1,498 |
| 1993 | 669 | 411 | 0 | 0 | 153 | 19 | 6 | 207 | 426 | 0 | 0 | 0 | 3 | 1,894 |
| 1994 | 542 | 253 | 0 | 0 | 371 | 9 | 3 | 184 | 397 | 0 | 0 | 0 | 17 | 1,776 |
| 1995 | 618 | 383 | 0 | 0 | 115 | 11 | 6 | 150 | 500 | 0 | 0 | 42 | 12 | 1,837 |
| 1996 | 527 | 347 | 0 | 0 | 315 | 15 | 3 | 176 | 517 | 0 | 0 | 72 | 11 | 1,983 |
| 1997 | 652 | 404 | 0 | 0 | 203 | 19 | 6 | 240 | 544 | 0 | 0 | 82 | 14 | 2,164 |
| 1998 | 714 | 311 | 167 | 20 | 115 | 0 | 16 | 80 | 365 | 0 | 3 | 148 | 19 | 1,958 |
| 1999 | 628 | 302 | 170 | 26 | 163 | 0 | 33 | 43 | 513 | 0 | 2 | 66 | 15 | 1,961 |
| 2000 | 562 | 310 | 230 | 34 | 127 | 0 | 25 | 56 | 402 | 1 | 6 | 69 | 24 | 1,846 |
| 2001 | 594 | 461 | 161 | 25 | 135 | 0 | 30 | 59 | 377 | 3 | 2 | 137 | 28 | 2,012 |
| 2002 | 660 | 406 | 2 | 45 | 167 | 0 | 29 | 548 | 748 | 1 | 4 | 164 | 49 | 2,823 |
| 2003 | 683 | 675 | 0 | 35 | 156 | 0 | 10 | 638 | 743 | 1 | 2 | 206 | 42 | 3,191 |
| 2004 | 844 | 629 | 0 | 38 | 230 | 0 | 3 | 526 | 797 | 4 | 0 | 143 | 34 | 3,248 |
| 2005 | 947 | 828 | 0 | 28 | 154 | 0 | 44 | 342 | 756 | 3 | 0 | 201 | 39 | 3,342 |
| Total | 8,980 | 6,116 | 730 | 251 | 2,614 | 97 | 226 | 3,504 | 7,346 | 13 | 19 | 1,330 | 307 | 31,533 |

SOURCE: Department of Defense.

**TABLE App-A-5** DoD Phase I Awards by Component, per Year (Percent of Annual Awards)

| | Phase I Awards (Percent of Annual Awards) | | | | | | | | | | | | | |
|---|---|---|---|---|---|---|---|---|---|---|---|---|---|---|
| Year | AF | ARMY | BMDO | CBD | DARPA | DSWA | DTRA | MDA | NAVY | NGA | NIMA | OSD | SOCOM | Total |
| 1992 | 21.6 | 21.0 | 0.0 | 0.0 | 15.9 | 2.3 | 0.0 | 19.6 | 19.6 | 0.0 | 0.0 | 0.0 | 0.0 | 100.0 |
| 1993 | 36.1 | 22.1 | 0.0 | 0.0 | 5.4 | 1.5 | 0.0 | 11.3 | 23.5 | 0.0 | 0.0 | 0.0 | 0.2 | 100.0 |
| 1994 | 28.5 | 12.8 | 0.0 | 0.0 | 23.9 | 0.7 | 0.0 | 11.8 | 21.2 | 0.0 | 0.0 | 0.0 | 1.2 | 100.0 |
| 1995 | 33.8 | 20.0 | 0.0 | 0.0 | 2.5 | 0.9 | 0.0 | 9.2 | 29.6 | 0.0 | 0.0 | 3.3 | 0.8 | 100.0 |
| 1996 | 22.6 | 18.0 | 0.0 | 0.0 | 17.3 | 1.1 | 0.0 | 10.2 | 26.3 | 0.0 | 0.0 | 4.1 | 0.4 | 100.0 |
| 1997 | 28.2 | 19.1 | 0.0 | 0.0 | 9.6 | 1.2 | 0.0 | 12.4 | 24.9 | 0.0 | 0.0 | 3.9 | 0.7 | 100.0 |
| 1998 | 36.6 | 15.6 | 13.0 | 1.6 | 3.4 | 0.0 | 0.7 | 0.9 | 17.8 | 0.0 | 0.2 | 8.9 | 1.3 | 100.0 |
| 1999 | 29.9 | 14.1 | 12.2 | 1.4 | 8.5 | 0.0 | 1.9 | 0.0 | 29.1 | 0.0 | 0.1 | 2.1 | 0.6 | 100.0 |
| 2000 | 30.1 | 16.2 | 18.9 | 2.1 | 6.7 | 0.0 | 1.6 | 0.0 | 17.7 | 0.0 | 0.5 | 4.6 | 1.6 | 100.0 |
| 2001 | 28.5 | 21.5 | 12.3 | 1.4 | 6.3 | 0.0 | 1.8 | 0.0 | 18.4 | 0.0 | 0.2 | 8.2 | 1.5 | 100.0 |
| 2002 | 19.8 | 13.0 | 0.1 | 1.6 | 5.6 | 0.0 | 1.1 | 24.1 | 26.7 | 0.0 | 0.2 | 5.9 | 2.0 | 100.0 |
| 2003 | 21.2 | 16.7 | 0.0 | 1.2 | 4.3 | 0.0 | 0.3 | 21.5 | 26.0 | 0.0 | 0.1 | 7.4 | 1.3 | 100.0 |
| 2004 | 25.4 | 17.2 | 0.0 | 1.3 | 7.5 | 0.0 | 0.0 | 15.2 | 28.2 | 0.1 | 0.0 | 4.0 | 1.2 | 100.0 |
| 2005 | 25.9 | 30.1 | 0.0 | 0.9 | 3.2 | 0.0 | 1.7 | 10.2 | 19.9 | 0.1 | 0.0 | 7.0 | 1.1 | 100.0 |
| Total | 27.0 | 18.6 | 3.3 | 0.9 | 8.0 | 0.4 | 0.7 | 11.5 | 23.8 | 0.0 | 0.1 | 4.6 | 1.1 | 100.0 |

SOURCE: Department of Defense.

**TABLE App-A-6** DoD Phase II Awards by Component, per Year (Percent of Annual Awards)

| | Phase II Awards (Percent of Annual Awards) | | | | | | | | | | | | | |
|---|---|---|---|---|---|---|---|---|---|---|---|---|---|---|
| Year | AF | ARMY | BMDO | CBD | DARPA | DSWA | DTRA | MDA | NAVY | NGA | NIMA | OSD | SOCOM | Total |
| 1992 | 25.4 | 39.7 | 0.0 | 0.0 | 9.5 | 0.0 | 2.8 | 10.6 | 12.0 | 0.0 | 0.0 | 0.0 | 0.0 | 100.0 |
| 1993 | 33.7 | 20.8 | 0.0 | 0.0 | 14.0 | 0.0 | 1.0 | 10.2 | 20.3 | 0.0 | 0.0 | 0.0 | 0.0 | 100.0 |
| 1994 | 37.4 | 19.2 | 0.0 | 0.0 | 10.6 | 0.0 | 0.7 | 5.7 | 26.4 | 0.0 | 0.0 | 0.0 | 0.0 | 100.0 |
| 1995 | 33.2 | 22.8 | 0.0 | 0.0 | 14.6 | 0.0 | 1.0 | 5.9 | 22.1 | 0.0 | 0.0 | 0.0 | 0.3 | 100.0 |
| 1996 | 35.5 | 16.4 | 0.0 | 0.0 | 12.8 | 0.0 | 0.5 | 5.9 | 25.5 | 0.0 | 0.0 | 2.6 | 0.8 | 100.0 |
| 1997 | 34.6 | 17.7 | 0.0 | 0.0 | 8.9 | 0.0 | 0.9 | 8.0 | 25.7 | 0.0 | 0.0 | 3.6 | 0.5 | 100.0 |
| 1998 | 36.2 | 16.5 | 0.0 | 0.0 | 10.6 | 0.0 | 1.0 | 10.3 | 20.2 | 0.0 | 0.0 | 4.9 | 0.3 | 100.0 |
| 1999 | 37.3 | 18.5 | 0.0 | 1.1 | 7.7 | 0.0 | 1.2 | 7.6 | 18.8 | 0.0 | 0.0 | 6.5 | 1.2 | 100.0 |
| 2000 | 31.2 | 17.9 | 0.0 | 1.3 | 7.2 | 0.0 | 0.8 | 8.9 | 29.7 | 0.2 | 0.0 | 2.1 | 0.8 | 100.0 |
| 2001 | 31.5 | 25.6 | 0.0 | 1.0 | 7.5 | 0.0 | 0.9 | 8.4 | 19.4 | 0.4 | 0.0 | 4.1 | 1.1 | 100.0 |
| 2002 | 35.2 | 18.8 | 0.0 | 1.7 | 7.1 | 0.0 | 0.9 | 4.2 | 25.7 | 0.2 | 0.0 | 5.4 | 0.8 | 100.0 |
| 2003 | 21.7 | 30.0 | 0.0 | 0.9 | 6.1 | 0.0 | 0.3 | 17.1 | 17.9 | 0.1 | 0.0 | 4.6 | 1.3 | 100.0 |
| 2004 | 27.0 | 23.3 | 0.0 | 0.9 | 6.4 | 0.0 | 0.3 | 18.0 | 18.1 | 0.2 | 0.0 | 5.1 | 0.8 | 100.0 |
| 2005 | 34.0 | 12.3 | 0.0 | 0.7 | 8.0 | 0.0 | 0.4 | 10.2 | 29.1 | 0.1 | 0.0 | 3.8 | 1.4 | 100.0 |
| Total | 31.7 | 21.2 | 0.0 | 0.6 | 8.9 | 0.0 | 0.8 | 10.3 | 22.2 | 0.1 | 0.0 | 3.4 | 0.8 | 100.0 |

SOURCE: Department of Defense.

**TABLE App-A-7** DoD Phase I and Phase II Awards by Component, per Year (Percent of Annual Awards)

| | Phase I and Phase II Awards (Percent of Annual Awards) | | | | | | | | | | | | | |
|---|---|---|---|---|---|---|---|---|---|---|---|---|---|---|
| Year | AF | ARMY | BMDO | CBD | DARPA | DSWA | DTRA | MDA | NAVY | NGA | NIMA | OSD | SOCOM | Total |
| 1992 | 22.7 | 26.4 | 0.0 | 0.0 | 14.0 | 1.6 | 0.8 | 17.0 | 17.4 | 0.0 | 0.0 | 0.0 | 0.0 | 100.0 |
| 1993 | 35.3 | 21.7 | 0.0 | 0.0 | 8.1 | 1.0 | 0.3 | 10.9 | 22.5 | 0.0 | 0.0 | 0.0 | 0.2 | 100.0 |
| 1994 | 30.5 | 14.2 | 0.0 | 0.0 | 20.9 | 0.5 | 0.2 | 10.4 | 22.4 | 0.0 | 0.0 | 0.0 | 1.0 | 100.0 |
| 1995 | 33.6 | 20.8 | 0.0 | 0.0 | 6.3 | 0.6 | 0.3 | 8.2 | 27.2 | 0.0 | 0.0 | 2.3 | 0.7 | 100.0 |
| 1996 | 26.6 | 17.5 | 0.0 | 0.0 | 15.9 | 0.8 | 0.2 | 8.9 | 26.1 | 0.0 | 0.0 | 3.6 | 0.6 | 100.0 |
| 1997 | 30.1 | 18.7 | 0.0 | 0.0 | 9.4 | 0.9 | 0.3 | 11.1 | 25.1 | 0.0 | 0.0 | 3.8 | 0.6 | 100.0 |
| 1998 | 36.5 | 15.9 | 8.5 | 1.0 | 5.9 | 0.0 | 0.8 | 4.1 | 18.6 | 0.0 | 0.2 | 7.6 | 1.0 | 100.0 |
| 1999 | 32.0 | 15.4 | 8.7 | 1.3 | 8.3 | 0.0 | 1.7 | 2.2 | 26.2 | 0.0 | 0.1 | 3.4 | 0.8 | 100.0 |
| 2000 | 30.4 | 16.8 | 12.5 | 1.8 | 6.9 | 0.0 | 1.4 | 3.0 | 21.8 | 0.1 | 0.3 | 3.7 | 1.3 | 100.0 |
| 2001 | 29.5 | 22.9 | 8.0 | 1.2 | 6.7 | 0.0 | 1.5 | 2.9 | 18.7 | 0.1 | 0.1 | 6.8 | 1.4 | 100.0 |
| 2002 | 23.4 | 14.4 | 0.1 | 1.6 | 5.9 | 0.0 | 1.0 | 19.4 | 26.5 | 0.0 | 0.1 | 5.8 | 1.7 | 100.0 |
| 2003 | 21.4 | 21.2 | 0.0 | 1.1 | 4.9 | 0.0 | 0.3 | 20.0 | 23.3 | 0.0 | 0.1 | 6.5 | 1.3 | 100.0 |
| 2004 | 26.0 | 19.4 | 0.0 | 1.2 | 7.1 | 0.0 | 0.1 | 16.2 | 24.5 | 0.1 | 0.0 | 4.4 | 1.0 | 100.0 |
| 2005 | 28.3 | 24.8 | 0.0 | 0.8 | 4.6 | 0.0 | 1.3 | 10.2 | 22.6 | 0.1 | 0.0 | 6.0 | 1.2 | 100.0 |
| Total | 28.5 | 19.4 | 2.3 | 0.8 | 8.3 | 0.3 | 0.7 | 11.1 | 23.3 | 0.0 | 0.1 | 4.2 | 1.0 | 100.0 |

SOURCE: Department of Defense.

**TABLE App-A-8** DoD Phase I and Phase II Awards, per State (1992–2005)

| Phase I Awards | | | Phase II Awards | | | Total Awards | | |
|---|---|---|---|---|---|---|---|---|
| State | Number | Percent | State | Number | Percent | State | Number | Percent |
| CA | 4,929 | 22.61 | CA | 2,198 | 22.59 | CA | 7,127 | 22.60 |
| MA | 3,212 | 14.73 | MA | 1,412 | 14.51 | MA | 4,624 | 14.66 |
| VA | 1,607 | 7.37 | VA | 741 | 7.61 | VA | 2,348 | 7.45 |
| MD | 1,047 | 4.80 | OH | 447 | 4.59 | MD | 1,478 | 4.69 |
| OH | 929 | 4.26 | CO | 435 | 4.47 | OH | 1,376 | 4.36 |
| CO | 928 | 4.26 | MD | 431 | 4.43 | CO | 1,363 | 4.32 |
| NY | 869 | 3.99 | PA | 413 | 4.24 | NY | 1,269 | 4.02 |
| TX | 848 | 3.89 | NY | 400 | 4.11 | PA | 1,250 | 3.96 |
| PA | 837 | 3.84 | TX | 346 | 3.56 | TX | 1,194 | 3.79 |
| NJ | 687 | 3.15 | NJ | 309 | 3.18 | NJ | 996 | 3.16 |
| FL | 582 | 2.67 | FL | 240 | 2.47 | FL | 822 | 2.61 |
| AL | 513 | 2.35 | AL | 227 | 2.33 | AL | 740 | 2.35 |
| AZ | 461 | 2.11 | NM | 203 | 2.09 | NM | 659 | 2.09 |
| NM | 456 | 2.09 | MI | 194 | 1.99 | AZ | 616 | 1.95 |
| MI | 420 | 1.93 | CT | 191 | 1.96 | MI | 614 | 1.95 |
| CT | 392 | 1.80 | WA | 165 | 1.70 | CT | 583 | 1.85 |
| WA | 326 | 1.50 | AZ | 155 | 1.59 | WA | 491 | 1.56 |
| MN | 294 | 1.35 | MN | 127 | 1.30 | MN | 421 | 1.34 |
| NH | 248 | 1.14 | NH | 124 | 1.27 | NH | 372 | 1.18 |
| IL | 214 | 0.98 | IL | 94 | 0.97 | IL | 308 | 0.98 |
| GA | 202 | 0.93 | GA | 92 | 0.95 | GA | 294 | 0.93 |
| NC | 178 | 0.82 | NC | 79 | 0.81 | NC | 257 | 0.82 |
| UT | 151 | 0.69 | UT | 76 | 0.78 | UT | 227 | 0.72 |
| TN | 142 | 0.65 | TN | 69 | 0.71 | TN | 211 | 0.67 |
| OR | 140 | 0.64 | OR | 63 | 0.65 | OR | 203 | 0.64 |
| IN | 94 | 0.43 | NV | 48 | 0.49 | IN | 130 | 0.41 |

*continued*

**TABLE App-A-8** Continued

| Phase I Awards | | | Phase II Awards | | | Total Awards | | |
|---|---|---|---|---|---|---|---|---|
| State | Number | Percent | State | Number | Percent | State | Number | Percent |
| MO | 92 | 0.42 | WI | 46 | 0.47 | WI | 129 | 0.41 |
| OK | 84 | 0.39 | IN | 36 | 0.37 | MO | 122 | 0.39 |
| WI | 83 | 0.38 | RI | 32 | 0.33 | NV | 117 | 0.37 |
| RI | 73 | 0.33 | MO | 30 | 0.31 | OK | 113 | 0.36 |
| NV | 69 | 0.32 | OK | 29 | 0.30 | RI | 105 | 0.33 |
| ME | 63 | 0.29 | ME | 26 | 0.27 | ME | 89 | 0.28 |
| HI | 59 | 0.27 | WV | 25 | 0.26 | HI | 79 | 0.25 |
| LA | 57 | 0.26 | MT | 25 | 0.26 | DE | 77 | 0.24 |
| DE | 56 | 0.26 | KS | 22 | 0.23 | LA | 75 | 0.24 |
| SC | 56 | 0.26 | DE | 21 | 0.22 | SC | 75 | 0.24 |
| WV | 50 | 0.23 | HI | 20 | 0.21 | WV | 75 | 0.24 |
| DC | 49 | 0.22 | VT | 19 | 0.20 | MT | 72 | 0.23 |
| MT | 47 | 0.22 | SC | 19 | 0.20 | KS | 67 | 0.21 |
| KS | 45 | 0.21 | LA | 18 | 0.18 | DC | 62 | 0.20 |
| VT | 43 | 0.20 | ID | 13 | 0.13 | VT | 62 | 0.20 |
| ID | 36 | 0.17 | DC | 13 | 0.13 | ID | 49 | 0.16 |
| MS | 26 | 0.12 | NE | 11 | 0.11 | MS | 37 | 0.12 |
| IA | 21 | 0.10 | MS | 11 | 0.11 | IA | 30 | 0.10 |
| KY | 20 | 0.09 | IA | 9 | 0.09 | NE | 29 | 0.09 |
| NE | 18 | 0.08 | AR | 8 | 0.08 | KY | 27 | 0.09 |
| AR | 15 | 0.07 | KY | 7 | 0.07 | AR | 23 | 0.07 |
| WY | 10 | 0.05 | ND | 6 | 0.06 | ND | 15 | 0.05 |
| ND | 9 | 0.04 | WY | 3 | 0.03 | WY | 13 | 0.04 |
| AK | 7 | 0.03 | SD | 2 | 0.02 | AK | 9 | 0.03 |
| SD | 7 | 0.03 | AK | 2 | 0.02 | SD | 9 | 0.03 |
| Total | 21,801 | | Total | 9,732 | | Total | 31,533 | |

SOURCE: Department of Defense.

**TABLE App-A-9** DoD Awards by Demographics

DoD Awards to Woman-owned Firms

| Year | Number of Phase I Awards | Number of Phase II Awards | Total |
|---|---|---|---|
| 1992 | 89 | 29 | 118 |
| 1993 | 110 | 49 | 159 |
| 1994 | 118 | 36 | 154 |
| 1995 | 101 | 53 | 154 |
| 1996 | 134 | 51 | 185 |
| 1997 | 129 | 58 | 187 |
| 1998 | 108 | 62 | 170 |
| 1999 | 129 | 44 | 173 |
| 2000 | 110 | 62 | 172 |
| 2001 | 141 | 75 | 216 |
| 2002 | 219 | 60 | 279 |
| 2003 | 213 | 110 | 323 |
| 2004 | 206 | 117 | 323 |
| 2005 | 303 | 132 | 435 |
| Total | 2,110 | 938 | 3,048 |

DoD Awards to Minority-owned Firms

| Year | Number of Phase I Awards | Number of Phase II Awards | Total |
|---|---|---|---|
| 1992 | 127 | 67 | 194 |
| 1993 | 190 | 69 | 259 |
| 1994 | 199 | 43 | 242 |
| 1995 | 180 | 89 | 269 |
| 1996 | 197 | 86 | 283 |
| 1997 | 196 | 117 | 313 |
| 1998 | 137 | 107 | 244 |
| 1999 | 195 | 73 | 268 |
| 2000 | 143 | 74 | 217 |
| 2001 | 148 | 80 | 228 |
| 2002 | 249 | 63 | 312 |
| 2003 | 229 | 98 | 327 |
| 2004 | 200 | 102 | 302 |
| 2005 | 219 | 116 | 335 |
| Total | 2,609 | 1,184 | 3,793 |

DoD Awards to Minority Woman-owned Firms

| Year | Number of Phase I Awards | Number of Phase II Awards | Total |
|---|---|---|---|
| 1992 | 32 | 8 | 40 |
| 1993 | 31 | 7 | 38 |
| 1994 | 20 | 5 | 25 |
| 1995 | 25 | 12 | 37 |
| 1996 | 27 | 11 | 38 |
| 1997 | 31 | 18 | 49 |
| 1998 | 26 | 14 | 40 |
| 1999 | 31 | 8 | 39 |
| 2000 | 19 | 10 | 29 |
| 2001 | 28 | 9 | 37 |
| 2002 | 51 | 14 | 65 |
| 2003 | 34 | 19 | 53 |
| 2004 | 38 | 16 | 54 |
| 2005 | 43 | 23 | 66 |
| Total | 436 | 174 | 610 |

DoD Awards to Either Minority- or Woman-owned Firms

| Year | Number of Phase I Awards | Number of Phase II Awards | Total |
|---|---|---|---|
| 1992 | 184 | 88 | 272 |
| 1993 | 269 | 111 | 380 |
| 1994 | 297 | 74 | 371 |
| 1995 | 256 | 130 | 386 |
| 1996 | 304 | 126 | 430 |
| 1997 | 294 | 157 | 451 |
| 1998 | 219 | 155 | 374 |
| 1999 | 293 | 109 | 402 |
| 2000 | 234 | 126 | 360 |
| 2001 | 261 | 146 | 407 |
| 2002 | 417 | 109 | 526 |
| 2003 | 408 | 189 | 597 |
| 2004 | 368 | 203 | 571 |
| 2005 | 479 | 225 | 704 |
| Total | 4,283 | 1,948 | 6,231 |

SOURCE: Department of Defense.

**TABLE App-A-10** Award Distribution

| | Mean ($) | | Count | | Minimum ($) | | Maximum ($) | | Sum ($) | | Median ($) | |
|---|---|---|---|---|---|---|---|---|---|---|---|---|
| FY | Phase I | Phase II | Phase I | Phase II | Phase I | Phase II | Phase I | Phase II | Phase I | Phase II | Phase I | Phase II |
| 1992 | 51,952 | 502,886 | 1,065 | 433 | 23,980 | 77,464 | 150,000 | 6,190,970 | 55,328,849 | 217,749,677 | 49,985 | 496,424 |
| 1993 | 56,939 | 560,752 | 1,303 | 591 | 6,622 | 50,000 | 238,729 | 4,293,621 | 74,191,797 | 331,404,639 | 50,000 | 502,567 |
| 1994 | 74,454 | 626,667 | 1,370 | 406 | 25,698 | 94,995 | 118,086 | 1,975,000 | 102,001,625 | 254,426,790 | 69,969 | 646,025 |
| 1995 | 76,269 | 668,247 | 1,262 | 575 | 9,164 | 114,749 | 129,770 | 4,236,522 | 96,251,012 | 384,241,870 | 71,900 | 650,400 |
| 1996 | 81,695 | 692,883 | 1,372 | 611 | 34,150 | 69,977 | 150,803 | 5,776,851 | 112,084,899 | 423,351,421 | 70,757 | 710,501 |
| 1997 | 88,074 | 742,196 | 1,526 | 638 | 6,170 | 87,650 | 153,675 | 6,838,043 | 134,401,334 | 473,520,837 | 98,796 | 744,797 |
| 1998 | 89,242 | 748,204 | 1,286 | 672 | 5,388 | 69,673 | 163,805 | 2,382,173 | 114,765,110 | 502,792,826 | 99,114 | 748,237 |
| 1999 | 85,741 | 789,013 | 1,393 | 568 | 0 | 190,195 | 198,216 | 4,089,106 | 119,437,606 | 448,159,151 | 91,629 | 749,262 |
| 2000 | 89,587 | 817,288 | 1,220 | 626 | 0 | 99,304 | 179,968 | 3,562,762 | 109,296,293 | 511,622,538 | 99,000 | 749,110 |
| 2001 | 91,656 | 836,488 | 1,310 | 702 | 45,500 | 124,997 | 262,540 | 6,361,394 | 120,069,946 | 587,214,239 | 99,568 | 748,121 |
| 2002 | 88,049 | 844,539 | 2,162 | 661 | 40,002 | 65,000 | 303,996 | 7,674,976 | 190,362,208 | 558,240,048 | 98,449 | 749,731 |
| 2003 | 88,776 | 807,484 | 2,113 | 1,078 | 36,595 | 8,897 | 448,796 | 4,024,384 | 187,584,379 | 870,468,216 | 98,979 | 747,245 |
| 2004 | 95,224 | 772,776 | 2,075 | 1,173 | 25,000 | 104,211 | 597,999 | 4,829,998 | 197,590,105 | 906,466,318 | 99,820 | 747,989 |
| 2005 | 91,076 | 730,747 | 2,344 | 998 | 33,944 | 50,000 | 163,050 | 4,294,783 | 213,482,151 | 729,285,506 | 99,439 | 748,994 |

SOURCE: Department of Defense.

**TABLE App-A-11** Phase I–Extra-large Awards

| Fiscal Year | >$150,000 | Percent | n |
|---|---|---|---|
| 1992 | 0 | 0.0 | 1,065 |
| 1993 | 3 | 0.2 | 1,303 |
| 1994 | 0 | 0.0 | 1,370 |
| 1995 | 0 | 0.0 | 1,262 |
| 1996 | 1 | 0.1 | 1,372 |
| 1997 | 1 | 0.1 | 1,526 |
| 1998 | 1 | 0.1 | 1,286 |
| 1999 | 3 | 0.2 | 1,393 |
| 2000 | 1 | 0.1 | 1,220 |
| 2001 | 4 | 0.3 | 1,310 |
| 2002 | 7 | 0.3 | 2,162 |
| 2003 | 4 | 0.2 | 2,113 |
| 2004 | 4 | 0.2 | 2,075 |
| 2005 | 2 | 0.1 | 2,344 |

SOURCE: Department of Defense.

**TABLE App-A-12** Phase II Multiple-award Winners 1992–2005 at DoD

| Firm Name | Number of Phase I Awards | Number of Phase II Awards |
|---|---|---|
| FOSTER-MILLER, INC. | 361 | 140 |
| PHYSICAL OPTICS CORP. | 316 | 117 |
| PHYSICAL SCIENCES, INC. | 170 | 75 |
| MISSION RESEARCH CORP. | 126 | 69 |
| ALPHATECH, INC. | 117 | 68 |
| CREARE, INC. | 129 | 60 |
| CHARLES RIVER ANALYTICS, INC. | 112 | 60 |
| CFD RESEARCH CORP. | 107 | 56 |
| TRITON SYSTEMS, INC. | 125 | 55 |
| COHERENT TECHNOLOGIES, INC. | 101 | 53 |
| TECHNOLOGY SERVICE CORP. | 90 | 42 |
| CYBERNET SYSTEMS CORP. | 95 | 41 |
| SCIENTIFIC SYSTEMS CO., INC. | 91 | 38 |
| DIGITAL SYSTEM RESOURCES, INC. | 48 | 36 |
| STOTTLER HENKE ASSOC., INC. | 84 | 36 |
| TEXAS RESEARCH INSTITUTE AUSTIN, INC. | 77 | 36 |
| ORINCON CORP. | 97 | 36 |
| METROLASER, INC. | 66 | 35 |
| SYSTEMS & PROCESS ENGINEERING CO. | 69 | 35 |
| TOYON RESEARCH CORP. | 65 | 34 |
| Total and Average (conversion rate) | 2,446 | 1,122 |

SOURCE: DoD awards database.

**TABLE App-A-13** Phase II Awards

| Fiscal Year | Count | Mean ($) | Maximum ($) | Minimum ($) | Sum ($) | Median ($) |
|---|---|---|---|---|---|---|
| 1992 | 433 | 502,886 | 6,190,970 | 77,464 | 217,749,677 | 496,424 |
| 1993 | 591 | 560,752 | 4,293,621 | 50,000 | 331,404,639 | 502,567 |
| 1994 | 406 | 626,667 | 1,975,000 | 94,995 | 254,426,790 | 646,025 |
| 1995 | 575 | 668,247 | 4,236,522 | 114,749 | 384,241,870 | 650,400 |
| 1996 | 611 | 692,883 | 5,776,851 | 69,977 | 423,351,421 | 710,501 |
| 1997 | 638 | 742,196 | 6,838,043 | 87,650 | 473,520,837 | 744,797 |
| 1998 | 672 | 748,204 | 2,382,173 | 69,673 | 502,792,826 | 748,237 |
| 1999 | 568 | 789,013 | 4,089,106 | 190,195 | 448,159,151 | 749,262 |
| 2000 | 626 | 817,288 | 3,562,762 | 99,304 | 511,622,538 | 749,110 |
| 2001 | 702 | 836,488 | 6,361,394 | 124,997 | 587,214,239 | 748,121 |
| 2002 | 661 | 844,539 | 7,674,976 | 65,000 | 558,240,048 | 749,731 |
| 2003 | 1,078 | 807,484 | 4,024,384 | 8,897 | 870,468,216 | 747,245 |
| 2004 | 1,173 | 772,776 | 4,829,998 | 104,211 | 906,466,318 | 747,989 |
| 2005 | 998 | 730,747 | 4,294,783 | 50,000 | 729,285,506 | 748,994 |

SOURCE: Department of Defense.

**TABLE App-A-14** Phase II by Demographics—Woman-owned

| | Count | Mean ($) | Maximum ($) | Minimum ($) | Sum |
|---|---|---|---|---|---|
| 1992 | 29 | 502,363 | 748,583 | 200,000 | 14,568,532 |
| 1993 | 49 | 547,885 | 1,230,770 | 163,021 | 26,846,376 |
| 1994 | 36 | 580,652 | 852,668 | 150,000 | 20,903,470 |
| 1995 | 53 | 638,463 | 1,707,915 | 224,960 | 33,838,550 |
| 1996 | 51 | 623,435 | 996,288 | 363,449 | 31,795,210 |
| 1997 | 58 | 657,143 | 1,249,871 | 183,998 | 38,114,287 |
| 1998 | 62 | 777,291 | 1,958,009 | 268,238 | 48,192,039 |
| 1999 | 44 | 773,759 | 1,626,792 | 515,786 | 34,045,389 |
| 2000 | 62 | 789,916 | 2,017,000 | 199,958 | 48,974,781 |
| 2001 | 75 | 819,575 | 3,615,525 | 190,093 | 61,468,152 |
| 2002 | 60 | 860,196 | 2,335,074 | 65,000 | 51,611,735 |
| 2003 | 110 | 803,592 | 1,799,791 | 261,943 | 88,395,121 |
| 2004 | 117 | 735,810 | 1,519,149 | 369,632 | 86,089,731 |
| 2005 | 132 | 729,750 | 4,294,783 | 299,921 | 96,327,003 |

SOURCE: Department of Defense.

**TABLE App-A-15** Phase II by Demographics—Minority-owned

| | Count | Mean ($) | Maximum ($) | Minimum ($) | Sum ($) |
|---|---|---|---|---|---|
| 1992 | 67 | 489,965 | 1,109,000 | 127,471 | 32,827,629 |
| 1993 | 69 | 525,801 | 1,044,526 | 112,775 | 36,280,278 |
| 1994 | 43 | 586,361 | 950,000 | 197,920 | 25,213,518 |
| 1995 | 89 | 666,279 | 1,550,000 | 247,339 | 59,298,801 |
| 1996 | 86 | 636,184 | 1,070,892 | 99,736 | 54,711,843 |
| 1997 | 117 | 705,905 | 1,900,572 | 206,467 | 82,590,936 |
| 1998 | 107 | 776,878 | 1,958,009 | 200,000 | 83,125,959 |
| 1999 | 73 | 816,322 | 2,631,489 | 474,959 | 59,591,489 |
| 2000 | 74 | 856,165 | 3,319,713 | 374,912 | 63,356,204 |
| 2001 | 80 | 838,708 | 3,615,525 | 124,997 | 67,096,603 |
| 2002 | 63 | 797,885 | 1,470,483 | 298,762 | 50,266,775 |
| 2003 | 98 | 780,273 | 1,880,376 | 154,388 | 76,466,762 |
| 2004 | 102 | 767,082 | 1,500,000 | 399,974 | 78,242,363 |
| 2005 | 116 | 725,954 | 1,636,356 | 265,273 | 84,210,673 |

SOURCE: Department of Defense.

**TABLE App-A-16** Phase II by Demographics—Both Woman- and Minority-owned

| | Count | Mean ($) | Maximum ($) | Minimum ($) | Sum ($) |
|---|---|---|---|---|---|
| 1992 | 8 | 508,405 | 748,583 | 300,000 | 4,067,240 |
| 1993 | 7 | 533,298 | 1,044,526 | 247,707 | 3,733,085 |
| 1994 | 5 | 540,330 | 750,000 | 455,810 | 2,701,652 |
| 1995 | 12 | 614,381 | 822,583 | 371,191 | 7,372,572 |
| 1996 | 11 | 602,246 | 750,000 | 374,975 | 6,624,703 |
| 1997 | 18 | 613,276 | 813,216 | 206,467 | 11,038,971 |
| 1998 | 14 | 957,908 | 1,958,009 | 284,000 | 13,410,707 |
| 1999 | 8 | 851,092 | 1,626,792 | 598,865 | 6,808,736 |
| 2000 | 10 | 796,973 | 1,449,309 | 374,912 | 7,969,731 |
| 2001 | 9 | 1,073,370 | 3,615,525 | 190,093 | 9,660,331 |
| 2002 | 14 | 783,051 | 1,241,412 | 298,762 | 10,962,716 |
| 2003 | 19 | 869,871 | 1,799,791 | 598,405 | 16,527,544 |
| 2004 | 16 | 799,599 | 1,459,029 | 548,659 | 12,793,584 |
| 2005 | 23 | 774,584 | 1,501,563 | 399,847 | 17,815,442 |

SOURCE: Department of Defense.

**TABLE App-A-17** Phase II Awards by Unit

| | Number of Awards | | | | | | | | | | | | | | | Percent |
|---|---|---|---|---|---|---|---|---|---|---|---|---|---|---|---|---|
| Component | 1992 | 1993 | 1994 | 1995 | 1996 | 1997 | 1998 | 1999 | 2000 | 2001 | 2002 | 2003 | 2004 | 2005 | Awards | of Total |
| AF | 110 | 199 | 152 | 191 | 217 | 221 | 243 | 212 | 195 | 221 | 233 | 234 | 317 | 339 | 3,084 | 31.7 |
| ARMY | 172 | 123 | 78 | 131 | 100 | 113 | 111 | 105 | 112 | 180 | 124 | 323 | 273 | 123 | 2,068 | 21.2 |
| BMDO | 0 | 0 | 0 | 0 | 0 | 0 | 0 | 0 | 0 | 0 | 0 | 0 | 0 | 0 | 0 | 0.0 |
| CBD | 0 | 0 | 0 | 0 | 0 | 0 | 0 | 6 | 8 | 7 | 11 | 10 | 11 | 7 | 60 | 0.6 |
| DARPA | 41 | 83 | 43 | 84 | 78 | 57 | 71 | 44 | 45 | 53 | 47 | 66 | 75 | 80 | 867 | 8.9 |
| DSWA | 0 | 0 | 0 | 0 | 0 | 0 | 0 | 0 | 0 | 0 | 0 | 0 | 0 | 0 | 0 | 0.0 |
| DTRA | 12 | 6 | 3 | 6 | 3 | 6 | 7 | 7 | 5 | 6 | 6 | 3 | 3 | 4 | 77 | 0.8 |
| MDA | 46 | 60 | 23 | 34 | 36 | 51 | 69 | 43 | 56 | 59 | 28 | 184 | 211 | 102 | 1,002 | 10.3 |
| NAVY | 52 | 120 | 107 | 127 | 156 | 164 | 136 | 107 | 186 | 136 | 170 | 193 | 212 | 290 | 2,156 | 22.2 |
| NGA | 0 | 0 | 0 | 0 | 0 | 0 | 0 | 0 | 1 | 3 | 1 | 1 | 2 | 1 | 9 | 0.1 |
| NIMA | 0 | 0 | 0 | 0 | 0 | 0 | 0 | 0 | 0 | 0 | 0 | 0 | 0 | 0 | 0 | 0.0 |
| OSD | 0 | 0 | 0 | 0 | 16 | 23 | 33 | 37 | 13 | 29 | 36 | 50 | 60 | 38 | 335 | 3.4 |
| SOCOM | 0 | 0 | 0 | 2 | 5 | 3 | 2 | 7 | 5 | 8 | 5 | 14 | 9 | 14 | 74 | 0.8 |
| Total | 433 | 591 | 406 | 575 | 611 | 638 | 672 | 568 | 626 | 702 | 661 | 1,078 | 1,173 | 998 | 9,732 | 100 |

SOURCE: Department of Defense.

## OUTCOMES

**TABLE App-A-18** Project Status

**1. What is the current status of the project funded by the referenced SBIR award? Select the one best answer.**

Percentages are based on the 920 respondents who answered this question (#1).

| | Count | Percent | |
|---|---|---|---|
| | 43 | 4.7 | a. Project has not yet completed Phase II. |
| | 214 | 23.3 | b. Efforts at this company have been discontinued. No sales or additional funding resulted from this project. |
| | 92 | 10.0 | c. Efforts at this company have been discontinued. The project did result in sales, licensing of technology, or additional funding. |
| | 244 | 26.5 | d. Project is continuing post-Phase II technology development. |
| | 145 | 15.8 | e. Commercialization is underway. |
| | 182 | 19.8 | f. Products/Processes/Services are in use by target population/customer/consumers. |
| n=920 | 920 | 100.0 | |
| | 43 | 4.7 | a. |
| | 214 | 23.3 | b. |
| | 257 | 27.9 | Projects not completed or discontinued without sales. |
| | 214 | 23.3 | b. |
| | 92 | 10.0 | c. |
| *Next: Question 2* | 306 | 33.3 | Projects discontinued. |
| | 92 | 10.0 | c. |
| | 244 | 26.5 | d. |
| | 145 | 15.8 | e. |
| | 182 | 19.8 | f. |
| *Next: Question 3* | 663 | 72.1 | Projects that could answer commercialization questions. |

SOURCE: NRC Phase II Survey.

**TABLE App-A-19** Reasons for Discontinuing

**2. Did the reasons for discontinuing this project include any of the following?**
(PLEASE SELECT YES OR NO FOR EACH REASON AND NOTE THE ONE PRIMARY REASON)

*(see Question 1)* 300 of 306 projects answered this question (#2).
The percentages below are the percent of the discontinued projects that responded with the indicated response.

| | Yes | No | Primary Reason | Yes (Percent) | No (Percent) | Primary Reason (Percent) | |
|---|---|---|---|---|---|---|---|
| | 79 | 221 | 31 | 26 | 74 | 10 | a. Technical failure or difficulties |
| | 171 | 129 | 78 | 57 | 43 | 26 | b. Market demand too small |
| | 58 | 242 | 8 | 19 | 81 | 3 | c. Level of technical risk too high |
| | 169 | 131 | 59 | 56 | 44 | 20 | d. Not enough funding |
| | 95 | 205 | 21 | 32 | 68 | 7 | e. Company shifted priorities |
| | 37 | 263 | 7 | 12 | 88 | 2 | f. Principal investigator left |
| | 178 | 122 | 30 | 59 | 41 | 10 | g. Project goal was achieved (e.g. prototype delivered for federal agency use) |
| | 14 | 286 | 7 | 5 | 96 | 2 | h. Licensed to another company |
| | 71 | 229 | 10 | 24 | 76 | 3 | i. Product, process, or service not competitive |
| | 60 | 240 | 9 | 20 | 80 | 3 | j. Inadequate sales capability |
| n=300 | 65 | 235 | 40 | 22 | 78 | 13 | k. Other (please specify): |

SOURCE: NRC Phase II Survey.

## TABLE App-A-20 Sales

| | |
|---|---|
| | **3. Has your company and/or licensee had any actual sales of products, processes, services or other sales incorporating the technology developed during this project?** (Select all that apply.) This question was not answered for those projects still in Phase II (5 percent) or for projects, which were discontinued without sales or additional funding (23 percent). The denominator for the percentages below is all projects that answered the survey. |
| *(See Question 1)* | 660 of 663 projects answered this question (#3). |

| | Count | Percent | |
|---|---|---|---|
| *(See Question 8)* | 169 | 25.6 | a. No sales to date, but sales are expected |
| | 71 | 10.8 | b. No sales to date nor are sales expected |
| | 319 | 48.3 | c. Sales of product(s) |
| | 57 | 8.6 | d. Sales of process(es) |
| | 180 | 27.3 | e. Sales of services(s) |
| | 61 | 9.2 | f. Other sales (e.g. rights to technology, licensing, etc.) |
| n=660 | 857 | 129.8 | Multiple answers allowed. Question asked to select all that applied. |

| Count | Percent | | |
|---|---|---|---|
| 214 | | 1b | |
| 71 | | 3b | |
| 285 | 31.0 | | We can conclude that 31 percent of the projects have no sales to date, and expect none. |

SOURCE: NRC Phase II Survey.

## TABLE App-A-21 Project Status II

| | Percent | Count | |
|---|---|---|---|
| 1a | 4.7 | 43 | Phase II Project not completed yet |
| 1b | 23.3 | 214 | Projects abandoned, no sales/revenues |
| (1c,d,e,f)x(3a) | 17.6 | 162 | Projects in development, expecting sales |
| (1d,e,f)x(3b) | 4.2 | 39 | Projects in development, not expecting sales |
| (1c,d,e,f)x(3c,d,e,f) | 45.7 | 420 | Projects with some sales/revenues |
| (1c)x(3a) | 0.8 | 7 | (Discont. w/sales)x(no sales to date, sales expected) |
| (1c)x(3b) | 3.0 | 28 | (Discont. w/sales)x(no sales to date, none expected) |
| | 0.8 | 7 | Projects dropped out of survey before answering |
| | 100.0 | 920 | |

SOURCE: NRC Phase II Survey.

**TABLE App-A-22** Sales Dollars

(see sales)

Of the 420 projects that reported sales or licensing at time of survey,
399 projects reported the year for first sale (Question 4).
Of the 399 projects that reported the year for first sale,
378 projects reported company sales dollar amount > 0 (Question 4b company only)

| Count | Percent | |
|---|---|---|
| 246 | 65.1 | > 0 and <$1M |
| 94 | 24.9 | $1M to <$5M |
| 10 | 2.6 | $5M to <$10M |
| 25 | 6.6 | $10M to <$50M |
| 3 | 0.8 | $50M + |
| 378 | 100.0 | Total |

| *q_4_b_company* | |
|---|---|
| Mean | $2,894,834 |
| Standard Error | $506,081 |
| Median | $500,000 |
| Mode | $500,000 |
| Standard Deviation | $9,839,338 |
| Range | $121,999,999 |
| Minimum | $1 |
| Maximum | $122,000,000 |
| Sum | $1,094,247,315 |
| Count | 378 |

Of the 420 projects that reported sales or licensing at time of survey,
399 projects reported the year for first sale (Question 4).
Of the 399 projects that reported the year for first sale,
393 projects reported a combined sales dollar amount > 0 (Questions 4a + 4b)

| Count | Percent | |
|---|---|---|
| 235 | 59.8 | > 0 and <$1M |
| 106 | 27.0 | $1M to <5M |
| 49 | 12.5 | $5M to <50M |
| 3 | 0.8 | $50M + |
| 393 | 100.0 | Total |

| *4: total combined sales* | |
|---|---|
| Mean | $3,244,750 |
| Standard Error | $500,065 |
| Median | $599,000 |
| Mode | $1,000,000 |
| Standard Deviation | $9,913,396 |
| Range | $124,499,998 |
| Minimum | $2 |
| Maximum | $124,500,000 |
| Sum | $1,275,186,865 |
| Count | 393 |

Of the 420 projects that reported sales or licensing at time of survey,
142 projects reported the year for first sale by licensees (Question 4).
Of the 142 projects that reported the year for first sale by lincensees,
33 projects reported company sales dollar amount > 0 (Question 4b licensee)

| Count | Percent | |
|---|---|---|
| 17 | 4.3 | > 0 and <$1M |
| 10 | 2.5 | $1M to <5M |
| 6 | 1.5 | $5M to <50M |
| 0 | 0.0 | $50M + |
| 33 | 100.0 | Total |

| *q_4_b_licensee* | |
|---|---|
| Mean | $2,833,485 |
| Standard Error | $824,733 |
| Median | $500,000 |
| Mode | $100,000 |
| Standard Deviation | $4,737,728 |
| Range | $19,991,000 |
| Minimum | $9,000 |
| Maximum | $20,000,000 |
| Sum | $93,505,000 |
| Count | 33 |

SOURCE: NRC Phase II Survey.

**TABLE App-A-23** Sales by Licensees

*(see sales)*

Of the 420 projects that reported sales or licensing at time of survey,
142 projects reported the year for first sale (Question 4).
Of the 142 projects that reported the year for first sale,
33 projects reported company sales dollar amount > 0 (Question 4b company only)

| Count | Percent | | *q_4_b_licensee* | |
|---|---|---|---|---|
| 17 | 51.5 | > 0 and <$1M | Mean | $2,833,485 |
| 10 | 30.3 | $1M to <$5M | Median | $500,000 |
| 2 | 6.1 | $5M to <$10M | Mode | $100,000 |
| 4 | 12.1 | $10M to <$50M | Standard Deviation | $4,737,728 |
| 0 | 0.0 | $50M + | Range | $19,991,000 |
| | | | Minimum | $9,000 |
| 33 | 100.0 | Total | Maximum | $20,000,000 |
| | | | Sum | $93,505,000 |
| | | | Count | 33 |

SOURCE: NRC Phase II Survey.

## TABLE App-A-24 Products

| | | | |
|---|---|---|---|
| | **Question 3: These are projects that have received at least some revenue from sales or licensing.** Double counts were removed from question 3c, d, e, and f (Using the Phase 2 Survey). 660 projects answered question #3, 420 answered positive to some revenue. | | |
| | Count | Percent | |
| n=920 | 420 | 45.7 | of the 920 total projects surveyed |
| | Phase 2 projects not yet completed were removed. | | |
| | 920 | | Total Projects Surveyed |
| | 43 | 1a | Project has not yet completed Phase II |
| | 877 | | |
| | Count | Percent | |
| n=877 | 420 | 47.9 | |
| *Next: Sales Dollars* | Thus, 48 percent of completed Phase II projects had some revenue from sales or licensing at time of survey. | | |
| | **Question 7: Did a commercial product result from this Phase II project?** Of the possible 660 responses, only 390 answered | | |
| | Count | Percent | |
| | 254 | 65.1 | Reported "Yes" for a commercial product as result of this project? (Question 7) |
| | 136 | 34.9 | Reported "No" for a commercial product as result of this project? (Question 7) |
| n=390 | 390 | 100.0 | Total responses to Question 7. |
| | For the whole survey (n=920), only 27 percent reported a commercial product. This would assume that the 270 nonreponses to Q7 (660 – 390 = 270) did not have a commercial product. | | |
| n=920 | 254 | 27.6% | for (n=920) |

SOURCE: NRC Phase II Survey.

**TABLE App-A-25** Sales Forecasts

*(See Question 3)* From question 3a, 169 projects did not have sales, but expect some in the future.

168 responded to question 8: Expected year of sales?

| Year of Expected Sales | Number of Projects | Percent | Mean Sales Expectation ($) | Projects Reporting Expections | Percent |
|---|---|---|---|---|---|
| 2005 | 37 | 22.0 | 869,167 | 36 | 33.0 |
| 2006 | 70 | 41.7 | 1,873,156 | 64 | 58.7 |
| 2007 | 31 | 18.5 | 152,500 | 6 | 5.5 |
| 2008 | 18 | 10.7 | 550,000 | 2 | 1.8 |
| 2009 | 5 | 3.0 | | 0 | 0.0 |
| 2010 | 7 | 4.2 | 100,000 | 1 | 0.9 |
| | 168 | | | 109 | |

More than 80 percent of projects reporting a year of expected sales were within 3 years.
More the 90 percent of project reporting expected sales dollars were within 2 years.

SOURCE: NRC Phase II Survey.

**TABLE App-A-26** Sales Expectations

| | Count | Percent | | |
|---|---|---|---|---|
| *(See Question 1)* | 43 | | 1a | Phase II not yet complete |
| *(See Question 1)* | 214 | | 1b | Sales not expected |
| *(See Question 1)* | 663 | | 1c, d, e, f | Had sales or expect sales (See details below) |
| | 920 | | | Total Surveyed |
| *(See Sales)* | 420 | | | Reported sales |
| *(See Question 3)* | 169 | | 3a | Sales are expected |
| *(See Question 3)* | 71 | | 3b | Sales not expected |
| *(See Question 3)* | 3 | | | No response. 660 of 663 answered the question. |
| | 663 | | 1c, d, e, f | Had sales or expect sales |
| | 420 | 46 | | Reported sales |
| | 169 | 18 | | Sales expected |
| | 285 | 31 | 1b+3b | Sales not expected |
| | 43 | 5 | | Phase II not complete |
| | 3 | 0 | | No response |
| | 920 | 100 | | |

In the Phase II survey, the term "sales" was defined to inclued all sales of a product, process, or service, to federal or private sector customes resulting from the technology developed during the Phase II project. A sale also includes licensing, the sale of technology or rights, etc.

SOURCE: NRC Phase II Survey.

**TABLE App-A-27** Sales by Number of Employees at Time of Survey

Of the 420 projects that reported sales or licensing at time of survey,
only 399 projects reported the year for first sale (Question 4).
Of the 399 projects that reported the year for first sale,
393 projects reported a combined sales dollar amount > 0 (Questions 4a + 4b)
Of the 393 project that reported a combined sales dollar amount > 0,
379 projects reported their current number of of employees.

| Employees | <$100K | $100K to < $1M | $1M to <$5M | $5M to <$50M | >$50M | Total | Percent |
|---|---|---|---|---|---|---|---|
| 0–5 | 19 | 33 | 6 | 2 | | 60 | 15.8 |
| 6–10 | 12 | 24 | 9 | 6 | | 51 | 13.5 |
| 11–15 | 12 | 21 | 10 | 3 | | 46 | 12.1 |
| 16–25 | 8 | 19 | 24 | 2 | | 53 | 14.0 |
| 26–50 | 7 | 21 | 17 | 17 | | 62 | 16.4 |
| 51–100 | 6 | 12 | 20 | 8 | | 46 | 12.1 |
| 101–250 | 6 | 20 | 8 | 5 | | 39 | 10.3 |
| 251–500 | 1 | 5 | 6 | 5 | 2 | 17 | 5.0 |
| 500+ | 0 | 2 | 0 | 1 | | 3 | 0.8 |
| Missing | | | | | | | |
| | 71 | 157 | 100 | 49 | 2 | 379 | |
| Percent | 18.7% | 41.4% | 26.4% | 12.9% | 0.5% | | 100.0% |

The above table has multiple responses from some firms.
278 of the 379 projects are from unique firms (see below).

| Surveys per Firm | Number of Firms | Number of Surveys |
|---|---|---|
| 1 | 217 | 217 |
| 2 | 43 | 86 |
| 3 | 9 | 27 |
| 4 | 6 | 24 |
| 5 | 1 | 5 |
| 7 | 1 | 7 |
| 13 | 1 | 13 |
| | 278 | 379 |

*(see Sales by Firm Growth for the raw data)*

SOURCE: NRC Phase II Survey.

**TABLE App-A-28** Sales by Size of Current Employment

| Firm Size at Survey | Number with Sales | Percent with Sales | Number with No Sales | Percent with No Sales | Number of Firms | Percent of All Firms |
|---|---|---|---|---|---|---|
| 0–5 | 58 | 14.2 | 93 | 18.2 | 151 | 16.4 |
| 6–10 | 53 | 13.0 | 54 | 10.6 | 107 | 11.6 |
| 11–15 | 45 | 11.0 | 40 | 7.8 | 85 | 9.2 |
| 16–20 | 25 | 6.1 | 29 | 5.7 | 54 | 5.9 |
| 21–25 | 21 | 5.1 | 23 | 4.5 | 44 | 4.8 |
| 26–30 | 17 | 4.2 | 26 | 5.1 | 43 | 4.7 |
| 31–40 | 23 | 5.6 | 10 | 2.0 | 33 | 3.6 |
| 41–50 | 28 | 6.8 | 31 | 6.1 | 59 | 6.4 |
| 51–75 | 27 | 6.6 | 40 | 7.8 | 67 | 7.3 |
| 76–100 | 14 | 3.4 | 21 | 4.1 | 35 | 3.8 |
| 101–200 | 35 | 8.6 | 50 | 9.8 | 85 | 9.2 |
| 201–300 | 11 | 2.7 | 8 | 1.6 | 19 | 2.1 |
| 301–500 | 12 | 2.9 | 19 | 3.7 | 31 | 3.4 |
| 500+ | 3 | 0.7 | 3 | 0.6 | 6 | 0.7 |
| Missing | 37 | 9.0 | 64 | 12.5 | 101 | 11.0 |
| | 409 | 100.0 | 511 | 100.0 | 920 | 100.0 |

SOURCE: NRC Phase II Survey.

## TABLE App-A-29 Sales by Firm Growth

Of the 420 projects that reported sales or licensing at time of survey,
only 399 projects reported the year for first sale (Question 4).
Of the 399 projects that reported the year for first sale,
393 projects reported a combined sales dollar amount > 0 (Questions 4a + 4b)
Of the 393 project that reported a combined sales dollar amount > 0
379 projects reported their current number of of employees.

| Employee Change | <$100K | $100K to < $1M | $1M to <$5M | $5M to <$50M | >$50M | Total | Percent |
|---|---|---|---|---|---|---|---|
| –26 + | 1 | 4 | 0 | 1 | | 6 | 1.6 |
| –6 to –25 | 4 | 7 | 1 | 1 | | 13 | 3.4 |
| –1 to –5 | 9 | 13 | 1 | 0 | | 23 | 6.1 |
| 0–5 | 29 | 50 | 19 | 8 | | 106 | 28.0 |
| 6–10 | 10 | 27 | 20 | 4 | | 61 | 16.1 |
| 11–25 | 9 | 26 | 28 | 6 | | 69 | 18.2 |
| 26–50 | 7 | 13 | 20 | 17 | | 57 | 15.0 |
| 51–100 | 2 | 10 | 4 | 6 | | 22 | 5.8 |
| 101–250 | | 5 | 6 | 5 | 2 | 16 | 4.7 |
| 251–500 | | 1 | 1 | 0 | | 2 | 0.5 |
| 500+ | | 1 | 0 | 1 | | 2 | 0.5 |
| Missing | | | | | | | |
| Total | 71 | 157 | 100 | 49 | 2 | 379 | |
| Percent | 18.7% | 41.4% | 26.4% | 12.9% | 0.5% | | 100.0% |
| Mean | 7.0 | 24.0 | 31.3 | 60.2 | 224.0 | 28.5 | |

Mean change in employees by sales classification

Employee change = (16b – 16a) of the Phase II survey
16a Number of employees (if known) when Phase II proposal was submitted
16b Current number of employees

SOURCE: NRC Phase II Survey.

**TABLE App-A-30** Firm Growth by Year of Award

811 answered question 16a and 16b

| Year | Number of Projects | Growth Sum (Number of Employees) | Average Employment Growth per Project |
|---|---|---|---|
| 2001 | 135 | 1,588 | 11.8 |
| 2000 | 132 | 2,122 | 16.1 |
| 1999 | 89 | 2,151 | 24.2 |
| 1998 | 97 | 3,244 | 33.4 |
| 1997 | 69 | 1,438 | 20.8 |
| 1996 | 60 | 2,415 | 40.3 |
| 1995 | 73 | 1,452 | 19.9 |
| 1994 | 49 | 3,087 | 63.0 |
| 1993 | 65 | 1,608 | 24.7 |
| 1992 | 42 | 1,984 | 47.2 |
| | 811 | 21,089 | 26.0 |

NOTE: Firms that "closed" were not surveyed, thus the "true" average growth could be lower.

SOURCE: NRC Phase II Survey.

**TABLE App-A-31** Customers

5. To date, approximately what percent of total sales from the technology developed during this project have gone to the following customers? (If none enter 0 [zero]. Round percentages. Answers should add to about 100 percent). 920 firms responded to this question as to what percent of their sales went to each agency or sector

| | |
|---|---|
| Domestic private sector | 21% |
| Department of Defense (DoD) | 38% |
| Prime contractors for *DoD or NASA* | 12% |
| NASA | 1% |
| *Agency that awarded the Phase II* | —% |
| Other federal agencies *(Pull down)* | 1% |
| State or local governments | 1% |
| Export Markets | 11% |
| Other (Specify)____________ | 16% |

SOURCE: NRC Phase II Survey.

**TABLE App-A-32** Sales by Year of Award

| Year of Award | Number of Respondents | Number of Respondents with Sales | Total Reported Sales ($) | Average Reported Sales (n=378) ($) | Average Reported Sales (All Firms n=920) ($) |
|---|---|---|---|---|---|
| 1992 | 46 | 19 | 150,371,265 | 7,914,277 | 3,268,941 |
| 1993 | 75 | 24 | 89,398,693 | 3,886,900 | 1,191,983 |
| 1994 | 55 | 27 | 128,949,544 | 4,775,909 | 2,344,537 |
| 1995 | 82 | 31 | 92,889,225 | 2,996,427 | 1,132,795 |
| 1996 | 62 | 30 | 72,995,813 | 2,433,194 | 1,177,352 |
| 1997 | 78 | 30 | 84,807,417 | 2,826,914 | 1,087,275 |
| 1998 | 107 | 48 | 104,409,505 | 2,175,198 | 975,790 |
| 1999 | 94 | 49 | 76,685,396 | 1,565,008 | 815,802 |
| 2000 | 147 | 71 | 248,397,371 | 3,498,555 | 1,689,778 |
| 2001 | 174 | 49 | 44,331,086 | 904,716 | 254,776 |
| Total | 920 | 378 | 1,093,235,315 | 2,892,157 | 1,188,299 |

This table only counts firm sales from the NRC Phase II Survey, question 4a.
This table does not count reported sales by lincensee, or "other sales "4b for firms or lincensee.

SOURCE: NRC Phase II Survey.

**TABLE App-A-33** Commercialization Lags

| **Time elapsed between award and sales** | | | **Frequency by award year for companies still expecting sales** | | |
|---|---|---|---|---|---|
| Elapsed Years | Number of Projects | Percentage of Responding Projects | Year of Award | Projects Expecting Sales | Historical Success Percentage |
| –10 | 1 | 0.3 | 1992 | 3 | |
| –6 | 1 | 0.3 | 1993 | 4 | |
| –4 | 2 | 0.5 | 1994 | 4 | 0.3 |
| –3 | 1 | 0.3 | 1995 | 8 | 0.8 |
| –2 | 6 | 1.6 | 1996 | 6 | 0.3 |
| –1 | 14 | 3.7 | 1997 | 13 | 0.3 |
| 0 | 40 | 10.6 | 1998 | 18 | 2.4 |
| 1 | 65 | 17.2 | 1999 | 19 | 3.7 |
| 2 | 83 | 22.0 | 2000 | 32 | 4.8 |
| 3 | 80 | 21.2 | 2001 | 61 | 10.1 |
| 4 | 38 | 10.1 | | | |
| 5 | 18 | 4.8 | | | |
| 6 | 14 | 3.7 | | | |
| 7 | 9 | 2.4 | | | |
| 8 | 1 | 0.3 | | | |
| 9 | 1 | 0.3 | | | |
| 10 | 3 | 0.8 | | | |
| 11 | 1 | 0.3 | | | |
| | 378 | | | | |

SOURCE: NRC Phase II Survey.

## TABLE App-A-34 Type of Product

**Q10 How did you (or do you expect to) commercialize your SBIR award?**

| Number of Responses | Percentage | |
|---|---|---|
| 17 | 3.1 | No commercial product, process, or service was/is planned |
| 178 | 32.1 | As software |
| 334 | 60.3 | As hardware (final product, component, or intermediate hardware product) |
| 127 | 22.9 | As process technology |
| 102 | 18.4 | As new or improved service capability |
| 2 | 0.4 | As a drug |
| 2 | 0.4 | As a biologic |
| 86 | 15.5 | As a research tool |
| 11 | 2.0 | As educational materials |
| 54 | 9.7 | Other, please explain |
| 913 | | |

| | |
|---|---|
| n=554 | Responses to Question 10 of the Phase II survey |
| 233 | Respondents with multiple responses |
| 321 | Respondents with a single response |
| 554 | Total Responses |

**Most Common Paired Responses**

| | |
|---|---|
| 67 | Paired Responses to (c) hardware and (d) process technology |
| 53 | Paired Responses to (c) hardware and (b) software |
| 53 | Paired Responses to (c) hardware and (e) new or improved service capability |
| 48 | Paired Responses to (c) hardware and (h) research tool |
| 41 | Paired Responses to (d) process technology and (e) new or improved service |
| 35 | Paired Responses to (b) software and (e) new or improved service capability |
| 34 | Paired Responses to (b) software and (h) research tool |
| 34 | Paired Responses to (e) new or improved service capability and (h) research tool |
| 32 | Paired Responses to (b) software and (d) process technology |
| 27 | Paired Responses to (d) process technology and (h) research tool |

SOURCE: NRC Phase II Survey.

## TABLE App-A-35 Sales by Number of Previous SBIR Awards

**Q19 How many SBIR awards did your company receive prior to the Phase I that led to this Phase II?**

| Additional Prior SBIRs | Number of Responses by Reported Sales | | | | | | |
|---|---|---|---|---|---|---|---|
| | $0 | <$100K | $100K to < $1M | $1M to <$5M | $5M to <$50M | $50M> | Total |
| 0 | 118 | 34 | 57 | 35 | 11 | 1 | 256 |
| | 46.1% | 13.3% | 22.3% | 13.7% | 4.3% | 0.4% | |
| 1 | 48 | 7 | 10 | 8 | 3 | 0 | 76 |
| | 63.2% | 9.2% | 13.2% | 10.5% | 3.9% | 0.0% | |
| 2 | 39 | 8 | 16 | 12 | 2 | 0 | 77 |
| | 50.6% | 10.4% | 20.8% | 15.6% | 2.6% | 0.0% | |
| 3–5 | 68 | 10 | 30 | 17 | 6 | 2 | 133 |
| | 51.1% | 7.5% | 22.6% | 12.8% | 4.5% | 1.5% | |
| 6–10 | 51 | 7 | 14 | 13 | 5 | 0 | 90 |
| | 56.7% | 7.8% | 15.6% | 14.4% | 5.6% | 0.0% | |
| 11–15 | 18 | 1 | 4 | 5 | 5 | 0 | 33 |
| | 54.5% | 3.0% | 12.1% | 15.2% | 15.2% | 0.0% | |
| 16–25 | 23 | 1 | 7 | 5 | 5 | 0 | 41 |
| | 56.1% | 2.4% | 17.1% | 12.2% | 12.2% | 0.0% | |
| 26–50 | 24 | 6 | 7 | 7 | 3 | 0 | 47 |
| | 51.1% | 12.8% | 14.9% | 14.9% | 6.4% | 0.0% | |
| 51–100 | 22 | 1 | 4 | 1 | 0 | 0 | 28 |
| | 78.6% | 3.6% | 14.3% | 3.6% | 0.0% | 0.0% | |
| 101+ | 36 | 3 | 10 | 2 | 0 | 0 | 51 |
| | 70.6% | 5.9% | 19.6% | 3.9% | 0.0% | 0.0% | |
| Total | 447 | 78 | 159 | 105 | 40 | 3 | 832 |
| | 53.7% | 9.4% | 19.1% | 12.6% | 4.8% | 0.4% | |

832 answered either 4 or 19 (n=832)
88 did not answer 4 or 19 (due to not have Phase II done (43) or dropped out (45))
920

| Additional Prior SBIRs | Sales Reported (Percent) | | | | | | |
|---|---|---|---|---|---|---|---|
| | 0 | <$100K | $100K to < $1M | $1M to <$5M | $5M to <$50M | $50M> | Total Responses |
| 0 | 46.1 | 13.3 | 22.3 | 13.7 | 4.3 | 0.4 | 256 |
| 1 | 63.2 | 9.2 | 13.2 | 10.5 | 3.9 | 0.0 | 76 |
| 2 | 50.6 | 10.4 | 20.8 | 15.6 | 2.6 | 0.0 | 77 |
| 3–5 | 51.1 | 7.5 | 22.6 | 12.8 | 4.5 | 1.5 | 133 |
| 6–10 | 56.7 | 7.8 | 15.6 | 14.4 | 5.6 | 0.0 | 90 |
| 11–15 | 54.5 | 3.0 | 12.1 | 15.2 | 15.2 | 0.0 | 33 |
| 16–25 | 56.1 | 2.4 | 17.1 | 12.2 | 12.2 | 0.0 | 41 |
| 26–50 | 51.1 | 12.8 | 14.9 | 14.9 | 6.4 | 0.0 | 47 |
| 51–100 | 78.6 | 3.6 | 14.3 | 3.6 | 0.0 | 0.0 | 28 |
| 101+ | 70.6 | 5.9 | 19.6 | 3.9 | 0.0 | 0.0 | 51 |
| Total | 53.7 | 9.4 | 19.1 | 12.6 | 4.8 | 0.4 | 832 |

SOURCE: NRC Phase II Survey.

**TABLE App-A-36** Additional Investment

**Additonal Investment/funding other than SBIR**

| | Any Investment | | |
|---|---|---|---|
| | Responses | Percent | |
| Yes | 451 | 52.9 | 49.02174 |
| No | 402 | 47.1 | 43.69565 |
| Total | 853 | 100.0 | 920 |

| | Investment ≥ 1$M | |
|---|---|---|
| | Responses | Percent |
| Yes | 130 | 15.2 |
| No | 723 | 84.8 |
| Total | 853 | 100.0 |

**Further Investment in SBIR Projects**

| | | | |
|---|---|---|---|
| $50M+ | 2 | $106,700,000 | 0.2345 |
| $5M to <$50M | 32 | $324,151,193 | 3.7515 |
| $1M to <$5M | 96 | $202,819,919 | 11.254 |
| $100K to <$1M | 241 | $90,112,919 | 28.253 |
| <$100K | 80 | $3,291,603 | 9.3787 |
| None | 402 | $0 | 47.128 |
| Total Investments | 853 | $727,075,634 | |
| Percentage of all respondents | 52.9% | | |
| Average (all) | | $852,375 | |
| Average (with investment) | | $1,612,141 | |

**Any Additional Investments**

| | Any Investment | |
|---|---|---|
| | Responses | Percent |
| Yes | 801 | 93.8 |
| No | 53 | 6.2 |
| Total | 854 | 100.0 |

The above table looks for investment from questions 19, 20, 21, and 22

SOURCE: NRC Phase II Survey.

## TABLE App-A-37 Investment Sources

**Further Investments in SBIR Projects**

| Source of Investment | Total Investment ($) | Percent | Number of Investments | Percent | Average Investment ($) |
|---|---|---|---|---|---|
| Non-SBIR Federal Funds | 332,374,455 | 45.4 | 205 | 25.8 | 1,621,339 |
| Private Investment from U.S. Venture Capital | 155,768,006 | 21.3 | 30 | 3.8 | 5,192,267 |
| Your Own Company | 83,640,416 | 11.4 | 274 | 34.4 | 305,257 |
| Private Investment from Other Private Equity | 71,066,831 | 9.7 | 55 | 6.9 | 1,292,124 |
| Private Investment from Other Domestic Private Company | 52,602,991 | 7.2 | 105 | 13.2 | 500,981 |
| Private Investment from Foreign Investment | 15,404,973 | 2.1 | 19 | 2.4 | 810,788 |
| Personal Funds | 14,013,832 | 1.9 | 65 | 8.2 | 215,597 |
| State or Local Government | 5,536,863 | 0.8 | 31 | 3.9 | 178,608 |
| Colleges or Universities | 1,667,264 | 0.2 | 12 | 1.5 | 138,939 |

SOURCE: NRC Phase II Survey.

## TABLE App-A-38 Related SBIR Awards

**Additional SBIR Funding**

| Number of Phase II Awards | Number of Companies | Percent |
|---|---|---|
| 28 | 1 | 0.1 |
| 9 | 1 | 0.1 |
| 7 | 4 | 0.5 |
| 6 | 6 | 0.7 |
| 5 | 12 | 1.5 |
| 4 | 23 | 2.8 |
| 3 | 29 | 3.6 |
| 2 | 106 | 13.0 |
| 1 | 172 | 21.2 |
| | | |
| 65 | 354 | 43.5 |

Total answering the question = 813

SOURCE: NRC Phase II Survey.

**TABLE App-A-39** Equity-related Activities

| | **Equity-Related Activities** | | | | | | |
|---|---|---|---|---|---|---|---|
| | U.S. Companies/Investors | | | Foreign Companies/Investors | | | |
| Activities | Finalized (Percent) | Ongoing (Percent) | Total (Percent) | Finalized (Percent) | Ongoing (Percent) | Total (Percent) | All Companies/ Investors (Percent) |
| Licensing Agreement(s) | 16 | 16 | 32 | 3 | 5 | 8 | 33.4 |
| Sale of Company | 1 | 5 | 6 | 0 | 1 | 1 | 6.3 |
| Partial Sale of Company | 1 | 4 | 5 | 0 | 1 | 1 | 5.3 |
| Sale of Technology Rights | 4 | 10 | 14 | 1 | 3 | 4 | 15.3 |
| Company Merger | 0 | 3 | 3 | 0 | 1 | 1 | 3.6 |
| Joint Venture Agreement | 4 | 8 | 12 | 1 | 2 | 3 | 13.4 |
| Marketing/Distribution Agreement(s) | 11 | 9 | 20 | 5 | 4 | 9 | 22.5 |
| Manufacturing Agreement(s) | 3 | 9 | 12 | 3 | 2 | 5 | 15.2 |
| R&D Agreement(s) | 14 | 14 | 28 | 3 | 3 | 6 | 28.3 |
| Customer Alliance(s) | 14 | 14 | 28 | 5 | 3 | 7 | 27.9 |
| Other *Specify*____ | 2 | 2 | 4 | 0 | 1 | 1 | 3.9 |

n=620 Responses to Question 12 of the Phase II survey

**Projects with at least one other company**.

| | | |
|---|---|---|
| Yes | 67.0% | 415 |
| No | 33.0% | 204 |
| Total | | 619 |

| | |
|---|---|
| 214 | Projects discontinued with no sales (not asked Question 12) |
| 43 | Projects not finished with Phase II (not asked Question 12) |
| 44 | Projects dropped out of the survey before Question 12 |
| 920 | |

SOURCE: NRC Phase II Survey.

## TABLE App-A-40 The "Go" Decision

**Counterfactual: Greenlighting the Project**
In the absence of the SBIR award, would the project have been implemented?

| | Number of Responses | Percent | |
|---|---|---|---|
| a | 18 | 2.9 | Definitely Yes |
| b | 62 | 10.0 | Probably Yes |
| c | 107 | 17.3 | Uncertain |
| d | 201 | 32.5 | Probably Not |
| e | 230 | 37.2 | Definitely Not |
| | 618 | 100.0 | |

n=618 Responses to Question 13 of the Phase II survey

SOURCE: NRC Phase II Survey.

## TABLE App-A-41 Patenting Activities

**Patents and Patent Applications**

| | Projects with Patent Applications | | Projects with Patents Awards | |
|---|---|---|---|---|
| Yes | 281 | 34.4% | 205 | 25.1% |
| No | 535 | 65.6% | 611 | 74.9% |
| | 816 | | 816 | |

This data excludes respondents with projects not yet completing Phase II (Q1a)

**Number of Patents per Company, Responses (%)**

| | Responses | Percent |
|---|---|---|
| 1 patent | 129 | 63 |
| 2 patents | 43 | 21 |
| 3–5 patents | 25 | 12 |
| 6–10 patents | 4 | 2 |
| 11–20 patents | 4 | 2 |

SOURCE: NRC Phase II Survey.

## TABLE App-A-42 Patents and Marketing

**Q18 Results**

270 Projects with Some Patents, etc.
546 Projects with No Patents, etc.

816

346 Projects with Some Patent Attempts, etc.
470 Projects with No Patent Attempts, etc.

816

**Patents, Copyrights, Trademarks**
614 of the 816 responses question 18 responded to question 11

| | Counts | |
|---|---|---|
| 24.3 | 149 | No Activity + No Marketing |
| 36.2 | 222 | No Activity + Marketing |
| 7.7 | 47 | Activity + No Marketing |
| 31.9 | 196 | Activity + Marketing |
| | 614 | |

**Patents, Copyrights, Trademarks, or Pending Similar Items**
614 of the 816 responses question 18 responded to question 11

| | Counts | |
|---|---|---|
| 21.5 | 132 | No Pending + No Marketing |
| 28.0 | 172 | No Pending + Marketing |
| 10.4 | 64 | Pending + No Marketing |
| 40.1 | 246 | Pending + Marketing |
| | 614 | |

SOURCE: NRC Phase II Survey.

## TABLE App-A-43 Patents and Firm Size

| Firm Size at Survey | Responses | Patents | Ratio |
|---|---|---|---|
| 0–5 | 151 | 74 | 0.490066 |
| 6–10 | 107 | 57 | 0.53 |
| 11–15 | 85 | 31 | 0.36 |
| 16–20 | 54 | 20 | 0.37 |
| 21–25 | 50 | 25 | 0.50 |
| 26–30 | 37 | 24 | 0.65 |
| 31–40 | 33 | 23 | 0.70 |
| 41–50 | 59 | 34 | 0.58 |
| 51–75 | 67 | 69 | 1.03 |
| 76–100 | 35 | 7 | 0.20 |
| 101–200 | 85 | 18 | 0.21 |
| 201–300 | 19 | 4 | 0.21 |
| 301–500 | 31 | 12 | 0.39 |
| 500+ | 6 | 0 | 0.00 |

SOURCE: NRC Phase II Survey.

**TABLE App-A-44** Scientific Publications Developed as a Result of Phase II Project

| | Number Submitted | Number Published |
|---|---|---|
| Scientific Publications | 1,028 | 990 |

NOTE: Results for 816 respondents to survey question.
SOURCE: NRC Phase II Survey, Question 18.

**TABLE App-A-45** University Involvement

**Q31 Any involvement by universities faculty, graduate students, and/or university developed technologies?**

| | |
|---|---|
| Yes | 212 |
| No | 625 |
| Total | 837 |

**University Involvement in SBIR Projects**

| | |
|---|---|
| 1.3% | The Principal Investigator (PI) for this Phase II project was at the time of the project a faculty member. |
| 1.3% | The Principal Investigator (PI) for this Phase II project was at the time of the project an adjunct faculty member. |
| 13.6% | Faculty member(s) or adjunct faculty member(s) work on this Phase II project in a role other than PI, e.g., consultant. |
| 11.4% | Graduate students worked on this Phase II project. |
| 9.2% | University/college facilities and/or equipment were used on this Phase II project. |
| 2.2% | The technology for this project was licensed from a university or college. |
| 3.9% | The technology for this project was originally developed at a university or college by one of the participants in this Phase II project. |
| 12.5% | A university or college was a subcontractor on this Phase II project. |

n=837

SOURCE: NRC Phase II Survey.

# Appendix B

# NRC Phase II and Firm Surveys

The first section of this appendix describes the methodology used to survey Phase II SBIR awards (or contracts.) The second part presents the results—first of the awards and then of the firm survey. (Appendix C presents the Phase I survey.)

## ABOUT THE SURVEYS

### Starting Date and Coverage

The survey of SBIR Phase II awards was administered in 2005, and included awards made through 2001. This allowed most of the Phase II awarded projects (nominally two years) to be completed, and provided some time for commercialization. The selection of the end date of 2001 was consistent with a GAO study, which in 1991, surveyed awards made through 1987.

A start date of 1992 was selected. The year 1992 for the earliest Phase II project was considered a realistic starting date for the coverage, allowing inclusion of the same (1992) projects as the DoD 1996 survey, and of the 1992, and 1993 projects surveyed in 1998 for SBA. This adds to the longitudinal capacities of the study. The 10 years of Phase II coverage spanned the period of increased funding set-asides and the impact of the 1992 reauthorization. This time frame allowed for extended periods of commercialization and for a robust spectrum of economic conditions. Establishing 1992 as the cut-off date for starting the survey helped to avoid the problem that older awards suffer from several problems, including meager early data collection as well as potentially irredeemable data loss; the fact that some firms and PIs are no longer in place; and fading memories.

## Award Numbers

While adding the annual awards numbers of the five agencies would seem to define the larger sample, the process was more complicated. Agency reports usually involve some estimating and anticipation of successful negotiation of selected proposals. Agencies rarely correct reports after the fact. Setting limitations on the number of projects to be surveyed from each firm required knowing how many awards each firm had received from all five agencies. Thus, the first step was to obtain all of the award databases from each agency and combine them into a single database. Defining the database was further complicated by variations in firm identification, location, phone numbers, and points of contact within individual agency databases. Ultimately, we determined that 4,085 firms had been awarded 11,214 Phase II awards (an average of 2.7 Phase II awards per firm) by the five agencies during the 1992–2001 time frame. Using the most recent awards, the firm information was updated to the most current contact information for each firm.

## Sampling Approaches and Issues

The Phase II survey used an array of sampling techniques, to ensure adequate coverage of projects to address a wide range both of outcomes and potential explanatory variables, and also to address the problem of skew. That is, a relatively small percentage of funded projects typically account for a large percentage of commercial impact in the field of advanced, high-risk technologies.

- **Random samples.** After integrating the 11,214 awards into a single database, a random sample of approximately 20 percent was sampled. Then a random sample of 20 percent was ensured for each year; e.g., 20 percent of the 1992 awards, of the 1993 awards, etc. Verifying the total sample one year at a time allowed improved ability to adapt to changes in the program over time, as otherwise the increased number of awards made in recent years might dominate the sample.
- **Random sample by agency.** Surveyed awards were grouped by agency; additional respondents were randomly selected as required to ensure that at least 20 percent of each agency's awards were included in the sample.
- **Firm surveys.** After the random selection, 100 percent of the Phase IIs that went to firms with only one or two awards were polled. These are the hardest firms to find for older awards. Address information is highly perishable, particularly for earlier award years. For firms that had more than two awards, 20 percent were selected, but no less than two.
- **Top performers.** The problem of skew was dealt with by ensuring that all Phase IIs known to meet a specific commercialization threshold (total of $10 million in the sum of sales plus additional investment) were surveyed (derived from the DoD commercialization database). Since 56 percent of all awards were

in the random and firm samples described above, only 95 Phase IIs were added in this fashion.

- **Coding.** The project database tracks the survey sample, which corresponds with each response. For example, it is possible for a randomly sampled project from a firm that had only two awards to be a top performer. Thus, the response could be analyzed as a random sample for the program, a random sample for the awarding agency, a top performer, and as part of the sample of single or double winners. In addition, the database allows examination of the responses for the array of potential explanatory or demographic variables.
- **Total number of surveys.** The approach described above generated a sample of 6,410 projects, and 4,085 firm surveys—an average of 1.6 award surveys per firm. Each firm receiving at least one project survey also received a firm survey. Although this approach sampled more than 57 percent of the awards, multiple-award winners, on average, were asked to respond to surveys covering about 20 percent of their projects.

## Administration of the Survey

The questionnaire drew extensively from the one used in the 1999 National Research Council assessment of SBIR at the Department of Defense, *The Small Business Innovation Research Program: An Assessment of the Department of Defense Fast Track Initiative*.[1] That questionnaire in turn built upon the questionnaire for the 1991 GAO SBIR study. Twenty-four of the 29 questions on the earlier NRC study were incorporated. The researchers added 24 new questions to attempt to understand both commercial and noncommercial aspects, including knowledge base impacts, of SBIR, and to gain insight into impacts of program management. Potential questions were discussed with each agency, and their input was considered. In determining questions that should be in the survey, the research team also considered which issues and questions were best examined in the case studies and other research methodologies. Many of the resultant 33 Phase II Award survey questions and 15 Firm Survey questions had multiple parts.

The surveys were administered online, using a Web server. The formatting, encoding, and administration of the survey was subcontracted to BRTRC, Inc., of Fairfax, VA.

There are many advantages to online surveys (including cost, speed, and possibly response rates). Response rates become clear fairly quickly, and can rapidly indicate needed follow up for nonrespondents. Hyperlinks provide amplifying information, and built-in quality checks control the internal consistency of the responses. Finally, online surveys allow dynamic branching of question sets,

[1]National Research Council, *The Small Business Innovation Research Program: An Assessment of the Department of Defense Fast Track Initiative*, Charles W. Wessner, ed., Washington, DC: National Academy Press, 2000.

with some respondents answering selected subsets of questions but not others, depending on prior responses.

Prior to the survey, we recognized two significant advantages of a paper survey over an online one. For every firm (and thus every award), the agencies had provided a mailing address. Thus, surveys could be addressed to the firm president or CEO at that address. That senior official could then forward the survey to the correct official within the firm for completion. For an online survey we needed to know the email address of the correct official. Also, each firm needed a password to protect its answers. We had an SBIR Point of Contact (POC) and email address and password for every firm, which had submitted for a DoD SBIR 1999 survey. However, we had only limited email addresses and no passwords for the remainder of the firms. For many, the email addresses that we did have were those of Principal Investigators rather than an official of the firm. The decision to use an online survey meant that the first step of survey distribution was an outreach effort to establish contact with the firms.

## Outreach by Mail

This outreach phase began with the establishing a NAS registration Web site which allowed each firm to establish a POC, email address and password. Next, the Study Director, Dr. Charles Wessner, sent a letter to those firms for which email contacts were not available. Ultimately only 150 of the 2,080[2] firms provided POC/email after receipt of this letter. Six hundred fifty of those letters were returned by the post office as invalid addresses. Each returned letter required thorough research by calling the agency provided phone number for the firm, then using the Central Contractor Registration database, Business.com (powered by Google) and Switchboard.com to try to find correct address information. When an apparent match was found, the firm was called to verify that it was in fact the firm, which had completed the SBIR. Two hundred thirty-seven of the 650 missing firms were so located. Another ten firms were located which had gone out of business and had no POC.

Two months after the first mailing, a second letter from the Study Director went to firms whose first letter had not been returned, but which had not yet registered a POC. This letter also went to 176 firms, which had a POC email, but no password, and to the 237 newly corrected addresses. The large number of letters (277) from this second mailing that were returned by the postal service, indicated that there were more bad addresses in the first mailing than indicated by its returned mail. (If the initial letter was inadvertently delivered, it may have been thrown away.) Of the 277 returned second letters, 58 firms were located using the search methodology described above. These firms were asked on the

[2]The letter was also erroneously sent to an additional 43 firms that had received only STTR awards.

phone to go to the registration Web site to enter POC/email/password. A total of 93 firms provided POC/email/password on the registration site subsequent to the second mailing. Three additional firms were identified as out of business.

The final mailing, a week before survey, was sent to those firms that had not received either of the first two letters. It announced the study/survey and requested support of the 1,888 CEOs for which we had assumed good POC/email information from the DoD SBIR submission site. That letter asked the recipients to provide new contact information at the DoD submission site if the firm information had changed since their last submission. One hundred seventy-three of these letters were returned. We were able to find new addresses for 53 of these, and ask those firms to update their information. One hundred fifteen firms could not be found and five more were identified as out of business.

The three mailings had demonstrated that at least 1,100 (27 percent) of the mailing addresses were in error, 734 of which firms could not be found, and 18 were reported to be out of business.

## Outreach by Email

We began Internet contact by emailing the 1,888 DoD Points of Contact (POCs) to verify their email and give them opportunity to identify a new POC. Four hundred ninety-four of those emails bounced. The next email went to 788 email addresses that we had received from agencies as PI emails. We asked that the PI have the correct company POC identify themselves at the NAS Update registration site. One hundred eighty-eight of these emails bounced. After more detailed search of the list used by NIH to send out their survey, we identified 83 additional PIs and sent them the PI email discussed above. Email to the POCs not on the DoD Submission site resulted in 110 more POC/email/Password being registered on the NAS registration site.

We began the survey at the end of February with an email to 100 POCs as a beta test and followed that with another email to 2,041 POCs (total of 2,141) a week later.

## Survey Responses

By August 5, 2005, five months after release of the survey, 1,239 firms had begun and 1,149 firms had completed at least 14 of 15 questions on the firm survey. Project surveys were begun on 1,916 Phase II awards. Of the 4,085 firms that received Phase II SBIR awards from DoD, NIH, NASA, NSF, or DoE from 1992 to 2001, an additional seven firms were identified as out of business (total of 25) and no email addresses could be found for 893. For an additional 500 firms, the best email addresses that were found were also undeliverable. These 1,418 firms could not be contacted, thus had no opportunity to complete the surveys. Of these

**TABLE App-B-1** NRC Phase II Survey Responses by Agency, August 4, 2005

| Agency | Phase II Sample Size | Awards with Good Email Addresses | Percent of Sample Awards with Good Email Addresses | Answered Survey as of August 4, 2005 | Surveys as a Percent of Sample | Surveys as a Percent of Awards Contacted |
|---|---|---|---|---|---|---|
| DoD | 3,055 | 2,191 | 72 | 920 | 30 | 42 |
| NIH | 1,680 | 1,127 | 67 | 496 | 30 | 44 |
| NASA | 779 | 534 | 69 | 181 | 23 | 34 |
| NSF | 457 | 336 | 74 | 162 | 35 | 48 |
| DoE | 439 | 335 | 76 | 157 | 36 | 47 |
| Total | 6,408 | 4,523 | 70 | 1,916 | 30 | 42 |

firms, 585 had mailing addresses known to be bad. The 1,418 firms that could not be contacted were responsible for 1,885 of the individual awards in the sample.

Using the same methodology as the GAO had used in the 1992 report of their 1991 survey of SBIR, undeliverables and out-of-business firms were eliminated prior to determining the response rate. Although 4,085 firms were surveyed, 1,418 firms were eliminated as described. This left 2,667 firms, of which 1,239 responded, representing a 46 percent response rate by firms,[3] which could respond. Similarly when the awards, which were won by firms in the undeliverable category, were eliminated (6,408 minus 1,885), this left 4,523 projects, of which 1,916 responded, representing a 42 percent response rate. Table App-B-1 displays by agency the number of Phase II awards in the sample, the number of those awards, which by having good email addresses had the opportunity to respond, and the number that responded.[4] Percentages displayed are the percentage of awards with good addresses, the percentage of the sample that responded, and the responses as a percentage of awards with the opportunity to respond.

The NRC Methodology report had assumed a response rate of about 20 percent. Considering the length of the survey and its voluntary nature, the rate achieved was relatively high and reflects both the interest of the participants in the SBIR program and the extensive follow-up efforts. At the same time, the possibility of response biases that could significantly affect the survey results must be recognized. For example, it may be possible that some of the firms that could not be found have been unsuccessful and folded. It may also be possible that unsuccessful firms were less likely to respond to the survey.

[3]Firm information and response percentages are not displayed in Table App-B-1, which displays by agency, since many firms received awards from multiple agencies.

[4]The average firm size for awards, which responded, was 37 employees. Nonresponding awards came from firms that averaged 38 employees. Since responding Phase IIs were more generally more recent than nonresponding, and awards have gradually grown in size, the difference in average award size ($655,525 for responding and $649,715 for nonresponding) seems minor.

# NRC Phase II Survey Results for DoD

NOTE: SURVEY RESPONSES APPEAR IN BOLD, AND EXPLANATORY NOTES APPEAR IN TYPEWRITER FONT. FOR FURTHER DETAIL AND ANALYSIS OF SURVEY RESULTS, SEE APPENDIX A.

**Project Information** 920 respondents answered the first question. Since respondents are directed to skip certain questions based on prior answers, the number that responded varies by question. Also some respondents did not complete their surveys. 837 completed all applicable questions. For computation of averages, such as average sales, the denominator used was 920, the number of respondents who answered the first question. Where appropriate, the basis for calculations is provided in red after the question.

PROPOSAL TITLE:
AGENCY: DoD
TOPIC NUMBER:
PHASE II CONTRACT/GRANT NUMBER:

Part I. Current status of the Project

1. What is the current status of the project funded by the referenced SBIR award? *Select the one best answer.* Percentages are based on the 920 respondents who answered this question.
   a. **5%** Project has not yet completed Phase II. *Go to question 21*
   b. **23%** Efforts at this company have been discontinued. No sales or additional funding resulted from this project. *Go to question 2*
   c. **10%** Efforts at this company have been discontinued. The project did result in sales, licensing of technology, or additional funding. *Go to question 2*
   d. **27%** Project is continuing post-Phase II technology development. *Go to question 3*
   e. **16%** Commercialization is underway. *Go to question 3*
   f. **20%** Products/Processes/Services are in use by target population/customer/consumers. *Go to question 3*

2. Did the reasons for discontinuing this project include any of the following? *PLEASE SELECT YES OR NO FOR EACH REASON AND NOTE THE ONE PRIMARY REASON*
300 projects were discontinued. The % below are the percent of the discontinued projects that responded with the indicated response.

| | Yes | No | Primary Reason |
|---|---|---|---|
| a. Technical failure or difficulties | **26%** | **74%** | **10%** |
| b. Market demand too small | **57%** | **43%** | **26%** |
| c. Level of technical risk too high | **19%** | **81%** | **3%** |
| d. Not enough funding | **56%** | **44%** | **20%** |
| e. Company shifted priorities | **32%** | **68%** | **7%** |
| f. Principal investigator left | **12%** | **88%** | **2%** |
| g. Project goal was achieved (e.g., prototype delivered for federal agency use) | **59%** | **41%** | **10%** |
| h. Licensed to another company | **4%** | **96%** | **2%** |
| i. Product, process, or service not competitive | **24%** | **76%** | **3%** |
| j. Inadequate sales capability | **20%** | **80%** | **3%** |
| k. Other (please specify): ________________ | **22%** | **78%** | **13%** |

*The next question to be answered depends on the answer to question 1. If c, go to question 3. If b, skip to question 16.*

Part II. Commercialization activities and planning.

Questions 3–7 concern actual sales to date resulting from the technology developed during this project. **Sales** includes all sales of a product, process, or service, to federal or private sector customers resulting from the technology developed during this Phase II project. A sale also includes licensing, the sale of technology or rights, etc.

3. Has your company and/or licensee had any actual sales of products, processes, services or other sales incorporating the technology developed during this project? *Select all that apply.* This question was not answered for those projects still in Phase II (5%) or for projects, which were discontinued without sales or additional funding (23%). The denominator for the percentages below is all projects that answered the survey. Only 72% of all projects, which answered the survey, could respond to this question.

   a. **18%** No sales to date, but sales are expected *Skip to question 8.*
   b. **8%** No sales to date nor are sales expected *Skip to question 11.*
   c. **35%** Sales of product(s)
   d. **6%** Sales of process(es)
   e. **20%** Sales of services(s)
   f. **7%** Other sales (e.g., rights to technology, licensing, etc.)

From the combination of responses 1b, 3a, and 3b, we can conclude that 31% had no sales and expect none, and that 18% had no sales but expect sales.

4. For your company and/or your licensee(s), when did the first sale occur, and what is the approximate amount of total sales resulting from the technology developed during this project? If multiple SBIR awards contributed to the ultimate commercial outcome, report only the share of total sales appropriate to this SBIR project. *Enter the requested information for your company in the first column and, if applicable and if known, for your licensee(s) in the second column. Enter approximate dollars. If none, enter 0 (zero).*

| | Your Company | Licensee(s) |
|---|---|---|
| a. Year when first sale occurred. | ☐☐☐☐ | ☐☐☐☐ |

**41% reported a year of first sale. 63% of these first sales occurred in 2000 or later. 15% reported a licensee year of first sale. 64% of these first sales occurred in 2000 or later.**

| | Your Company | Licensee(s) |
|---|---|---|
| b. Total Sales Dollars of Product (s) Process(es) or Service(s) to date. (Average of 920 survey respondents) | **$1,316,573** | **$102,005** |

**Although 399 reported a year of first sale, only 378 reported sales >0. Their average sales were $3,204,358. Over half of the total sales dollars were due to 7 projects, each of which had $30,000,000 or more in sales. The highest reporting project had $99,000,000 in sales. Similarly of the 142 projects that reported a year of first licensee sale, only 34 reported actual licensee sales >0. Their reported average sales were $3,673,370. 37% of the total sales dollars was due to 2 projects, each of which had 15,000,000 or more licensee sales. The highest reporting project had 20,000,000 in licensee sales.**

| | Your Company | Licensee(s) |
|---|---|---|
| c. Other Total Sales Dollars (e.g., Rights to technology, Sale of spin-off company, etc.) to date. (Average of 920 survey respondents) | **$60,917** | **$33,750** |

**Combining the responses for b and c, the average for each of the 920 projects that responded to the survey is thus sales of nearly $1.4 million by the SBIR company and over $135,000 in sales by licensees.**

---

Display this box for Q 4 & 5 if project commercialization is known.

Your company reported sales information to DoD as a part of an SBIR proposal or to NAS as a result of an earlier NAS request. This information may be useful in answering the prior question or the next question. You reported as of *(date):* DoD sales *($ amount),* Other federal sales *($ amount),* Export sales *($ amount),* Private-sector sales *($ amount),* and other sales *($ amount).*

---

5. To date, approximately what percent of total sales from the technology developed during this project have gone to the following customers? *If none enter 0 (zero). Round percentages. Answers should add to about 100%.*[5] 920 firms responded to this question as to what percent of their sales went to each agency or sector.

| | |
|---|---|
| Domestic private sector | **21%** |
| Department of Defense (DoD) | **38%** |
| Prime contractors for *DoD or NASA* | **12%** |
| NASA | **1%** |
| *Agency that awarded the Phase II* | **–%** |
| Other federal agencies *(Pull down)* | **1%** |
| State or local governments | **1%** |
| Export Markets | **11%** |
| Other (Specify)____________ | **16%** |

---

The following questions identify the product, process, or service resulting from the project supported by the referenced SBIR award, including its use in a fielded federal system or a federal acquisition program.

---

6. Is a federal system or acquisition program using the technology from this Phase II?
   If yes, please provide the name of the federal system or acquisition program that is using the technology. **12% reported use in a federal system or acquisition program.**

7. Did a commercial product result from this Phase II project? **27% reported a commercial product.**

8. If you have had no sales to date resulting from the technology developed during this project, what year do you expect the first sales for your company or its licensee? Only firms that had no sales but answered that they expected sales got this question.

   **18% expected sales.** The year of expected first sale is ☐☐☐☐
   **82% of those expecting sales expected sales to occur before 2008**

9. For your company and/or your licensee, what is the approximate amount of total sales expected between now and the end of 2006 resulting from the technology developed during this project? *(If none, enter 0 [zero].)* This

[5]Please note: If a NASA SBIR award, the prime contractors line will state "Prime contractors for NASA." The "Agency that awarded the Phase II" will only appear if it is not DoD or NASA. The name of the actual awarding agency will appear.

question was seen by those who already had sales and those w/o sales who reported expecting sales; however, averages are computed for all who took the survey since all could have expected sales.

a. Total sales dollars of product(s), process(es) or services(s) expected between now and the end of 2006. (Average of 920 projects) **$900,280**

b. Other Total Sales Dollars (e.g., rights to technology, sale of spin-off company, etc.) expected between now and the end of 2006. (Average of 920 projects) **$129,372**

c. Basis of expected sales estimate. *Select all that apply.*
   - a. **18%** Market research
   - b. **31%** Ongoing negotiations
   - c. **33%** Projection from current sales
   - d. **4%** Consultant estimate
   - e. **31%** Past experience
   - f. **34%** Educated guess

10. How did you (or do you expect to) commercialize your SBIR award?
   - a. **3%** No commercial product, process, or service was/is planned.
   - b. **32%** As software
   - c. **60%** As hardware (final product, component, or intermediate hardware product)
   - d. **23%** As process technology
   - e. **18%** As new or improved service capability
   - f. **0%** As a drug
   - g. **0%** As a biologic
   - h. **15%** As a research tool
   - i. **2%** As educational materials
   - j. **10%** Other, please explain ______________________

11. Which of the following, if any, describes the type and status of marketing activities by your company and/or your licensee for this project? *Select one for each marketing activity.* This question answered by 620 firms, which completed Phase II and have not discontinued the project, w/o sales or additional funding.

| | Marketing activity | Planned | Need Assistance | Underway | Completed | Not Needed |
|---|---|---|---|---|---|---|
| a. | Preparation of marketing plan | **11%** | **7%** | **21%** | **28%** | **33%** |
| b. | Hiring of marketing staff | **10%** | **5%** | **9%** | **19%** | **57%** |
| c. | Publicity/advertising | **12%** | **8%** | **22%** | **17%** | **41%** |
| d. | Test marketing | **10%** | **6%** | **16%** | **13%** | **54%** |
| e. | Market Research | **9%** | **9%** | **23%** | **21%** | **37%** |
| f. | Other *(Specify)* | **1%** | **2%** | **1%** | **1%** | **37%** |

Part III. Other outcomes

12. As a result of the technology developed during this project, which of the following describes your company's activities with other companies and investors? *(Select all that apply.)* Percentage of the 620 who answered this question.

| | | U.S. Companies/Investors | | Foreign Companies/Investors | |
|---|---|---|---|---|---|
| | Activities | Finalized Agreements | Ongoing Negotiations | Finalized Agreements | Ongoing Negotiations |
| a. | Licensing Agreement(s) | **16%** | **16%** | **3%** | **5%** |
| b. | Sale of Company | **1%** | **5%** | **0%** | **1%** |
| c. | Partial sale of Company | **1%** | **4%** | **0%** | **1%** |
| d. | Sale of technology rights | **4%** | **10%** | **1%** | **3%** |
| e. | Company merger | **0%** | **3%** | **0%** | **1%** |
| f. | Joint Venture agreement | **4%** | **8%** | **1%** | **2%** |
| g. | Marketing/distribution agreement(s) | **11%** | **9%** | **5%** | **4%** |
| h. | Manufacturing agreement(s) | **3%** | **9%** | **3%** | **2%** |
| i. | R&D agreement(s) | **14%** | **14%** | **3%** | **3%** |
| j. | Customer alliance(s) | **14%** | **14%** | **5%** | **3%** |
| k. | Other *Specify*___________ | **2%** | **2%** | **0%** | **1%** |

13. In your opinion, in the absence of this SBIR award, would your company have undertaken this project? *Select one.*
Percentage of the 618 who answered this question.
    a. **3%** Definitely yes
    b. **10%** Probably yes *If selected a or b, go to question 14.*
    c. **17%** Uncertain
    d. **33%** Probably not
    e. **37%** Definitely not *If c, d, or e, skip to question 16.*

14. If you had undertaken this project in the absence of SBIR, this project would have been Questions 14 and 15 were answered only by the 13% who responded that they definitely or probably would have undertaken this project in the absence of SBIR.

a. **7%** Broader in scope
b. **44%** Similar in scope
c. **50%** Narrower in scope

15. In the absence of SBIR funding, *(Please provide your best estimate of the impact.)*
    a. The start of this project would have been delayed about **an average of 11** months.
    **62% of the 81 firms expected the project would have been delayed. 49% (40 firms) expected the delay would be at least 12 months. 22% anticipated a delay of at least 24 months.**
    b. The expected duration/time to completion would have been
        1) **78%** Longer
        2) **10%** The same
        3) **0%** Shorter
        **12%** No response
    c. In achieving similar goals and milestones, the project would be
        1) **0%** Ahead
        2) **11%** The same place
        3) **69%** Behind
        **20%** No response

16. Employee information. *Enter number of employees. You may enter fractions of full-time effort ( e.g., 1.2 employees). Please include both part-time and full-time employees, and consultants, in your calculation.*

| | |
|---|---|
| Number of employees (if known) when Phase II proposal was submitted | **Ave = 35**<br>**8% report 0**<br>**29% report 1-5**<br>**30% report 6-20**<br>**13% report 21-50**<br>**8% report >100** |
| Current number of employees | **Ave = 60**<br>**1% report 0**<br>**17% report 1-5**<br>**30% report 6-20**<br>**22% report 21-50**<br>**18% report >100** |

| | |
|---|---|
| Number of current employees who were hired as a result of the technology developed during this Phase II project. | **Ave = 2.5**<br>**48% report 0**<br>**42% report 1-5**<br>**5% report 6-20**<br>**2% report >20** |
| Number of current employees who were retained as a result of the technology developed during this Phase II project | **Ave = 2.3**<br>**44% report 0**<br>**47% report 1-5**<br>**5% report 6-20**<br>**2% report >20** |

17. The Principal Investigator for this Phase II Award was a (check all that apply)
    a. **4%** Woman
    b. **11%** Minority
    c. **86%** Neither a woman or minority

18. Please give the number of patents, copyrights, trademarks, and/or scientific publications for the technology developed as a result of this project. *Enter numbers. If none, enter 0 (zero).* Results are for 816 respondents to this question.

| Number Applied For/ Submitted | | Number Received/ Published |
|---|---|---|
| **836** | Patents | **398** |
| **71** | Copyrights | **62** |
| **211** | Trademarks | **176** |
| **1,028** | Scientific Publications | **990** |

Part IV. Other SBIR funding

19. How many SBIR awards did your company receive prior to the Phase I that led to this Phase II?
    a. Number of previous Phase I awards. **Average of 22. 29% had no prior Phase I and another 42% had 5 or less prior Phase I.**
    b. Number of previous Phase II awards. **Average of 8. 46% had no prior Phase II and another 36% had 5 or less prior Phase II.**

20. How many SBIR awards has your company received that are related to the project/technology supported by this Phase II award?
    a. Number of related Phase I awards. **Average of two awards. 46% had no prior related Phase I and another 48% had 5 or less prior related Phase I.**
    b. Number of related Phase II awards. **Average of one. 56.5% had no prior related Phase II and another 42.1% had 5 or less prior related Phase II.**

Part V. Funding and other assistance

21. Prior to this SBIR Phase II award, did your company receive funds for research or development of the technology in this project from any of the following sources? Of 854 respondents.
    a. **22%** Prior SBIR *Excluding the Phase I, which proceeded this Phase II.*
    b. **12%** Prior non-SBIR federal R&D
    c. **2%** Venture Capital
    d. **8%** Other private company
    e. **6%** Private investor
    f. **27%** Internal company investment (including borrowed money)
    g. **2%** State or local government
    h. **1%** College or University
    i. **5%** Other *Specify* _________

---

Commercialization of the results of an SBIR project normally requires additional developmental funding. Questions 22 and 23 address additional funding. Additional Developmental Funds include non-SBIR funds from federal or private sector sources, or from your own company, used for further development and/or commercialization of the technology developed during this Phase II project.

---

22. Have you received or invested any additional developmental funding in this project?
    a. **54%** Yes *Continue*
    b. **46%** No *Skip to question 24.*

23. To date, what has been the total additional developmental funding for the technology developed during this project? Any entries in the **Reported** column are based on information previously reported by your firm to DoD or NAS. They are provided to assist you in completing the **Developmental Funding** column. Previously reported information did not include investment by your company or personal investment. *Please update this information to include breaking out Private investment and Other investment by subcategory. Enter dollars provided by each of the listed sources. If none, enter 0 (zero).* The dollars shown are determined by dividing the total

funding in that category by the 920 respondents who started the survey to determine an average funding. Only 462 of these respondents reported any additional funding.

| Source | Reported | Developmental Funding |
|---|---|---|
| a. Non-SBIR federal funds | $_ _, _ _ _, _ _ _ | **$ 361,277** |
| b. Private Investment | $_ _, _ _ _, _ _ _ | |
| (1) U.S. venture capital | | **$169,313** |
| (2) Foreign investment | | **$16,744** |
| (3) Other private equity | | **$77,246** |
| (4) Other domestic private company | | **$57,177** |
| c. Other sources | $_ _, _ _ _, _ _ _ | |
| (1) State or local governments | | **$6,018** |
| (2) College or universities | | **$1,812** |
| d. Not previously reported | | |
| (1) Your own company (Including money you have borrowed) | | **$90,913** |
| (2) Personal funds | | **$15,232** |
| Total average additional developmental funding, all sources, per award | | $795,734 |

24. Did this award identify matching funds or other types of cost sharing in the Phase II Proposal?[6]
    a. **81%** No matching funds/co-investment/cost sharing were identified in the proposal. *If a, skip to question 26.*
    b. **16%** Although not a DoD Fast Track, matching funds/co-investment/cost sharing were identified in the proposal.
    c. **3%** Yes. This was a DoD Fast Track proposal.

25. Regarding sources of matching or co-investment funding that were proposed for Phase II, check all that apply. The percentages below are computed for those 161 projects, which reported matching funds.
    a. **40%** Our own company provided funding (includes borrowed funds)
    b. **20%** A federal agency provided non-SBIR funds
    c. **43%** Another company provided funding
    d. **14%** An angel or other private investment source provided funding
    e. **12%** Venture Capital provided funding

[6]The words underlined appear only for DoD awards.

26. Did you experience a gap between the end of Phase I and the start of Phase II?
    a. **69%** Yes *Continue*
    b. **31%** No *Skip to question 29.*
    The average gap reported by 584 respondents was 5 months. 3% of the respondents reported a gap of two or more years.

27. Project history. Please fill in for all dates that have occurred. This information is meaningless in aggregate. It has to be examined project by project in conjunction with the date of the Phase I end and the date of the Phase II award to calculate the gaps.

    Date Phase I ended *Month/year*

    Date Phase II proposal submitted *Month/year*

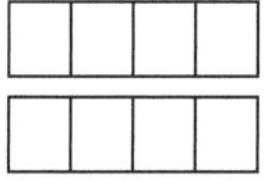

28. If you experienced funding gap between Phase I and Phase II for this award, *(select all answers that apply)*
    a. **58%** Stopped work on this project during funding gap.
    b. **34%** Continued work at reduced pace during funding gap.
    c. **4%** Continued work at pace equal to or greater than Phase I pace during funding gap.
    d. **7%** Received bridge funding between Phase I and II.
    e. **2%** Company ceased all operations during funding gap.

29. Did you receive assistance in Phase I or Phase II proposal preparation for this award? Of 791 repondents.
    a. **2%** State agency provided assistance.
    b. **2%** Mentor company provided assistance.
    c. **0%** Regional association provided assistance.
    d. **3%** University provided assistance.
    e. **93%** We received no assistance in proposal preparation.

    Was this assistance useful?
    a. **58%** Very Useful
    b. **40%** Somewhat Useful
    c. **2%** Not Useful

30. In executing this award, was there any involvement by university faculty, graduate students, and/or university developed technologies? Of 837 respondents.
    **25%** Yes
    **75%** No

31. This question addresses any relationships between your firm's efforts on this Phase II project and any University (ies) or College (s). The percentages are computed against the 837 who answered question 30, not just those who answered yes to question 30. ***Select all that apply.***
    a. **1%** The Principal Investigator (PI) for this Phase II project was at the time of the project a faculty member.
    b. **1%** The Principal Investigator (PI) for this Phase II project was at the time of the project an adjunct faculty member.
    c. **14%** Faculty member(s) or adjunct faculty member(s) work on this Phase II project in a role other than PI, e.g., consultant.
    d. **11%** Graduate students worked on this Phase II project.
    e. **9%** University/College facilities and/or equipment were used on this Phase II project.
    f. **2%** The technology for this project was licensed from a University or College.
    g. **4%** The technology for this project was originally developed at a University or College by one of the percipients in this Phase II project.
    h. **13%** A University or College was a subcontractor on this Phase II project.

In remarks enter the name of the University or College that is referred to in any blocks that are checked above. If more than one institution is referred to, briefly indicate the name and role of each.

32. Did commercialization of the results of your SBIR award require FDA approval? Yes **2%**

    In what stage of the approval process are you for commercializing this SBIR award?
    a. **0%** Applied for approval
    b. **1%** Review ongoing
    c. **0%** Approved
    d. **0%** Not Approved
    e. **1%** IND: Clinical trials
    f. **0%** Other

# NRC Firm Survey Results

**NOTE: ALL RESULTS APPEAR IN BOLD. RESULTS ARE REPORTED FOR ALL 5 AGENCIES (DoD, NIH, NSF, DoE, AND NASA).**

**1,239 firms began the survey. 1,149 completed through question 14. 1,108 completed all questions.**

---

**If your firm is registered in the DoD SBIR/STTR Submission Web site, the information filled in below is based on your latest update as of September 2004 on that site. Since you may have entered this information many months ago, you may edit this information to make it correct. In conjunction with that information, the following additional information will help us understand how the SBIR program is contributing to the formation of new small businesses active in federal R&D and how they impact the economy. Questions A–G are autofilled from Firm database, when available.**

---

A. Company Name: ______________________________
B. Street Address: ______________________________
C. City: ______________________ State: ____ Zip: ________
D. Company Point of Contact: ______________________
E. Company Point of Contact Email: __________________
F. Company Point of Contact Phone: (___) ___ - ____ Ext: __________
G. The year your company was founded: __________

1. Was your company founded because of the SBIR Program?
   a. **79%** No
   b. **8%** Yes
   c. **13%** Yes, In part

2. Information on company founders. *Please enter zeros or the correct number in each pair of blocks.*

   a. Number of founders. [ ][ ]
      **5%** unknown
      **40%** 1
      **30%** 2
      **13%** 3
      **8%** 4
      **2%** 5
      **2%** >5
      **Average = 2 founders/firm**

b. Number of other companies started by one or more of the founders. ☐☐

**5%** Unknown
**46%** Started no other firms
**23%** Started 1 other firm
**13%** Started 2 other firms
**7%** Started 3 other firms
**3%** Started 4 other firms
**3%** Started 5 or more other firms
**Average number of other firms founded is one.**

c. Number of founders who have a business background. ☐☐

**5%** Unknown
**50%** No founder known to have business background
**30%** One founder with business background
**14%** More than one founder with business background

d. Number of founders who have an academic background ☐☐

**5%** Unknown
**29%** No founder known to have academic background
**38%** One founder with academic background
**28%** More than one founder with academic background

3. What was the most recent employment of the company founders prior to founding this company? *Select all that apply.* **Total >100% since many companies had more than one founder.**
   a. **65%** Other private company
   b. **36%** College or University
   c. **9%** Government
   d. **10%** Other

4. How many SBIR and/or STTR awards has your firm received from the federal government?
   a. Phase I: ________ **Average number of Phase I reported was 14.**
      **13%** **1 Phase I**
      **34%** **2 to 5 Phase I**
      **24%** **6 to 10 Phase I**
      **14%** **11 to 20 Phase I**
      **11%** **21 to 50 Phase I**
      **3%** **51 to 100 Phase I**
      **2%** **>100 Phase I Five firms reported >300 Phase I**

What year did you receive your first Phase I Award? _______

**3% reported 1983 or sooner**
**33% reported 1984 to 1992**
**40% reported 1993 to 1997**
**24% reported 1998 or later**

b. Phase II: ________ **Average number of Phase II reported was 7**

**27% 1 Phase II**
**44% 2 to 5 Phase II**
**15% 6 to 10 Phase II**
**8% 11 to 20 Phase II**
**5% 21 to 50 Phase II**
**1% >50 Phase II Four firms reported >100 Phase II**

What year did you receive your first Phase II Award? _______

**3% reported 1983 or sooner**
**22% reported 1984 to 1992**
**35% reported 1993 to 1997**
**41% reported 1998 or later**

5. What percentage of your company's growth would you attribute to the SBIR program after receiving its first SBIR award?

a. **31%** Less than 25%
b. **25%** 25% to 50%
c. **20%** 51% to 75%
d. **24%** More than 75%

6. Number of company employees (including all affiliates):

a. At the time of your company's first Phase II Award: ____

**56% 5 or less**
**28% 6 to 20**
**9% 21 to 50**
**8% > 50 Fourteen firms (1.3%( had greater than 200 employees at time of first Phase II.**

b. Currently: ______

**29% 5 or less**
**37% 6 to 20**
**17% 21 to 50**
**13% 51 to 200**
**5% > 200 Eleven firms report over 500 current employees.**

7. What percentage of your total R&D effort (Man-hours of Scientists and Engineers) was devoted to SBIR activities during the most recent fiscal year?___%
   **22%** **0% of R&D was SBIR during most recent fiscal year.**
   **16%** **1% to 10% of R&D was SBIR during most recent fiscal year.**
   **11%** **11% to 25% of R&D was SBIR during most recent fiscal year.**
   **18%** **26% to 50% of R&D was SBIR during most recent fiscal year.**
   **14%** **51% to 75% of R&D was SBIR during most recent fiscal year.**
   **19%** **>75% of R&D was SBIR during most recent fiscal year.**

8. What was your company's total revenue for the last fiscal year?
   a. **10%** <$100,000
   b. **18%** $100,000–$499,999
   c. **16%** $500,000–$999,999
   d. **33%** $1,000,000–$4,999,999
   e. **14%** $5,000,000–$19,999,999
   f. **6%** $20,000,000–$99,999,999
   g. **1%** $100,000,000 +
   h. **0.4%** Proprietary information

9. What percentage of your company's revenues during its last fiscal year is federal SBIR and/or STTR funding (Phase I and/or Phase II)? ___________
   **30%** **0% of revenue was SBIR (Phase I or II) during most recent fiscal year.**
   **17%** **1% to 10% of revenue was SBIR (Phase I or II) during most recent fiscal year.**
   **11%** **11% to 25% of revenue was SBIR (Phase I or II) during most recent fiscal year.**
   **13%** **26% to 50% of revenue was SBIR (Phase I or II) during most recent fiscal year.**
   **13%** **51% to 75% of revenue was SBIR (Phase I or II) during most recent fiscal year.**
   **13%** **76% to 99% of revenue was SBIR (Phase I or II) during most recent fiscal year.**
   **4%** **100% of revenue was SBIR (Phase I or II) during most recent fiscal year.**

10. **This question eliminated from the survey as redundant.**

11. Which, if any, of the following has your company experienced as a result of the SBIR Program? *Select all that apply.*

    a. **Fifteen** firms made an initial public stock offering in calendar year
    **Seven reported prior to 2000; two in 2000; four in 2004; and one in both 2006 and 2007**

    b. **Six** planned an initial public stock offering for 2005/2006.

    c. **14%** Established one or more spin-off companies.

    How many spin-off companies?
    **242** Spin-off companies were formed.

    d. **84%** reported None of the above.

12. How many patents have resulted, at least in part, from your company's SBIR and/or STTR awards?
    **43% reported no patents resulting from SBIR/STTR.**
    **16% reported one patent resulting from SBIR/STTR.**
    **27% reported 2 to 5 patents resulting from SBIR/STTR.**
    **13% reported 6 to 25 patents resulting from SBIR/STTR.**
    **1% reported >25 patents resulting from SBIR/STTR.**

**A total of over 3,350 patents were reported; an average of almost 3 per firm**

---

The remaining questions address how market analysis and sales of the commercial results of SBIR are accomplished at your company.

---

13. This company normally first determines the potential commercial market for an SBIR product, process or service
    a. **66%** Prior to submitting the Phase I proposal
    b. **21%** Prior to submitting the Phase II proposal
    c. **9%** During Phase II
    d. **3%** After Phase II

14. Market research/analysis at this company is accomplished by: *(Select all that apply.)*
    a. **28%** The Director of Marketing or similar corporate position
    b. **7%** One or more employees as their primary job
    c. **41%** One or more employees as an additional duty
    d. **23%** Consultants
    e. **53%** The Principal Investigator
    f. **67%** The company president or CEO
    g. **1%** None of the Above

15. Sales of the product(s), process(es) or service(s) that result from commercialising an SBIR award at this company are accomplished by: *Select all that apply.*
    a. **35%** An in-house sales force
    b. **52%** Corporate officers
    c. **30%** Other employees
    d. **30%** Independent distributors or other company(ies) with which we have marketing alliances
    e. **26%** Other company(ies), which incorporate our product into their own
    f. **9%** Spin-off company(ies)
    g. **26%** Licensing to another company
    h. **11%** None of the above

# Appendix C

# NRC Phase I Survey

## SURVEY DESCRIPTION

This section describes a survey of Phase I SBIR awards over the period 1992–2001. The intent of the survey was to obtain information on those which did not proceed to Phase II, although most of the firms that did receive a Phase II were also surveyed.

Over that period the five agencies (DoD, DoE, NIH, NASA, and NSF) made 27,978 Phase I awards. Of the total number for the five agencies, 7,940 Phase I awards could be linked to one of the 11,214 Phase II awards made from 1992–2001. To avoid putting an unreasonable burden on the firms that had many awards, we identified all firms that had over 10 Phase I awards that apparently had not received a Phase II. For those firms, we did not survey any Phase I awards that also received a Phase II. This meant that 1,679 Phase Is were not surveyed.

We chose to survey the Principal Investigator (PI) rather than the firm to reduce the number of surveys that any one person would have to complete. In addition, if the Phase I did not result in a Phase II, the PI was more likely to have a better memory of it than firm officials. There were no PI email addresses for 5,030 Phase I awardees. This reduced the number of surveys sent since the survey was conducted by email.

Thus there were 21,269 surveys (27,978 minus 1,679 minus 5,030 = 21,269) emailed to 9,184 PIs). Many PIs had received multiple Phase I awards. Of these surveys, 6,770 were undeliverable. This left possible responses of 14,499. Of these, there were 2,746 responses received. The responses received represented 9.8 percent of all Phase I awards for the five agencies, or 12.9 percent of all surveys emailed, and 18.9 percent of all possible responses.

The agency breakdown, including Phase I survey results, is given in Table App-C-1.

**TABLE App-C-1** Agency Breakdown for NRC Phase I Survey

| Phase I Project Surveys by Agency | Phase I Awards, 1992–2001 | Answered Survey (Number) | Answered Survey (%) |
|---|---|---|---|
| DoD | 13,103 | 1,198 | 9 |
| DoE | 2,005 | 281 | 14 |
| NASA | 3,363 | 303 | 9 |
| NIH | 7,049 | 716 | 10 |
| NSF | 2,458 | 248 | 10 |
| TOTAL | 27,978 | 2,746 | 10 |

## SURVEY PREFACE

This survey is an important part of a major study commissioned by the U.S. Congress to review the SBIR program as it is operated at various federal agencies. The assessment, by the National Research Council (NRC), seeks to determine both the extent to which the SBIR programs meet their mandated objectives, and to investigate ways in which the programs could be improved. Over 1,200 firms have participated earlier this year in extensive survey efforts related to firm dynamics and Phase II awards. This survey attempts to determine the impact of Phase I awards that do not go on to Phase II. We need your help in this assessment. We believe that you were the PI on the listed Phase I.

We anticipate that the survey will take about 5-10 minutes of your time. If this Phase I resulted in a Phase II, this survey has only 3 questions; if there was not a Phase II, there are 14 questions. Where dollar figures are requested (sales or funding), please give your best estimate. Responses will be aggregated for statistical analysis and not attributed to the responding firm/PI, without the subsequent explicit permission of the firm.

Since you have been the PI on more than one Phase I from 1992 to 2001, you will receive additional surveys. These are not duplicates. Please complete as many surveys for those Phase Is that did not result in a Phase II as you deem to be reasonable.

Further information on the study can be found at *<http://www7.nationalacademies.org/sbir>*. BRTRC, Inc., is administering this survey for the NRC. If you need assistance in completing the survey, call 877-270-5392. If you have questions about the assessment more broadly, please contact Dr. Charles Wessner, Study Director, NRC.

**Project Information**
**Proposal Title:**
**Agency:**
**Firm Name:**
**Phase I Contract / Grant Number:**

# NRC Phase I Survey Results

**NOTE: RESULTS APPEAR IN BOLD. RESULTS ARE REPORTED FOR ALL 5 AGENCIES (DoD, NIH, NSF, DoE, AND NASA). EXPLANATORY NOTES ARE IN TYPEWRITER FONT.**

`2,746 responded to the survey. Of these 1,380 received the follow on Phase II. 1,366 received only a Phase I.`

1. Did you receive assistance in preparation for this Phase I proposal?

| Phase I only | | | Received Phase II | |
|---|---|---|---|---|
| **95%** | No | *Skip to Question 3.* | **93%** | No |
| **5%** | Yes | *Go to Question 2.* | **7%** | Yes |

2. If you received assistance in preparation for this Phase I proposal, put an X in the first column for any sources that assisted and in the second column for the most useful source of assistance. Check all that apply. `Answered by 74 Phase I only and 91 Phase II who received assistance.`

| | **Phase I only**<br>Assisted/Most Useful | **Received Phase II**<br>Assisted/Most Useful |
|---|---|---|
| State agency provided assistance | **10/3** | **11/10** |
| Mentor company provided assistance | **15/9** | **21/15** |
| University provided assistance | **31/17** | **34/22** |
| Federal agency SBIR program managers or technical representatives provided assistance | **16/8** | **25/19** |

3. Did you receive a Phase II award as a sequential direct follow-on to this Phase I award? *If yes, please check yes. Your survey would have been automatically submitted with the HTML format. Using this Word format, you are done after answering this question. Please email this as an attachment to jcahill@brtrc.com, or fax to Joe Cahill 703 204 9447. Thank you for you participation.* `2,746 responses`

**50%** No. We did not receive a follow-on Phase II after this Phase I.
**50%** Yes. We did receive the follow-on Phase II after this Phase I.

4. Which statement correctly describes why you did not receive the Phase II award after completion of your Phase I effort. *Select best answer.* All questions which follow were answered by those 1,366 who did not receive the follow-on Phase II. % based on 1,366 responses.

**33%** The company did not apply for a Phase II. *Go to question 5.*
**63%** The company applied, but was not selected for a Phase II. *Skip to question 6.*
**1%** The company was selected for a Phase II, but negotiations with the government failed to result in a grant or contract. *Skip to question 6.*
**3%** Did not respond to question 4.

5. The company did not apply for a Phase II because: *Select all that apply.* % based on 446 who answered "The company did not apply for a Phase II" in question 4.

**38%** Phase I did not demonstrate sufficient technical promise.
**11%** Phase II was not expected to have sufficient commercial promise.
**6%** The research goals were met by Phase I. No Phase II was required.
**34%** The agency did not invite a Phase II proposal.
**3%** Preparation of a Phase II proposal was considered too difficult to be cost effective.
**1%** The company did not want to undergo the audit process.
**8%** The company shifted priorities.
**5%** The PI was no longer available.
**6%** The government indicated it was not interested in a Phase II.
**13%** Other—explain:

6. Did this Phase I produce a noncommercial benefit? Check all responses that apply. % based on 1,366.

**59%** The awarding agency obtained useful information.
**83%** The firm improved its knowledge of this technology.
**27%** The firm hired or retained one or more valuable employees.
**17%** The public directly benefited or will benefit from the results of this Phase I. *Briefly explain benefit.*
**13%** This Phase I was essential to founding the firm or to keeping the firm in business.
**8%** No

7. Although no Phase II was awarded, did your company continue to pursue the technology examined in this Phase I? *Select all that apply.* % based on 1,366.

**46%** The company did not pursue this effort further.
**22%** The company received at least one subsequent Phase I SBIR award in this technology.
**14%** Although the company did not receive the direct follow-on Phase II to the this Phase I, the company did receive at least one other subsequent Phase II SBIR award in this technology.
**12%** The company received subsequent federal non-SBIR contracts or grants in this technology.
**9%** The company commercialized the technology from this Phase I.
**2%** The company licensed or sold its rights in the technology developed in this Phase I.
**16%** The company pursued the technology after Phase I, but it did not result in subsequent grants, contracts, licensing or sales.

## Part II. Commercialization

8. How did you, or do you, expect to commercialize your SBIR award? *Select all that apply.* % based on 1,366.

**33%** No commercial product, process, or service was/is planned.
**16%** As software
**32%** As hardware (final product component or intermediate hardware product)
**20%** As process technology
**11%** As new or improved service capability
**15%** As a research tool
**4%** As a drug or biologic
**3%** As educational materials

9. Has your company had any actual sales of products, processes, services or other sales incorporating the technology developed during this Phase I? *Select all that apply.* % based on 1,366.

**5%** Although there are no sales to date, the outcome of this Phase I is in use by the intended target population.
**65%** No sales to date, nor are sales expected. *Go to question 11.*
**15%** No sales to date, but sales are expected. *Go to question 11.*
**9%** Sales of product(s)
**1%** Sales of process(es)

**6%** Sales of services(s)
**2%** Other sales (e.g., rights to technology, sale of spin-off company, etc.)
**2%** Licensing fees

10. For your company and/or your licensee(s), when did the first sale occur, and what is the approximate amount of total sales resulting from the technology developed during this Phase I? If other SBIR awards contributed to the ultimate commercial outcome, estimate only the share of total sales appropriate to this Phase I project. (Enter the requested information for your company in the first column and, if applicable and if known, for your licensee(s) in the second column. Enter dollars. If none, enter 0 (zero); leave blank if unknown.)

| | Your Company | Licensee(s) |
|---|---|---|
| a. Year when first sale occurred | **89 of 147**<br>**after 1999** | **11 of 13**<br>**after 1999** |
| b. Total Sales Dollars of Product(s), Process(es), or Service(s) to date | | |
| *(Sale Averages)* | **$84,735** | **$3,947** |
| Top 5 Sales<br>Accounts for 43% of all sales | 1. **$20,000,000**<br>2. **$15,000,000**<br>3. **$5,600,000**<br>4. **$5,000,000**<br>5. **$4,200,000** | |
| c. Other Total Sales Dollars (e.g., Rights to technology, Sale of spin-off company, etc.) to date | | |
| *(Sale Averages)* | **$1,878** | **$0** |

Sale averages determined by dividing totals by 1,366 responders.

11. If applicable, please give the number of patents, copyrights, trademarks, and/or scientific publications for the technology developed as a result of Phase I. (Enter numbers. If none, enter 0 [zero]; leave blank if unknown.)

# Applied For or Submitted / # Received/Published

**319 / 251** Patent(s)
**50 / 42** Copyright(s)
**52 / 47** Trademark(s)
**521 / 472** Scientific Publication(s)

12. In your opinion, in the absence of this Phase I award, would your company have undertaken this Phase I research? Select only one lettered response. If you select c, and the research, absent the SBIR award, would have been different in scope or duration, check all appopriate boxes. Unless otherwise stated, % are based on 1,366.

**5%** Definitely yes
**7%** Probably yes, similiar scope and duration
**16%** Probably yes, but the research would have been different in the following way
% based on 218 who responded probably yes, but research would have . . .
**75%** Reduced scope
**4%** Increased scope
**21%** No Response to scope
**5%** Faster completion
**51%** Slower completion
**44%** No Response to completion rate
**14%** Uncertain
**40%** Probably not
**16%** Definitely not
**4%** No Response to question 12

## Part III. Funding and Other Assistance

Commercialization of the results of an SBIR project normally requires additional developmental funding. Questions 13 and 14 address additional funding. Additional Developmental Funds include non-SBIR funds from federal or private sector sources, or from your own company, used for further development and/or commercialization of the technology developed during this Phase I project.

13. Have you received or invested any additional developmental funding in this Phase I? % based on 1,366.

**25%** Yes. Go to question 14.
**72%** No. Skip question 14 and submit the survey.
**3%** No response to question 13.

14. To date, what has been the approximate total additional developmental funding for the technology developed during this Phase I? (Enter numbers. If none, enter 0 [zero]; leave blank if unknown).

| Source | # Reporting that source | Developmental Funding (Average Funding) |
|---|---|---|
| a. Non-SBIR federal funds | **79** | **$72,697** |
| b. Private Investment | | |
| (1) U.S. Venture Capital | **13** | **$4,114** |
| (2) Foreign investment | **8** | **$4,288** |
| (3) Other private equity | **20** | **$7,605** |
| (4) Other domestic private company | **39** | **$8,522** |
| c. Other sources | | |
| (1) State or local governments | **20** | **$1,672** |
| (2) College or Universities | **6** | **$293** |
| d. Your own company (Including money you have borrowed) | **149** | **$21,548** |
| e. Personal funds of company owners | **54** | **$4,955** |

Average funding determined by dividing totals by 1,366 responders.

# Appendix D

# Case Studies

# 3e Technologies International (formerly known as AEPTEC)

*Zoltan Acs*
*University of Baltimore*

## FIRM OVERVIEW

3e Technologies International (3eTI), formerly known as AEPTEC, specializes in secure wireless network applications and wireless condition-based maintenance solutions. Condition-based maintenance refers to the use of advanced technologies to determine equipment condition, and potentially predict failure. The company's main service is providing wireless infrastructure and specialized applications of wireless technologies.

Steven Chen founded 3eTI in 1996. Chen originally sought to develop hard disk drives, but shifted the company's focus to secure wireless technology and condition-based maintenance because of product development opportunities made available through the SBIR program.

As a more flexible small business, 3eTI is able to economically customize solutions for the government market, gaining an advantage over larger companies that are focused primarily on mass market products. 3eTI specializes in customized, high-tech total solutions designed to meet a customer's specification, while still destined for a future in applications for the commercial market. For example, the company's wireless access point products (devices that connect computing platforms together to form a wireless network) compete with products from vendors such as Cisco and Motorola. 3eTI's customized solutions capabilities were demonstrated in a prior contract, where the company developed a robust one-box wireless sensor networking product, which compares to four-box solutions from competitors such as Cisco.

3eTI/AEPTEC was named the 5th fastest growing company in the Washington, DC metropolitan area in 2003. The company has also received the Navy SBIR Success Story Award, the Tibbetts award in 2002, and was #62 on the list of top Department of Defense contractors in 2003.

## THE ROLE OF SBIR

3eTI has been the recipient of 28 Phase I and 10 Phase II awards. Approximately eight percent of the company's revenue comes from SBIR Phase I and II awards.

The SBIR awards have been integral to the firm's current position. They allowed the company to keep its technology fresh and on the leading edge, which is imperative for a small company. 3eTI considers SBIR as a leading source of funding and the company anticipates applying for further SBIR awards.

3eTI has continued with its focus—determined mostly by early SBIR awards—on wireless technologies. Within that area, the company explored various possibilities and partnerships. For example, while the majority of the company's work has been geared towards U.S. Navy ships, 3eTI has expanded into wireless technologies that provide for antiterrorism and force protection at military bases and other government facilities.

The majority of the firm's commercialization activities are due to SBIR. Phase III awards have allowed the company to commercialize their technology by providing seed funding. It allowed 3eTI to build a reputation and further develop its products, leading to 3eTI's acquisition by EFJ, Inc.—a leading wireless telecommunications solutions company—in 2006.

3eTI has used SBIR's sole source justification to gain a competitive advantage over large systems integrators, like Lockheed Martin, in federal procurement. Under this provision, the procurement advantage remains even if the small business enters a partnership with, or is bought out by, a large publicly traded company.

SBIR also informs small firms about the technical direction federal agencies are taking, and allows them to provide input for those technical solutions. SBIR affords small companies like 3eTI an opportunity to understand what direction federal agencies are heading and what their expectations are in terms of technology and product development. It would be difficult for a small company to get that kind of information outside the program.

In 1998, when the company first started receiving SBIR awards, it had less than 20 employees. Currently, 3eTI has approximately 95 employees. 3eTI has a ratio of four research personnel to one manufacturing personnel. This ratio is due to the fact that the company has historically manufactured only prototypes or small orders of units, and outsourced larger volume manufacturing of finished products.

The firm has sold products resulting from the SBIR projects to both the federal and private sector. 3eTI has also filed several patents and has published several scientific papers. 3eTI trademarks include AirGuard, InfoMatics, and Virtual Perimeter Monitoring System (VPMS).

## IMPROVING SBIR

### Surviving Funding Gaps

Concern has been expressed regarding the delay between Phase I and Phase II awards. Funding gaps may be fatal to commercialization opportunities for small companies. When a company is counting on an award that is being dragged out, it may lose an important window of opportunity and customers that have their own timelines to follow. If a commercialization opportunity is lost, companies often have to start the award process all over with a new project. Unfortunately, agency contract officers often do not consider the real-world limitations their awardees face.

Such delays may occur because contracts must pass through several levels of bureaucracy, and delays accumulate at each level. The preferred solution to this problem includes establishing a timeline for the government to follow in administering awards, and requiring the agency to advance the company a small portion of the award after making the Phase II decision, but before the contract is finalized.

3eTI has survived such delays (sometimes as long as one year) only because the company was fortunate enough to receive overlapping SBIR awards and conducted successful commercialization activities. Although to some extent delay is built into the budget and manpower process, it cannot be sustained without winning multiple awards. In fact, the company very rarely stops work when waiting for the funding to arrive. While it has encountered delays in receiving Phase II money, it knew the award was approved. Nevertheless, such delays are hurdles for both the firm and the customer.

## Commercialization Programs

The company has participated in meetings organized by the SBIR, including meetings where various federal agencies participated as commercialization customers. 3eTI finds these events very helpful.

### Award Size

The amount of the Phase I award is seen as being adequate, considering an award period of six months. Phase I awards can be utilized well by companies that have a good idea formulated, giving the company an opportunity to make its plan and state its case. However, 3eTI believes that the size of Phase II awards should be increased. The standard award, approximately $750,000, is small considering today's market standards, and it has not changed in the past 5 years. The price of doing business increases as the project grows and diversifies. The ideal award would be a minimum of $1,000,000, with another $250,000+ available for Phase II Options. Also, companies need additional funding because it takes a long time to get Phase II awards to commercialization and transform them into finished products.

## Procurement

Individuals charged with procurement require extensive training. They must understand their legal responsibilities in the procurement process. According to the law, products developed under SBIR should receive priority consideration in purchase decisions. Sometimes procurement specialists fail to provide that consideration. In such instances, 3eTI has protested and educated those involved.

A lack of knowledge about the procurement process may inhibit a firm's participation in SBIR. When companies begin writing proposals, without understanding exactly the customer's needs, they invest valuable time and resources in a process doomed to failure.

## Venture Capital Participation

Small publicly traded companies should be allowed to participate in the SBIR program, even if they have access to private sources of funding, because the SBIR's objective is to promote innovation. Additionally, if small publicly traded companies are excluded from the program, the government will have to pay a higher price for their product by acquiring it commercially.

## Enlarging SBIR

The SBIR program should be larger and invest in its administration and customer service. The government benefits greatly from the program, because for $750,000, it gets a better and more cost-efficient product compared to the commercial sector alternative. Large commercial companies can have greater expenses, compared to small companies, and may reflect that in their prices.

# Advanced Ceramics Research[1]

*Irwin Feller*
*American Association for the Advancement of Science*

Advanced Ceramics Research (ACR) was originally incorporated as a start-up, self-financed firm in 1989 by Anthony Mulligan who had recently graduated in mechanical engineering from the University of Arizona, and Mark Angier who was still a student in mechanical engineering, also at the University of Arizona. Shortly after they were joined by Dr. Donald Uhlmann, a professor at the University of Arizona, and Kevin Stuffle, a chemical engineer previously employed at Ceramatec Corporation, Salt Lake City, Utah. In late 1996 Dr. Daniel Albrecht, retired CEO of Buehler Corporation, joined as a shareholder and officer until 2000. Since 2000, Angier and Mulligan have remained as the only shareholders and are active in the management of the company.

From its inception, ACR sought to become a product development company, capable of manufacturing products for a diverse set of industries based on its technological developments. Although its competitive advantage has been in its advanced technology, it has sought to avoid being limited to being a contract R&D house. Over its history, the relative emphasis on R&D, product development, and manufacturing has varied, being primarily shaped by market demand conditions for its end-user products. The firm has both an extended set of collaborative, network relationships with university researchers, who conduct basic research on materials, and "downstream" customers for its products.

Also, from its early inception, the firm knew about the SBIR program, but viewed its profit ceiling margins, placed at 5–7 percent, as too low to warrant much attention. Only commercial products were seen as yielding an adequate profit margin. Over time though, it has participated in the SBIR program of several federal agencies, including DoD, NASA, Department of Energy, and the National Science Foundation.

ACR's initial 2 products were PVA-SIC grinding stones and Polyurethane friction drive belts for the aluminum memory disk manufacturing industry. These two products were a direct result of a NASA Phase I SBIR program entitled "Laser Induced Thermal Micro-cracking for Ductile Regime Grinding of Large Optical Surfaces." While the program did not go on to Phase II, the commercial sales generated from the first two products was significant for the growth of the company.

The firm also saw market potential in developing products from advanced ceramics. The attractiveness of the SBIR program was that it would underwrite concept development. Firm representatives had several discussions with DoD

[1]Draft based on interview with Dr. Ranji Vaidyanthan, May 3, 2005, at the Navy Opportunity Forum and publicly available information.

**ADVANCED CERAMICS RESEARCH :**
**COMPANY FACTS AT A GLANCE**

Address: 3292 E. Hemisphere Loop, Tucson, AZ 85706
Phone: 520-573-6300; <*http://www.acrtucson.com*>

Year Started: 1989

Ownership: Private

Annual Sales:
FY2002: $5 million
FY2003: $8.3 million
FY2004: $11.5 million
FY2005: $20+ million

Number of Employees: 83

3-Year Sales Growth Rate: 250 percent
4-Year Sales Growth Rate: 400 percent

SIC:

Technology Focus: Advanced composite materials; rapid prototyping, UAV's, sensors

Number of SBIR Awards—Phase I
(DoD Phase I): 75
Number of SBIR Awards—Phase II
(DoD Phase II): 18

Awards: 2002 R&D 100 Awards (Fibrous monolith wear-resistant components that increased the wear life of mining drill bits), 2001 R&D 100 Award for water-soluble composite tooling material, 2000 R&D 100 Award for water-soluble rapid prototyping support material

officials about the SBIR program, but the catalytic event was a meeting with a DARPA program officer, Bill Coblenz. Coblenz already held a patent (issued in 1988) on ceramic materials. He was interested in supporting "far out" ideas related to the development of low-cost production processes on advanced ceramics, based on the technique of rapid prototyping. DARPA already was supporting research at the University of Michigan.

ACR was encouraged to begin work on low-cost production techniques. It did this under a series of DARPA awards and SBIR awards, although never concentrating on SBIR. Drawing in part on the advanced research being done at the

University of Michigan and drawing on its expertise in both advanced ceramics and manufacturing, ACR developed a general purpose technology of being able to convert autoCAD drawings into machine readable code, then to direct generation of ceramic, composite, and metal parts.

Initially SBIR awards accounted for nearly all of ACR's revenues. By 1993 the firm had transitioned to nearly 50 percent of its revenues from the commercial sector and about 25 percent of its revenues from non-SBIR government R&D funding, with the remaining 25 percent as SBIR revenues. For 2005 the company projects about $20 million in sales with about 15-20 percent of the revenues coming from STTR/SBIR Phase I and Phase II programs. The firm's R&D also has been underwritten by revenues generated by its manufacturing operations. Its primary use of SBIR awards was to develop specific application technologies based about its core technology.

One market that it saw as having considerable potential was that of developing and manufacturing "flexible carriers for hard-disk drives" for the electronics industry. After aggressively "knocking on doors" to gain customers, it soon became a major supplier to firms such as SpeedFam Corporation, Komag, Seagate, and IBM. ACR's competitive advantage rested in its ability to make prototypes accurately, quickly, and at competitive prices. Demand for this product line grew rapidly, enabling the firm to go to a 3-shift 7-day-a-week operation. In addition, ACR developed ancillary products related to testing and quality control tied to this product line.

The firm financed its expansion through a combination of retained earnings and license revenues, primarily from Smith Tools International, an oil and rock drilling company, and Kyocera, a Japanese based firm, which specialized in ceramics for communications applications, which licensed its Fibrous Monolith technology. ACR also reports receiving approximately $100,000 in the form of a bridge loan between a Phase I and Phase II award from a short-lived Arizona's state economic development program, funded from state lottery revenues. It reports no venture capital financing. It remains a privately held firm.

Demand for ACR's electronic products seemed to be on an upward trajectory through the 1990s. In response to demands from its primary customers for an increase in output from 5,000 to 60,000 units monthly, ACR built a new 30,000 square foot plant. The electronics market for ACR's products however declined abruptly in 1997, when 2 of its major customers—Seagate and Komag, two of the largest producers of hard disk drives, shifted production to Asia. This move represented both the shift from 8 inch to 5 inch and then 3.5 inch disks, and lower production costs, which drove down the price of the carrier components they produced from $16 to $1.50 per unit. The loss of its carrier business was a major reversal for the firm. Heavy layoff resulted, with employment declining to low of about 28 employees in 1998.

1998–1999 are described as years of reinvention for survival for ACR. The firm's R&D division, which formerly had been losing money, was now seen as having to become its primary source of revenue. The explicit policy was to undertake only that R&D which had discernible profit margins and the opportunity for near term commercialization. Previously, ACR had conducted a small number of Phase I SBIR awards, but had not actively pursued Phase II awards unless it could readily see the commercial product that was likely to flow from this research or it had a commercial partner.

ACR reports several outcomes from its participation in the SBIR program. As of 2005, it has received 75 Phase I and 21 Phase II awards. The larger number of awards have been from DoD, followed by NASA, with a few from the other agencies such as NSF and DoE. Products based on SBIR awards received from DARPA and NASA have had commercial sales of approximately $14 million.

ACR is now actively engaged in development and marketing of Silver Fox, a small unmanned aerial vehicle (UAV). R&D for the Silver Fox has been supported by awards under DoD's STTR program, and involves collaboration between ACR and researchers at the University of Arizona, University of California-Berkeley, the University of California-Los Angeles, and MIT.

The genesis of the project highlights the multiple uses of technological innovations. In 2000, while in DC to discuss projects with Office of Naval Research (ONR) program managers, ACR representatives also had a chance meeting with program manager for the Navy interested in small SWARM unmanned air vehicles (UAVs). At the time, ONR expressed and interest and eventually provided funding for developing a new low-cost small UAV as a means to engage in whale watching around Hawaii, with the objective of avoiding damage to the Navy's underwater sonic activities. Once developed however, the UAV's value as a more general purpose battlefield surveillance technology soon became apparent and ONR provided additional funding to further refine the UAV for warfighter use in Operation Iraqi Freedom.

ACR has a bonus compensation plan that rewards employees for invention disclosures, patents, licenses, and presentations at professional meetings. These incentives are seen as fostering outcome from SBIR awards (as with all other company activities).

ACR owns a 49 percent stake in a joint venture manufacturing company called Advanced Ceramics Manufacturing, LLC, which is located on the Tohono O'Odham Reservation south of Tucson, Arizona. Fifty-one percent is owned by Tribal Land Alotees. The company, which employs about 10 people who manufacture ceramic products in a multimillion dollar facility (15,000 square feet), is expected to do about $2.5 million in sales revenues over the next 12 months.

ACR also has also recently opened 2,500 square feet of laboratory and office space in Arlington, VA, where it is basing its new Sensors Division and providing customer support to its military customers with an initial staff of 8 persons.

Funding delays between Phase I and Phase II awards have been handled primarily through a process of shared decision making, leading to consensus-based reallocations of firm resources and staff assignments. ACR typically has several R&D projects occurring simultaneously. When delays occur, researchers are assembled to determine whether the firm's internal funds, including its IR&D funds, will be used to continue a specific project.

DoD's SBIR review and award procedures are seen as fair and timely. The dollar amounts of Phase I and Phase II awards and SBIR "paperwork" requirements likewise are seen as reasonable.

The Navy is seen as especially good in the speed with which it handles the selection process. It has reduced the length of time to make awards from 3 to 4 months to 2 months; NSF, by way of contrast, takes 6 months.

The length of the selection process across federal agencies does influence ACR's decisions. It is more likely to pursue Phase I awards from agencies such as DoD that have short selection cycles than those with long(er) ones.

The company has seen great benefit in accelerating commercialization of its SBIR/STTR programs through participation of the Navy's Technology Assistance Program (TAP). ACR first participated in the TAP program for its Water Soluble Tooling Technology, its Fibrous Monolith Technology, and its UAV technology. ACR's diligent following to what it learned in the Navy's TAP program has assisted it in receiving 3 separate Indefinite Deliverables, Indefinite Quantities (ID/IQ) Phase III contracts totaling $75 million. Each of the three technologies has received a $25 million ID/IQ contract to facilitate continued government use of the technology.

# Applied Signal Technology

*Irwin Feller*
*American Association for the Advancement of Science*

Applied Signal Technology (AST) was founded in 1984 by Gary Yancey, John Trieichler, Jim Collins, and Jane Sanchez. The four founders had been employed by Argo Systems, a California-based defense contractor, specializing in signal technology related to strategic intelligence, but left to start their own firm. Each of them invested their own funds and deferred their salaries for a year to provide the firm's initial working capital. The firm subsequently received several rounds of private financing prior to going public. AST went public in 1993, and is currently listed on the NASDAQ as ASPG.

The firm's initial revenues were derived largely from a 1-year consulting contract with Lockheed, augmented by smaller size contracts from DoD. The firm grew steadily throughout much of the 1990s, reaching a peak level of employment (650) and revenues ($115 million) in about 1999. The fall of the Berlin Wall, and the subsequent large reductions in expenditures by U.S. intelligence agencies led

**APPLIED SIGNAL TECHNOLOGY:**
**COMPANY FACTS AT A GLANCE**

Address: 400 West California Avenue
Sunnyvale, CA 94086
Phone: 408-749-1888

Year Started: 1984

Ownership: Publicly traded equity; NASDAQ

Annual Sales (FY2004) $142 million

Number of employees: 500

Sales Growth Rate: Doubling between 2002–2004

SIC Code: 3669

Technology Focus: Advanced digital signal processing products, systems, and services; signal intelligence

Number of SBIR Awards—Phase I: 3
Number of SBIR Awards—Phase II: 1
Number of Patents: 4

to a sharp fall in revenues and employment. The upsurge in national expenditures for intelligence following 9/11 however has led to the firm's renewed growth; firm revenues increased from $76 million to $143 million between 2002 and 2004. Current employment levels now hover about 640 (110 new employees were added with the acquisition of Dynamic Technologies, on July 1, 2005); the firm is no longer eligible for the SBIR program.

AST located in Sunnyvale in part because the founders lived in the Bay Area, but also because at the time of its founding, Silicon Valley was the place to be for cutting-edge technologies, especially those related to miniaturization, a key factor in expanding the use of digital signal processing to the needs of signal intelligence organizations.

From its inception, reflecting the expertise and orientation of its 3 founders, AST has sought to integrate R&D and manufacturing. Its activities thus extended from the design of digital microwave receivers to the manufacture of the relevant hardware.

AST's initial customers were U.S. intelligence agencies. Beginning about 1994, its line of business began to extend to DoD. In the interim period, the firm had tried unsuccessfully to develop product lines, such as in HDTV receivers, for the civilian market. Although its products were regarded as technical successes, the firm was not able to produce these products in an economically profitable manner. (It notes that the fundamental economics of producing for the DoD and intelligence customers is different from that of the domestic sector. For the former, engineering hours are a source of income; for the latter, they are a cost, with income not being generated until downstream sales of the final product occur.)

AST's involvement with the DoD SBIR program stems from a purposeful search by the firm to diversify its customer base. It systematically scanned the DoD Web site for topics that matched its technical capabilities, and soon found one that provided a "glove fit." It describes its initial interaction with the cognate DoD technical program officer as providing DoD with a "surprising" opportunity for a quantum leap in technology. AST eventually submitted 6 Phase I proposals, of which 3 were funded; 1 of which eventually led to a Phase II award.

AST's work on the SBIR project increased its opportunities to work with DoD program managers about the generation of future solicitation topics. This interaction however proved a mixed blessing. In one case, a topic generated by discussions with one program manager was, in the firm's view, redirected to benefit another firm; subsequently, performance on the award proved technically unsuccessful.

The DoD SBIR awards are described as also having yielded significant benefits to the firm. The Phase II award led to a "foundation product" for one of AST's division. SBIR also served to "validate" AST's marketing efforts in becoming a new, credible supplier to DoD Services. Successful performance of an SBIR award is seen as providing proof to DoD program managers that AST was capable of producing technically valuable products under given contract terms. Had it not been for SBIR, AST's business with the Special Operations Command would not have developed; Special Operations likely would have stayed with their pre-existing sources of supply. Program managers are described as being too busy with multiple contracts to search out or to respond attentively to new sources of technology. Their orientation is to hire a contractor to solve problems, not to necessarily seek out the most competitive performer. SBIR, by way of contrast, requires that they become involved with small firms, to look at technical options, and to allow for increased competition in the selection of R&D performers. It allows and encourages program managers to scan, and expand, the contractor base.

AST has received four patents over the course of its history.[2] Overall, firm strategy is not to actively pursue patents, relying instead on trade secrets and know-how to protect its intellectual property. In part, it follows this strategy because patents provide little net economic benefit. Its dominant customer is the U.S. government, which, under government contracts is entitled to royalty-free use. Also, by not patenting, the firm avoids the costs of patenting and the associated public disclosure of technological knowledge. Its patents relate to its earlier efforts to enter domestic commercial markets.

AST did not submit proposals under DoD's 2004 SBIR solicitation because it had 485 employees as of August, 2004, and projected that its employment level would exceed 500 at the time it might receive an award. The uncertainties of the Defense market however induce a margin of error to such estimates, especially for firms close to the ceiling level of eligibility. Thus, it recommends that eligibility for SBIR be based on a firm's employment level at the time it submits a proposal rather than this level at the time of an award.

AST experienced no difficulties with the gap in funding between its Phase I and Phase II award. Indeed, it completed its Phase I award early, (an outcome that its DoD program manager said had not happened before), and received accelerated funding to begin its Phase II work. The problem it encountered with SBIR processes related primarily to the slow pace of the initial review process. This process was described as having taken 4–6 months, which makes it exceedingly difficult for the firm to maintain the core staff that is expected to work on the project. An increase in the dollar amount of Phase I awards to the range of $150,000–$200,000 is recommended.

[2]One patent was sold.

# Bihrle Applied Research

*Irwin Feller*
*American Association for the Advancement of Science*

## FIRM HISTORY

Bihrle Applied Research was started in 1973, as a 1-person engineering consulting operation by William Bihrle, an aeronautical engineer, formerly with Grumman Aircraft, which had major facilities on Long Island, New York. Billy Barnhart, also a former Grumman aeronautical engineer, joined Bihrle in 1976 when the firm was preparing for its first NASA contract. Bihrle and Barnhart were each seeking to do more exciting research than they found possible while working with Grumman. The firm's operations were self-financed, with its very

**BIHRLE APPLIED RESEARCH:**
**COMPANY FACTS AT A GLANCE**

400 Jericho Turnpike
Jericho, NY 11753
Phone: 757-766-2416

Year Started: 1973
Ownership: Private
Annual Sales (FY2004/2005): $3,492,554
Number of Employees: 18
3-Year Sales Growth Rate: 47 percent
SIC: 8748
Technology Focus: Aeronautical engineering; software development; simulation and testing
Number of SBIR Awards
Phase I: 12
(DoD Phase I Awards): 10
Number of SBIR Awards
Phase II: 9
(DoD Phase II Awards): 8
Number of Patents: 0
Number of Publications: 20
Number of Presentations: 33
Awards: 4

early years described as being essentially focused on survival. Bihrle remains a privately owned firm. It has not sought venture capital.

Reflecting its initial formation, Bihrle has continued to locate its headquarters in Long Island. The firm's first major contract was with NASA's Langley Research Center, and involved work on the stall and spin properties of general aviation aircraft. NASA was a major customer for the firm for many years. Consequently, Bihrle has located the bulk of its engineering operations in Hampton, Virginia, near the NASA Langley Research Center.

The firm started to add employees by the late 1970s, as it secured new contracts from NASA and the Navy. Bihrle's work consists primarily of aeronautical research, specialized wind tunnel testing, and development of simulation techniques related to aeronautics, including the development of specialized software. Its primary customers are DoD, NASA, and airframe manufacturers.

## EXPERIENCES UNDER SBIR

Bihrle received its first SBIR award in 1986 from the Air Force. The firm had read and heard about the program, and as an ongoing performer of DoD research perceived that its technical competencies could be applied to topics identified in DoD's SBIR solicitations. Bihrle discovered that program managers, both in DoD and NASA, found the firm's research to be useful in achieving mission objectives, but at times could not justify funding it under existing program constraints, even those of a 6.2 (applied research) character. SBIR thus gave agency program managers greater flexibility to support relevant "general purpose" technologies that otherwise could not be supported under targeted R&D programs. SBIR was described as a "gold mine" to DoD program managers. Bihrle also found it useful to meet with DoD officials about mutually interesting areas of research, with a view toward having these interests reflected in the development of solicitation topics.

Bihrle's work under SBIR awards has centered mostly about the development of new testing techniques for wind tunnel testing and the development of new simulation techniques. These new technologies are broadly applicable across several DoD programs, including the joint strike fighter and the Navy's V-22 program, as well as for NASA. The firm also credits SBIR awards as having contributed significantly to its current portfolio of tools and methods.

The importance of SBIR as a source of the firm's revenue has varied over the years, but in several years has been a significant portion of total revenues, approaching 25–30 percent. Bihrle's ability to compete however for SBIR awards also has varied. In some years, it might be able to identify three–four topics in DoD's solicitations to which it could respond by submitting a proposal; in other years, there would be no topic that matched the firm's competencies.

DoD's administration of the SBIR program is credited with proceeding as advertised. Awards are made on a timely basis. Birhle also reports good success—approximately 75 percent—in competing for Phase II awards, based on its Phase I work. The impact on the firm of gaps in funding between the two awards varies widely depending on its overall pace of activity. On some occasions, the firm will use internal funds to keep a project moving; on other occasions, it will let the project sit until an award is made.

## PARTICIPATION IN STATE GOVERNMENT PROGRAMS

On one occasion, Bihrle was able to obtain interim funding from a New York state program that was designed to provide gap funding. The state program though was described as "strange," in that the firm had to add something "extra" to its pending proposal in order to qualify for funding.

## RECOMMENDATIONS ON SBIR PROGRAM

The firm views the current size of Phase I and Phase II awards as reasonable. It notes though that supporting a Phase II proposal with a budget below the maximum amount appears to appeal to program managers. Proposals with below-ceiling budgets thus seem to have a competitive advantage.

SBIR is seen as an important program both for DoD and the firm. It should be continued.

# Brimrose

*Zoltan Acs*
*University of Baltimore*

## FIRM OVERVIEW

Brimrose is a high-tech R&D company that uses SBIR awards to develop and commercialize new a technology development vehicle. The company manufactures, markets, and services optical process control spectrometer systems for the pharmaceutical, refinery, and chemical industries. These near infrared optical spectrometers operate in the harshest of environments around the world, from the arctic to equatorial climates, to the severest of vibration in the most dusty, abrasive, and dirtiest of the chemical plants.

SBIR is credited with providing tremendous growth opportunities for research and development of the company. Brimrose has progressed from a small, unknown company to a respectable high-tech company known for providing innovative technological solutions. Currently Brimrose is one of the world leaders in process control equipment. It is also known to various government and research laboratories for providing them with innovative photonic components and technical assistance.

## A UNIVERSITY START-UP—THE ROLE OF SBIR

The firm was founded in 1979 by Dr. Ronald G. Rosemeier. After receiving his Ph.D. in Materials Science from the Johns Hopkins University, he began working as a post-doctoral student at the University of Maryland. There, he started writing proposals for SBIR awards, finding eventual success with four Phase I awards. Based on this funding, which approximated $200,000, he founded his firm and started hiring his first employees. He also applied for commercial bank loans, but found that banks were not willing to give him loans backed by the SBIR awards. As a result, Dr. Rosemeier collected over $100,000 in credit card debt. Six months later, he wrote Phase II proposals and received 3 awards, approximating $1,500,000—which meant real money. He hired additional employees. At that time 10 percent of the business was commercial (selling X-ray imaging at tradeshows) and 90 percent SBIR. As the company started commercializing new products, however, this percentage shifted to 80 percent commercial revenue and 20 percent SBIR revenue.

Over its history, Brimrose has garnered 65 Phase I and 28 Phase II SBIR awards. A few of the SBIR awards have directly resulted into commercial products. However, most awards have helped improve the company's products through development of new technical concepts and improved manufacturing techniques.

Brimrose began operations with 6 employees and today averages approximately 60 employees. Most of the Research & Development team and few of the support staff were hired under SBIR-related activities.

## COMMERCIAL APPLICATIONS

The firm's commercialization strategy came out of, and has been greatly enhanced through, its participation in SBIR. The Phase I and Phase II SBIR funding has allowed the company to determine the feasibility of new technology and develop it to the point of prototype development without allocation of significant internal resources. Following prototype development, Brimrose uses internal funds from previous commercial sales to bring the technology to the point of commercial availability. Thus, the SBIR funds lower the company's financial burden by decreasing the risks associated with new technology development. As a small company, this help is vital to its commercialization strategy.

While the company conducts applied research in areas such as industrial process control spectroscopy in the pharmaceutical and petrochemical industries, nondestructive testing and evaluation and novel opto-electronics devices, the focus of the company is on products that can be commercialized.

For example, Brimrose has developed a near infrared optical process control spectrometer that is operating in Germany at the OMV Refinery. This instrument performs the final quality control inspection of aviation jet fuel that is directly dispensed into commercial aircraft at the Munich International Airport. With built-in online modem support and on-board Brimrose sensor technology, this system has operated 24 hours a day for more than 5 years without a single failure.

Its optical spectrometers are also used in drug manufacturing. Having passed GMP (Good Manufacturing Practices) requirements for hardware and software, the product has been certified by the FDA for drug manufacturing. AstraZeneca has chosen Brimrose systems exclusively for all their plants world-wide as the only process control spectrometer that will be used for manufacturing of their pharmaceutical drugs.

The company commercially also manufactures components such as fiber optic coupled diode pumped solid-state green lasers for the biological instrumentation market, fiber optic collimators and focusers, fiber optic coupled modulators and tunable IR filters, and frequency shifters for the telecom industry. In all of its fiber optic coupled products, the assembly and alignment fixturing is critical to the product specification requirements. Brimrose also has extensive experience in packaging design, which includes epoxy-free optical path configurations and hermetically seam-welded designs.

The company also sells products to the agencies that made the SBIR awards. About 15 percent of its outcomes go back to the agencies. Additionally they provide technology and support to non-SBIR government programs. Their government contracts presently deal with the development of extremely fast optical radar receivers, ultra-high precision, interferometer-less profilers/distance gauges for IC fabrication applications, room temperature solid state mid IR lasers for counter measures applications, manufacturing and development of hydrogen plasma assisted MOCVD reactors for laser wafer, diamond film, gallium nitride and silicon carbide film materials development. It has developed space-qualified components for optical spectrometer based on acousto-optic tunable filters.

A the projected market size for Brimrose' acousto-optics device-based technology product is $5,000,000. Its major competitors are Foss and Brucker. In the area of vibrometer and displacement measurement devices using Laser Doppler Velocimeter, the company's main competitors are MetroLasers in the United States and Poly-Tec, in Europe.

## KNOWLEDGE GENERATION

The company is a 1982 recipient of the IR 100 Award. In addition, the company has published over 150 scientific papers. Brimrose has also filed two patents, with four U.S. patents pending:

- System for visualization of solid-liquid interface during crystal growth U.S. Patent Number 5037621, August 6, 1991.
- "Device and method for optical path length measurement," Chen-Chia Wang, S. Trivedi, and J. Khurgin, U.S. Patent Number 6,600,564.
- "Apparatus for optical difference frequency measurement using pulsed light sources and an optical frequency sensor," Chen-Chia Wang, Sudhir Trivedi, Feng Jin, Ponciano Rodriguez, and Serguei Stepanov, U.S. Patent Pending.
- "Bi-Wavelength Optical Intensity Modulators using Materials with Saturable Absorptions," Chen-Chia Wang and Sudhir Trivedi, U.S. Patent Pending.
- "Multi-junction solar cell with improved conversion efficiency," G.V. Jagannathan, Feng Jin, and Sudhir B. Trivedi. U.S. Patent Pending.
- "High speed optical gain flattener," Jolanta I. Rosemeier, Ronald G. Rosemeier, and Feng Jin. U.S. Patent Pending.

## IMPROVING SBIR

### Surviving Funding Gaps

Brimrose notes that it is extremely committed to the development of its technology. During the gap between the Phase I and Phase II funding, it uses internal resources to continue technology development. By doing this, the company believes that it is even more prepared to continue with the second phase of the work, which benefits both the government agency that is supporting the work as well as Brimrose. By continuing technology development during the Phase I to Phase II gap, Brimrose is also better placed to procure other forms of funding should Phase II SBIR funding fail to materialize.

While commercial products have always been the main objective of Brimrose, these funding gaps nonetheless hurt the company's ability to function seamlessly.

### Award Size

Flexibility is important as the cost of developing new technologies varies from topic to topic. For some technologies, $100,000 is enough. A laser generation system itself costs $300,000.

### Award Selection

While Brimrose views the award selection process as being, overall, fair, it finds that the proposal process is very different from agency to agency. For instance, submitting proposals to the DoD is relatively straightforward and simple. However, other agencies such as NSF are quite the opposite.

Submission of the proposals is very complicated and time consuming, which just adds to the stress of whole proposal writing process. NSF also uses its own format for the commercialization report. Therefore, a great deal of effort goes into reformatting the information. The paperwork involved is found to the relatively the same for each agency, though the level of this paperwork has decreased as reporting becomes increasingly electronic.

Also, agencies focus the proposals on different areas. For instance, NSF puts a great deal of emphasis on the commercial aspects of the technology. Brimrose has put more effort into preparing the commercialization plan for an NSF proposal than it put into the technical parts of many proposals combined. As a small company, Brimrose found this to be a major diversion of its resources. The commercialization report can be used as an indication of its abilities to commercialize products.

Brimrose believes that there should be a standardized proposal submission process, proposal acceptance process and reporting process for all of the agencies. There should be awards or matching funds available for Phase III projects. Phase II programs should be open, not by invitation only (as it has been in the last three years). All Phase I awardees should have the opportunity to submit a Phase II proposal especially in the instance where the technical monitor has not been involved in the Phase I research effort. Finally, Brimrose recommends that the duration of the Phase I and Phase II efforts should be extended to 9–12 months and 30 months respectively.

### Agency Interaction

Some agencies provide personal technical monitoring where there is a great deal of interaction between the firm's principal investigators and the funding agency. This scenario usually leads to a far more productive program. Some agencies, on the other hand, use one technical monitor for dozens of Phase I programs. In this case, there is little personal interaction and it is very hard for the technical monitor to understand the full benefits of the innovative technology that the company is able to provide.

SBIR program management is not funded in some agencies, which results in poor follow-up and communication with the companies. In some cases, program managers are not conscientious about their responsibilities and are unresponsive to attempts to communicate with them. Finally, program administration is not transparent enough to monitor accountability.

These suggestions for improvement notwithstanding, Brimrose believes that the SBIR program has inherent strengths.

- The SBIR program gives financial resources to companies that may otherwise not be able to pursue their high-tech research.
- The SBIR award greatly improves the research capabilities of a firm. The award provides the financial resources to conduct a feasibility study and then to develop a prototype.
- SBIR awards provide funding to more high-risk technology than most other sources of funding.

# CFD Research Corporation

*Zoltan Acs*
*University of Baltimore*

## FIRM OVERVIEW

CFD Research Corporation (CFDRC) specializes in engineering simulations and innovative designs. CFDRC's software and expertise allow coupled multiscale, multiphysics simulations of fluid, thermal, chemical, biological, electrical, and mechanical phenomena for real-world applications from aerospace, biomedical, defense, materials, energy, and other industries. Such simulations provide clearer insights into complex systems and thus enable *Better Decisions* and facilitate *Better Products* with lower risk, reduced cost, and less time.

CFDRC has developed state-of-the-art simulation methodologies under SBIR awards from several agencies. The resulting software has been adapted and applied to a wide range of problems, mostly during commercialization phases. Using its software and experimental facilities, CFDRC develops new hardware concepts, innovative designs, and prototypes. CFDRC is a two-time (1996 and 2006) winner of the prestigious Tibbetts Award for innovation. CFDRC is rated by DoD in the top 10 percent of small business for technology commercialization, nationwide.

About half of CFD Research Corporation's work is research and development, the other half focused on application and transfer of developed technologies. Computational Fluid Dynamics (CFD) enables computer simulations of fluid flow, heat transfer and chemical reaction processes. The simulations (virtual prototyping) facilitate engineering design and reduce time-to-market for new advanced programs. Some examples include:

- Biomedical studies for natural and prosthetic heart valves; vestibular (inner ear) mechanics; and ultrasound blood flow meter.
- Design of low NOx combustors and fuel nozzles for aircraft engines.
- Optimization of diesel fuel injectors and climate control systems for automotive companies and Chemical Vapor Deposition(CVD) reactors for semiconductor manufacturers.
- Analysis of pilot ejection seats, Space Shuttle Main Engine, and Air Turbo Rocket (ATR) propulsion system components.

## COMPANY HISTORY

Founded in 1987 by Dr. Ashok K. Singhal, CFDRC has developed simulation software for submarine applications, and aircraft escape systems for the U.S.

Navy. CFDRC was also invited to be part of the redesign teams after the accidents of the shuttle Challenger, in 1986, and Columbia in 2002. In parallel, CFDRC has also worked with private companies like Motorola, Applied materials, Samsung, GE, P&W, RR, Baxter, J&J, P&G, Chrysler, and Caterpillar.

During the slow down of the aerospace markets, the company diversified heavily into electronics and expanded its services to the global market. In 1992, CFDRC embraced the idea of commercializing simulation software. The company created a product development department that focused on customer friendly interfaces, support, and marketing.

The company's commercial division expanded in parallel with its R&D activities. CFD Research Corporation became a well known leader in the domain of multiphysics and multidisciplinary simulations. The domain of multiphysics includes fluid flow, heat transfer, chemical reactions, biochemistry, electrical phenomena, optics, photonics, etc. The simulation software allowed customers to conduct numerical experiments in conjunction with very complex geometries. CFDRC services were used in conjunction with a variety of products in the field of mechanical engineering: aircrafts, cars, moving body problems, etc. In human physiology area, the developed technology can work at the tissue level, bacteria level, or drug development and delivery devises.

Refocusing its operations on advanced developments for aerospace and R&D in emerging areas such as renewable energy sources , nanotechnology and biotechnology, CFDRC sold its commercial software products division in 2004. After downsizing and refocusing the company, CFDRC began a new growth strategy that is focussed on application-specific software and hardware developments. This is also more aligned with the missions of the DoD and federal agencies. This approach has already resulted in faster growth over last 3 years. Presently, CFDRC has three business units: (1) Aerospace & Defense, (2) Biomedical & life Sciences, and (3) Nano Materials & Processes. The company's projects address challenging problems of National importance including: Chem-Bio Protection, Traumatic Brain Injury (TBI), Alternative Energy Sources, Low Emissions, Space Launch Systems and Missile Propulsion.

## THE ROLE OF SBIR

Since its founding, the company received approximately 125 Phase I awards and about 70 Phase II awards, garnering its first SBIR Phase I award in 1987. The first Phase II was received in 1988. Over the years, the average SBIR funding has been 30–40 percent of total revenue, with additional funding obtained through commercial contracts and Broad Agency Announcements. Currently CFDRC is the recipient of more than 10 Phase I awards from a variety of agencies, including NIH, NSF, and DARPA.

SBIR has helped CFDRC to recruit world class talent. Most employees are recruited from universities in the Northeast and California. CFDRC grew from 5 to over 125 employees by 2003. In January of 2004, the company peaked at 135 employees. Of these about 70 were Ph.D.s.

SBIR awards are also credited with playing a critical role in developing CFDRC's hallmark technological innovations, long-term research, and advanced applications.

## COMMERCIALIZATION

Based on its innovative technolgy, CFDRC won a Broad Agency Announcement against primes like Lockheed and Boeing in 1995. While the Air Force initially hesitated to award $1.5 million, 3-year contract to a company that never made an aircraft, they were nevertheless satisfied with the result. In 1999, after finishing the program sucessfully, the Air Force officials noted that large companies have not been able to provide the agency with such extensive development and services for such limited funds. It was a good confirmation of the value and ability of small businesses.

CFCRC is also building and marketing software capabilities in new areas, such as biotechnology and nanotechnology. In the past 10 years, the company increased investment in both, its engineering and bio labs. CFDRC's commercial software products (CFD-ACE and CFD-Fastran) are licensed by over 600 organizations, worldwide. Customers include Fortune 100 companies, start-up companies, government labs, and universities. In 2004, CFDRC sold its software product division to an international software company.CFDRC, however, retains royalty-free use of this technology and continues to develop the base product.

## KNOWLEDGE GENERATION

According to CFDRC, funding from SBIR has helped to pioneer the use of advanced CFD and multiphysics simulations in biotechnology and semiconductor equipment companies. It has also led to novel designs for lower emission fuel nozzles, and novel biochips for genomics and proteomics.

### Licenses and Patents

Presently CFDRC has over 25 patents, awarded or pending in the areas of biotechnology, combustion, propulsion, and materials. While the company's product division was a clean sale, CFDRC enjoys perpetual rights to customize the product for the purposes of customers like the DoD. However the company cannot sell the product in competition.

The company can now pursue patents even more aggressively, because after selling the product division there's no need to invest effort in marketing of software.

CFDRC tries to licence its products to large global companies, to ensure greater market coverage. Many designs and prototypes were not patented, however, CFDRC still has the data rights. Holding data rights and IP are important to the company. Patents are used as credentials in negotiations with larger companies.

### Publications and Trademarks

CFCRC has published or presented over 500 papers in international journals and conferences. In addition, it held the trademarks for 7 software product names, which were transferred with the sale of software division.

## IMPROVING SBIR

### Surviving Funding Gaps

Although funding gaps cause great inconveniences, the company has been able to manage the delay between Phase I and II by diversifying its base, by investing internal funds to keep the momentum, and by temorarily diverting staff to other projects.

### Award Size

CFCRC believes that Phase I awards in the $100,000–$125,000 range are more commensurate with current economy. The upper limit of the awards should be increased to primarily offset the cost of living. In 20 years there was only one increase from $50,000 to $75,000. In that spirit, the awards are due for one more adjustment. For Phase II awards, the award amounts should be in the $800,000–$1,200,000 range. Phase II awards size need be increased in order to produce more tangible developments. The 2-year time limit for Phase II awards is fine.

### Project Selection

Most agencies aapear to have developed good project selection procedures. The selection process is, by and large, fair and timely. There were few situations when CFDRC was discouraged by agencies that applied certain unwritten rules to their processes.

For example, some large agencies followed the practice of not awarding more than two contracts per company each year. Such decisions should be based on merit and potential for innovation. Otherwise, they penalize more established companies with a greater number of ideas and capacity in favor of newly established start-up companies with fewer ideas and submissions. Some other organizations attempt to limit the number of allowed proposals. per company. Although agencies may have a good reason for implementing such measures, they tempt companies to "play games" like dividing into 2–3 pieces, changing names, etc. Such practices interfere with overall innovation and commercialization goals.

Agencies should avoid limits on the number of awards or proposals and exceptions for VCs as this will hamper overall success of the SBIR program. Funding of venture-backed firms is fair only if a company has already proved success, has shown discipline in creating profit, developed some technological edges, or is ready to commercialize. For such companies, help from venture funds or large corporations will bring more regular management processes and financial due diligence to the table.

## Commercialialization

Different agencies view "commercialization" differently, which is natural and expected. However all reviewers are not educated or coached properly about this process. NASA and DoD are more centralized, and agencies like NIH and NSF have their own peer-reviewed processes. The company feels that the original intent of SBIR was innovation and commercialization. Agencies sometimes append supplementary considerations to the award process. While these additions are well meant, they interfere with commercialization success.

Federal agencies such as DoD, DoE, and NASA should provide incentives for prime contractors to partner with small businesses. It is very beneficial for small companies to pursue SBIR awards in partnership with primes. Finally, it ultimately benefits the government to leave data rights with small businesses.

# Ciencia

*Irwin Feller*
*American Association for the Advancement of Science*

## FIRM HISTORY

Ciencia was started in 1989 by Dr. Salvador Fernandez as a firm specializing in contract R&D for the Strategic Defense Initiative. Its early emphasis was on multispectral imaging for remote sensing. As the U.S. geopolitical situation changed, with less interest on space-based interceptors, the firm has shifted its emphasis. Its core technology remains centered about the development and application of photonic sensors and instrumentation, and it continues to conduct R&D on defense-related topics. Increasingly though it has directed its efforts towards biotechnology, biomedicine, and environmental monitoring. Ciencia's sales, as of 2005, were estimated at $2 million, distributed approximately equally between DoD and other federal government agents and sales to the private sector. In addition to DoD, it has conducted R&D for the Department of Energy and the National Institute of Mental Health.

Prior to founding Ciencia, Dr. Fernandez had worked for SRA, Glastonbury, CT, a defense-oriented R&D firm. Differences in the priority attached to Dr. Fernandez's research interests led him to leave the firm and start his own. Also, Dr. Fernandez had previous experience in starting a firm, having earlier launched one, funded by external investors.

As it has begun to concentrate more on development of products for the

**CIENCIA: COMPANY FACTS AT A GLANCE**

111 Roberts Street
East Hartford, CT 06108
Phone: 860-528-9737
Fax: 860-528-5658
*<http://www.ciencia.com>*

Employment: 12
Revenues: $2 million
Number of SBIR Awards
Phase I: 38
Phase II: 22
Number of Patents: 3

commercial sector, Ciencia has begun to address the issue of how to couple its contract R&D orientation—essentially providing a service—with that necessary for a producer of products, or goods. At present, it has handled its moves towards more "downstream," product orientation by partnering with Oriel Instruments, Inc., Stratford, CT. Under a license agreement with Ciencia, Oriel manufactures and markets a fluorescence lifetime measurement device technology based on Ciencia's technology.

## CAPITAL FORMATION

Ciencia is a privately held firm. Its initial capital came from several sources. The new firm was able to use the equipment infrastructure remaining from Dr. Fernandez's previous firm. It also received funding from the Connecticut Innovation Fund. Some of its early DoD contracts permitted it to purchase specialized equipment. Retained earnings were the final source of internally generated growth.

## INTELLECTUAL PROPERTY

Intellectual property protection is of increasing importance to the firm, especially as it shifts its emphasis to commercial markets. It has received 3 patents to date, and has several patents pending.

## SBIR EXPERIENCES

Ciencia's experiences with the SBIR program began early in the firm's history. It received its first Phase I SBIR award in 1992 from the then Ballistic Missile Defense Organization to develop acousto-optic tunable filters for separating and resolving light into different colors. This award was followed by a Phase II award in 1993. The firm has successfully competed for 38 Phase I awards, primarily from DoD and NASA, followed by 22 Phase II awards.

SBIR projects have played a tactical role in the firm's business strategy and development. Ciencia is market focused: Given its core technical expertise, it looks for market opportunities or, what it describes as, problems that need to be solved. It then looks to SBIR as a potential source of R&D funding to extend and adapt its technological competencies to address this need.

Despite its successes in these competitions, the firm is disappointed that its technical advances have not been more widely adopted by the sponsoring agencies. It notes that even as its work under the SBIR awards advance a technology, the stage of development at the end of a Phase II award may still not mean that a "practical" technology exists.

The process of technological innovation is seen as requiring additional time and support than is provided by Phase II awards. However, it has not succeeded in attracting Phase III awards from DoD or NASA. Ciencia also notes that the SBIR programs of NIH and NSF, 2 other sources of SBIR funding for the firm, do not provide Phase III awards.

The disjuncture between the gestation requirements for new technologies and the SBIR funding practices of federal agencies is highlighted for Ciencia by its experiences with one of its NASA projects. In this case, a NASA program manager was interested in the embryonic technology being developed by Ciencia under a Phase II award, but lacked program funds for a Phase III award. To advance the technology further, additional work on the project was funded under a new cycle of Phase I and Phase II awards.

Ciencia sees the SBIR competitive selection process as fair. However, feedback from the review process is highly variable among agencies. NIH, which employs a peer review system in proposal selection, is described as offering the best review. NSF's feedback procedures, which provide scientific comments, are also highly rated. By way of contrast, DoD's and NASA's procedures for providing feedback to firms are described as virtually useless. They are described as cursory, and as offering no information to a firm on why its proposals may have been funded or not.

## RECOMMENDATIONS FOR SBIR

DoD needs to address the lack of adequate Phase III funding for technologies advancing through the Phase II process. In part, this lack of support follows from the way selected DoD units are seen as perceiving the SBIR program. According to Ciencia, some units see SBIR as a mandated set-aside: They want their funds to provide some tangible outcomes, but do not treat SBIR projects as a necessary part of their mainstream R&D programs or necessarily tied to "end user" needs. Thus although the technology developed under Phase II may work, it may not fit into any existing "socket." Some topics may lead to Phase III awards; others not tied to mission needs, fall into a "nether world."

## SUMMARY

The SBIR program played an important role in the initial launching of Ciencia, and continues to be an important source of its new R&D funding as the firm has sought to widen the range of government and domestic markets to which its core technological expertise can be applied. Improvements in coupling the topic selection process with mission and end-user needs are seen as needed to increase and accelerate the transition of technologies emerging from Phase II projects into operational deployment.

# Custom Manufacturing & Engineering

*Irwin Feller*
*American Association for the Advancement of Science*

## FIRM HISTORY

Custom Manufacturing & Engineering (CME) was established in 1997 by Dr. Nancy Crews and 2 former co-workers as a spin-off from Lockheed Martin's Specialty Component Division, Largo, Florida. The spin-off was the culmination of a series of events affecting operation of a Department of Energy nuclear weapons plant as a government-owned, contractor-operated facility. Between 1957 and 1992, the complex was managed by General Electric. In 1992, Martin-Marietta won the contract that then merged into Lockheed Martin in 1995. Following easing of cold war tensions, the facility was closed in 1997.

Under a Defense Conversion Initiative, Custom Manufacturing & Engineering was formed to continue engineering design and manufacturing work on selected aspects of the facility's operations. The Defense Conversion effort allowed CME to purchase some of the equipment. Lockheed Martin Corporation novated some contracts, which provided an initial source of revenue. Dr. Crews had initially joined Lockheed Martin as senior manager of marketing and long-

**CUSTOM MANUFACTURING & ENGINEERING: COMPANY FACTS AT A GLANCE**

2904 44th Avenue North
St. Petersburg, FL 33714

Phone: 727-547-9799
Fax: 727-541-8822

Revenues: Proprietary information
Employment: 140

Number of SBIR Awards
- Phase I: 9 (+3 STTR awards)
- Phase II: 7 (+1 pending)
- Phase III: 4

Number of Patents: 1 provisional; 2 pending

Awards: National Tibbetts Award

range planning. Previously, she had worked for Eastman Kodak as a Program Manager and a Marketing Director.

CME's core technology centers about electrical power controls and integrated sensors. Its initial customers were DoD services and agencies, including the Army, Navy, and Air Force. Military sales continue to be the larger portion of the firm's revenues. Its business strategy has been to expand to the commercial sector by developing intelligent power system management and control technologies. Its current customers include several major defense and aerospace firms, other federal agencies, and a diversified set of commercial markets.

## CAPITAL FORMATION

CME started with a grant from Lockheed Martin Corporation and a personal investment by Dr. Crews. This capital infusion, coupled with acquisition of physical facilities and equipment from Lockheed Martin enabled CME to begin with an established physical plant and an ongoing contract. These arrangements enabled the firm to begin operations without recourse to external sources of capital. CME remains a privately held firm. It has not received funding from angel or venture capitalists.

## EXPERIENCES UNDER SBIR

Starting operations with an existing contract, CME's credits its growth during the first 5 years of operation to its success in winning SBIR contracts, primarily from DoD. These awards enabled CME to hire personnel with technological expertise that expanded its R&D capabilities, which in turn led to expanded applications of its core technologies, and its entry into new product markets and customers.

CME's success in winning DoD Phase III contracts also are described as direct outcomes of the R&D performed under its earlier SBIR awards. However, its success in securing Phase III contracts has been less than warranted given its technological advances, according to the firm. The difficulties it has encountered in transitioning Phase I and Phase II awards into Phase III awards or subsequent acquisition contracts is seen by the firm as caused by the disconnect between DoD's management of the SBIR program and its acquisition programs. The two are organizationally separate, with SBIR and acquisition programs in effect having different missions. SBIR seeks to generate new technologies, but acquisition programs are viewed as unresponsive to receiving new technologies.

SBIR's procedures for reviewing proposals are seen as timely and fair. In general, debriefings are viewed as providing useful information. One source of concern to the firm is the lack of specificity about the criteria to be applied in specific competitions. In one recent case, when CME inquired about why its proposal had not been funded, it was informed that it was because the firm lacked experience in manufacturing. Manufacturing experience, however, was not part of the stated criteria, according to CME.

CME has also encountered serious problems, what was described as a nightmare, with the delays between Phase I and Phase II awards. The delays have led to the loss of key personnel, necessitate the redeployment of staff from other projects to maintain work on a project, and require investments of internal funds without assurance about if or when Phase II funding will be awarded. In one recent experience with the Navy, the firm did not receive funding for an option year under its Phase I award, but then was asked to prepare a Phase II proposal, which, in effect, required it to continue work for almost 7 months without funding in order to keep the people employed and the subcontractors interested.

## INTELLECTUAL PROPERTY

Intellectual property protection is an important part of the firm's strategy, especially as a means of protecting its future. The SBIR program is of special value in following this strategy, as it helps underwrite the development of new technologies, which can then be patented. The costs of filing patents for a small firm however have limited the extent to which it can pursue patent protection on all its new technology.

## INVOLVEMENT WITH STATE GOVERNMENT PROGRAMS

CME has received grants from two programs administered by Enterprise Florida, the state's economic development program. One of these grants was intended to assist the firm commercialize a technology being developed under an SBIR award.

## RECOMMENDATIONS

The SBIR program needs to improve coordination between its SBIR and acquisitions programs to facilitate the transition to Phase III awards, and more generally, to DoD's acquisition programs. To accomplish this, acquisition program officers need to be involved in a systematic manner in the generation of SBIR topics. This participation would increase the likelihood that R&D conducted on SBIR awards addressed operational needs. Additionally, the specification of topics, whether oriented towards 6.1, 6.2, or 6.4, need to be better defined. Firms essentially are "blind" about the downstream, operational use of their R&D after they complete Phase II work.

The dollar ceilings on Phase I and Phase II awards also need to be increased. The current level of Phase I awards needs to be increased to the $100,000–$150,000 range; the current ceiling of $75,000 is too low to accomplish much.

Phase II awards should be increased up to $1 million (with an option to go above that amount) to permit firms to build hardware prototypes.

## SUMMARY

Established about a cluster of technologies and customers connected to its origins as a firm created out of a closed division of a DoE nuclear weapons facility, Custom Manufacturing & Engineering employed SBIR awards to diversify its technology base, and thus to widen its set of products, markets, and customers. SBIR awards served as an important source of revenue during the firm's first 5 years of operation, and as the basis for some of its recent Phase III awards from DoD. The firm's strategy however is to avoid becoming dependent on defense-related R&D contracts, and is actively seeking to enter additional commercial markets.

# Cybernet Systems Corporation

*Irwin Feller*
*American Association for the Advancement of Science*

## FIRM HISTORY

Cybernet was established by Heidi Jacobus in 1988. The firm's establishment is an outgrowth of her educational, professional, and family experiences, which Jacobus describes as a female version of a Horatio Alger story. Cybernet's establishment also is a distinctive example of the contribution of the DoD SBIR program in launching a firm.

Jacobus is the first member of her extended family to graduate from high school, much less college. Her early education was distinguished by academic scholarships from community organizations that enabled her to attend private preparatory schools. She then had a wide-ranging undergraduate education at Trinity College, Connecticut, majoring in psychology but taking courses in several fields. Previously, while in a high-school level preparatory school, she had taken summer courses in a regional college, which enabled her to graduate in 3½ years.

In 1973, while considering her future plans and awaiting a Spring graduation, she sought employment by posting her availability in buildings at the nearby University of Connecticut's Farmington campus. She was soon hired by the dean of the campus Medical School as a "girl Friday," laboratory assistant. The dean was an early pioneer in efforts to employ computer-assisted instruction for medical students, using the PLATO system that had recently been developed

**CYBERNET SYSTEMS CORPORATION: COMPANY FACTS AT A GLANCE**

727 Airport Boulevard
Ann Arbor, MI 48108
Phone: 734-668-2567, Fax: 734-668-8780
*<http://www.cybernet.com>*

Annual Sales (2004): $5 million
Employment: 50
Number of SBIR Awards
  Phase I: 89
  Phase II: 44
Number of Patents: 25

at the University of Illinois. Jacobus describes her 1½-year experiences in this position as being in on the ground floor of the development of a new technology, and as focusing her broad based intellectual interests on human interfaces with computer technology.

Her work experiences also brought her into ongoing contact with faculty at the University of Illinois, which in turn led to an offer of a graduate teaching assistantship from the university's computer science department. Jacobus had had no previous background in computer science, but by taking both undergraduate and graduate courses concurrently was able to earn an M.S. degree. She then advanced to Ph.D. level work, passing all courses and the Ph.D. prelims. However, in about 1978–1979, as she began to work on her dissertation, her dissertation advisor left for a 1-year sabbatical leave in Belgium. The then difficulties of communicating effectively and rapidly with an advisor located at such a distance would have caused a hiatus in Jacobus's dissertation work. Adding to the complexity of her choice was that during this period, Jacobus had met her future husband, Dr. Charles Jacobus, a Ph.D. in electrical engineering, while working in the university's computer laboratories. As her husband was completing his degree and about to enter the labor market, Jacobus had to decide between moving with her husband or waiting at the university until her advisor returned.

The decision was to move. She and her husband took positions with Texas Instruments, Dallas, Texas, where he worked on the design of semiconductors. He was also put on TI's "fast track," and was provided opportunities by the firm to obtain an Executive MBA from Southern Methodist University.

Jacobus worked in TI's central research laboratories (where she was the only female engineer). TI's research laboratory was described as the firm's "pie-in-the-sky" operation; researchers were given broad flexibility to select their projects; Jacobus chose to extend the work she had been doing on her dissertation, which focused on the ergonomics of human-computer interaction. She stopped her career however to have children. (At this time, mid-1980s, the absence of corporate leave policies and paucity of a company-based infrastructure to support working parents, especially mothers, limited the possibilities of combining parenting and work.)

In the late 1980s, Jacobus's husband was recruited by ITI, Ann Arbor, Michigan, which was in the midst of a major expansion. Ann Arbor was described as offering a quite different world from Dallas for an educated, career-oriented woman, especially a mother. Whereas in the Dallas area, Jacobus found her employment and career orientation an exception, in Ann Arbor, she found many peers who were actively also engaged outside of the home in careers or volunteer work. Her children now in nursery school, Jacobus joined the Junior League, where she worked with the area's Children's Science Museum, both as a docent and as the chair of a major fundraising drive.

This "reentry" into professional life led Jacobus to think afresh about completing her Ph.D. At the time, the 7-year deadline between passage of exams and defense of a dissertation was nearing, but she requested and received an extension from the University of Illinois. (Had she received her degree, Jacobus would have been the 4th female graduate student to receive a Ph.D. in Computer Science from the University.) She was also permitted to have an advisor in absentia, and re-started work on her dissertation, again focusing on human-computer interaction, under the supervision of a University of Michigan faculty member in industrial engineering. However, she soon experienced difficulties with the advisor about the thesis proposal she submitted to him.

At the same time, Jacobus had been working on a part-time basis in the University of Michigan's library on a project to develop an indexed reference book on the SBIR program. Contemporary practice then was for each federal agency to post its SBIR program and fundable topics of interest in separate documents and to gather them together in a single volume, but without a topical technology index. UM's library had a contract to prepare such an index.

While working on the project, Jacobus became aware of the SBIR program. Soon after, she saw her thesis topic, which had become the source of disagreement between her advisor and her, listed as a DARPA topic of interest. She distilled her thesis proposal into an SBIR proposal, noting also that she did not then have a firm, but would start one if she received an award. Subsequent to the submission of the proposal, she received a telephone call from a DARPA official stating that her proposal was "best" he had ever read. The award was for making graphic displays on airplane consoles compatible with the ergonomics of human perception. (In the early period of technical advances in computer graphics, an unknown was the human factors of man-machine interaction, specifically the ability of humans to effectively process and react to varying combinations of color, motion, and shape.)

Coincidentally, her husband required major surgery and was seriously ill, and as a result became unemployed. The contrast between the difficulties she had been encountering with working with a faculty member about the suitability of her thesis proposal and the accolades she received from DARPA, and her husband's illness required she become the primary family income earner, led her to say, "Good-bye University, hello DARPA."

The feedback from DARPA was the motivating event that gave Jacobus the courage to found Cybernet. She submitted SBIR proposals to other agencies, receiving awards from NASA and the Army, followed by the award from DARPA that had catalyzed the firm's founding. Cybernet was a bootstrap operation in every sense of the word; rather than a garage operation, it was literally a child's bedroom operation, housed in her daughter's bedroom. Jacobus had to learn the basics of government contract and accounting procedures, such as overhead rates, allowable expenditures, and related provisions. She managed this from reading manuals obtained at the regional SBA office; through purchasing technical assistance from local consultants, such as a retired, former DCAA contracting officer; and though assistance in understanding federal government contract procedures provided by the regional Small Business Development Center, located in a community college in Livonia, Michigan. As described by Jacobus, one of the side benefits of the SBIR program is that it forces "business discipline" on technologists. Cybernet started with 3 people. It then moved to rented, shared space in a complex of small offices carved out of the nearby facilities of the former Bendix Aerospace Laboratory.

Cybernet's growth initially was based on matching the technical competencies of a network of friends and colleagues with engineering backgrounds in the Ann Arbor region to lists of posted SBIR topics. Its core technology is the development and application of robotics technology solutions to human-machine interaction. Its expertise is centered about centered about distributed simulation and training, software intelligence, network connectivity, robotics, and man-machine interaction, which it seeks to apply to a diverse set of defense and nondefense industries. Cybernet thus describes itself as a "brain company with hands."

Based on the serendipitous events shaping its founding, Cybernet sees limited value to the type of formal strategic planning customarily associated with the launch process of start-up firms: Who could have predicted that from part-time work in a library one would have built a firm with 50 employees and current annual revenues of $5 million. Instead, it has an orientation, or culture, of going for it, that is of pursuing opportunities as they arise.

Undergirding this approach has been the firm's ability to integrate human factors expertise with the design and manufacture of functioning technologies. Here the firm has drawn upon the rich tradition and ready availability of robotics and related manufacturing expertise in the Ann Arbor region. As noted below, it has since expanded into several technological and market areas.

Cybernet's current revenues are derived approximately 70 percent from the federal government and 30 percent from the civilian sector. The firm remains active in the SBIR programs of several agencies. Employees are encouraged to seek out SBIR topics congruent with the firm's core technology, to form research teams, and to submit proposals. Employees receive monetary bonuses for successful proposals; so too do employees whose proposals are not funded.

## LOCATION

Cybernet's location in Ann Arbor offered both advantages and disadvantages for Cybernet. As noted, the firm was able to tap the region's pool of consultants-especially in contracting and intellectual property law—to address core business needs. Especially helpful in the firm's early days was that several of these consultants offered their services at below market rates, in effect taking Cybernet under their wing. The area's pool of engineers and technologists also provided a supply of individuals interested in pursuing new research and career opportunities via the competitive proposal route. It also provided a skilled and/or readily trainable workforce, as the firm expanded into new technological areas. Finally, the firm was able to draw upon the region's existing industrial base for hardware components; this access has proven especially important as the firm has sought to respond to the rapidly changing and pressing needs of U.S. military forces currently located in Kuwait.

A major shortcoming of its location in Michigan however was the difficulty it encountered in obtaining short-term, working capital from the region's banks. Even though it had an SBIR award, Cybernet initially encountered difficulty in obtaining a line of credit for $20,000 for payroll and related operating expenses. Regional banks did not consider an SBIR award as "bankable," because U.S. government contracts do not allow for assignment of claims in case of a firm's default. (According to the firm, it would have been easier to obtain the line of credit if their contract had been with K-Mart to produce T-shirts). Jacobus ended up having to sign a personal note, with her house as collateral, to obtain the loan.

Overall, a definite negative in obtaining bank credit is viewed as existing in the Midwest. Banks in other states, such as Virginia, Massachusetts, and California, are viewed as being far more understanding of and receptive to making loans on "intangible," research-oriented enterprises.

## CAPITAL FORMATION

The firm remains a privately held enterprise, although it has received selected infusions of investor capital, principally from a rippling process in which first firms and then members of the boards of these firms saw profit potential in Cybernet's embryonic technologies. Cybernet approached 3 original equipment manufacturers to manufacture a prototype of a medical device it was developing. One firm, Sparton Corporation, a 100-year-old electronics firm that did mid-tech production became fascinated with Cybernet's invention and decided to become an investor. A member of Sparton's board of directors, a retired investment banker who ran a boutique investment firm, arranged for several of his investors to invest in Cybernet. Together these 2 corporate groups of investors plus Ampex Corporation have invested $5m in Cybernet for a 20 percent ownership stake.

Bringing in outside investors, however, required Cybernet to elect becoming a C Corporation, whereas formerly it had been an S Corporation. The shift was made to satisfy the needs of the external investors that they not be taxed personally on Cybernet's profits. However, it resulted in Cybernet's internal stockholders now being responsible for both corporate and personal taxes.

These investments permitted Cybernet to free itself from reliance on government contracts and the SBIR program. However, its ventures into the private sector market have been limited by high marketing costs. For example, it has been able to place its Linux server in Best Buy stores. It describes the process of selling high-tech consumer products through mass distributors as akin to selling groceries: Suppliers must compete for and pay for shelf space and accept returns as unsold inventory.

Cybernet has had periodic discussions with venture capital firms about going public, but to date with no agreement. The firm sees itself as falling into the gray size area of VC interest. At its current annual sales volume of $5 million, it is too large for a true zero stage launch and too small ($10 million–$15 million being seen as the next threshold level) to warrant a near-term major public offering. Aligned with the difficulty that it initially had in securing bank loans, Cybernet also notes that although there are some venture capital firms in Michigan, these firms tend to focus on biotech firms or to invest out of state. They are not seen as focused on firms working on high-tech defense technologies.

In combination, the difficulties that it has experienced in securing venture capital while simultaneously developing new products for the DoD, as exemplified by its recent delivery of an automated tactical ammunition classification system in Kuwait for the Army, leads it to question the thrust of federal agency SBIR program managers to interpret the language in SBIR regulations on commercialization to mean venture capital investment whereas the appropriate interpretation would allow for non-SBIR government sales, such as the sale of their special purposed machinery for the Army in Kuwait.

Another source of capital for the firm's expansion was partial liquidation of its equity holdings in a collaborative venture with Immersion, a San Jose, California-based firm. The collaboration was based on a pooling of the respective patents each firm held on robotic sticks, a key technology in the production and marketing of computer games. Each firm found itself in discussions with Microsoft about access to their respective patents, patents that had the potential to challenge a core technological component of Microsoft's games. Immersion, unlike Cybernet, had easy access to venture capital, with investments from Intel, Apple, and other Silicon Valley sources, and was planning an IPO. (Immersion also was located nearby to these firms, facilitating both access to the venture capital and ongoing advice). Cybernet exchanged ownership of its patents for a 10 percent stake in Immersion. The public offering was highly successful, supported in part, by Microsoft's licensing of the robotic stick technology. Immersion stock rise and fell in part with the dot-com phenomena and in part with Microsoft's use of its technology. Partial liquidation of Cybernet's holdings in Immersion, however, provided new capital.

## INTELLECTUAL PROPERTY

As described above, intellectual property protection has been an important part of Cybernet's strategy. The firm currently has 25 patents, primarily centered about the design and construction of robotic sticks. These sticks, in turn, are used in a wide variety of end uses, including computer games, training of astronauts, and manufacturing control processes.

Attention to IP protection began early in the firm's history, in part precipitated by a negative experience in which it found that an employee at a federal agency technology transfer center had claimed a technology that Cybernet had described in one of its proposals. The first that the firm learned of this was when it saw the technology described in a technology transfer packet offered to the public.

Patent protection serves to enhance Cybernet's ability to market its technologies effectively as well as to generate revenues. More generally, protection of IP is seen as an important but demanding task for small high-tech firms. The challenge is seen as especially formidable for such firms under the SBIR program as their technologies course through Phase I and Phase II stages and begin to enter Phase III production. Large firms, including major DoD prime contractors, are seen as aggressive in seeking to trespass on the intellectual property of small firms. Large firms are viewed as treating small firms as "Kelly girls," that is the provider of services, not as independent producers. Immersion, for example, has recently been involved in suits against Microsoft and Sony for infringing on the firm's patents.

## INVOLVEMENT WITH STATE GOVERNMENT PROGRAMS

Over a 15-year period, Cybernet was somewhat successful in several efforts to secure financial support from several of Michigan's state technology developments, receiving several awards between $25,000 and $50,000. Michigan is seen as having an extensive array of technology development programs, but so weighted to supporting biotech start-ups that high-tech manufacturing firms receive limited attention. Last year, Michigan began to support SBIR companies' projects in designated technology areas with commitments for supplemental funds. Jacobus applauds such a program because it enhances the value Michigan companies can provide to their federal customer.

## EXPERIENCES WITH THE SBIR PROGRAM

The SBIR program is seen as one of the federal government's most competitive and fairly run programs. New topics and topic authors appear in each new round of SBIR solicitations; the proposal review process involves changing sets of agency reviewers, and each new round of solicitations attracts new sets of competitors—a distinctive and desirable aspect of the program.

Winners are selected on the basis of technical merit, relative to the stated topic. Technical merit, in the firm's view, should remain the primary criterion. One implication of adhering to this criterion is that there should be no quotas or ceilings established on the number of SBIR awards a single firm may receive, or exclusive focus on commercial product outcomes, as some have proposed. If a firm's proposals have the highest technical merit, they should be funded. The primary measure of SBIR program success should be cost effective solutions to Government needs and requirements. Alternative, nontechnical measures of success such as nongovernment sourced investment, "commercialization," job growth, or other quotas, if used to measure success would lead to much lower success rates. For example, according to the firm, since approximately only 10 percent of venture funded companies become successful growth companies leading to IPO or buy-out—using these measures as performance measures would set a relatively low success rate towards which the SBIR program might aspire.

The timeliness of the DoD SBIR review process has improved considerably over the past 15 years, especially as it has shifted from hard copy submissions and processing to computerized systems.

Cybernet's early experiences with SBIR included transition-funding gaps as long as 18 months. Improved processing of awards and increased attention to the difficulties faced by firms because of the gap has led DoD to act to shorten them considerably. In general, Cybernet handles gaps between the 2 phases by scraping together funds from different contracts to redeploy its personnel. It is able to do so because it has often had a portfolio of DoD SBIR awards from different Services, each of whom operates on a different time schedule in issuing solicitations and making awards. The gap between phases though is seen as a serious problem for small firms with few contracts. Not only does the gap create liquidity problems for the firm, but they also run the risk of losing their key technical personnel.

The firm though notes a downside to recent DoD efforts to shorten the transition gap. These efforts can lead agencies to ask Phase I awardees to submit Phase II proposals very early, as short as 4 months, in their Phase I work. This period of time can be too short for the firm to show results. Given that some DoD competitions now involve multiple Phase I awardees followed by down selection for Phase II awards, differences in the pace of work during a Phase I contract, not the realized outcomes at the end of the full award period, may lead to erroneous judgments about the technical importance of competitive projects. A gap of some duration may be a necessary part of a fully informed Phase II review.

Recent actions by DoD and other federal agencies, such as NIH, to permit increases in the size of Phase I awards to $75,000–$100,000 has been helpful since awards of this level are often necessary to demonstrate the feasibility of the technological approach being advanced in a proposal. From Cybernet's perspective though, it is not clear whether these increases are in response to the stated needs of firms for larger initial awards or a proxy endeavor to adjust for inflation as it is an attempt by selected agencies to reduce the excessive workload on SBIR managers. DoD SBIR contract monitors are seen as now having to handle too large a number of proposals. An undesirable consequence of this workload is that it reduces the appeal to them of working with small firms.

## RECOMMENDATIONS

SBIR should institute debriefings for firms that received Phase I awards but which either are not asked to propose for Phase II competitions or propose and then are not selected. At present, firms that do not receive Phase II invitations to propose are, at best, simply notified of that outcome. Compounding the problem, some DoD services only provide debriefing invitations to invited companies, leaving those who are not notified to wonder whether their Phase II proposal is actually "disinvited" or if a letter is lost in the mail. By way of contrast, in accord with federal contracting requirements, debriefings are mandated following Phase I competitions. Phase II solicitations currently are seen as extensions of Phase I work; they thus are not treated as full and open competitions subject to debriefing requirements.

Debriefings, which at times can contain paraphrases of reviewer comments, can be of considerable value to the firm as they can contain information that can guide the modification or correction of proposals in subsequent solicitation rounds. (Not helpful though are debriefings that simply note that the proposal was not sufficiently innovative, or at the other extreme, overly ambitious).

Attention to DoD's procedures of selecting Phase II awardees is needed. Some services are moving to issuing multiple Phase I awards on a topic and then down selecting for Phase II awards. However, rather than having more choices for the Phase II awards, some services have been pre-selecting invitees for the down selection stage. Issuing invitations benefits the service in reducing the number of proposals it needs to review, and indeed may be of some benefit to firms in saving them from expending effort on writing proposals that have little chance of winning. However, pre-selection is unfair to firms that continue to see their Phase I projects as competitive for Phase II awards.

Finally, many Phase I and Phase II proposal receive high technical ratings, but are not selected due to funding limitations. It would be helpful to provide an overall proposal ranking against those that were selected for each list that is funding limited. Many Phase I debriefings already provide this information ("xx proposals were received, yy were selected, your proposal was ranked zz of the xx received").

DoD's SBIR program needs to strengthen the relevance of its topic selection process to the "end users." SBIR awardees tend to work only with their service's technical monitor, who may provide no clue about the "bigger picture" of how a topic or technology fits into larger systems or how it will be used. This lack of information detracts from the utility of the SBIR project, and thus of the SBIR program.

Increased attention to building "hands across the water" between the SBIR program and DoD's management of major weapons systems is needed. Prime contractors have little motivation or incentive to work with small firms or to incorporate technologies developed by small firms into the weapons systems they are developing. Prime contractors would prefer to have add-on contracts to their work to develop the technologies listed in SBIR solicitations.

One recommended approach to improving links between the SBIR program and procurement and acquisitions programs would be to have SBIR program managers report directly to DoD major weapons/systems program managers. Another would be to route SBIR funding back through the program offices from which it is derived.

## SUMMARY

Cybernet's establishment was directly linked to these SBIR awards. Expressions of interest in Heidi Jacobus's research by a DARPA program manager catalyzed her willingness to leave a Ph.D. program and launch a new firm. The firm describes itself as a "brain company with hands," integrating emerging research findings on man-machine interactions with Michigan's traditional manufacturing capabilities. Cybernet also is an example of how SBIR funding bridges a gap between projects focused on basic research at major universities and major commercial and/or weapons systems development work performed by prime or major corporations. Starting with contracts from NASA and DoD, the firm has developed a broadened, diverse set of technologies and markets, ranging, extending from federal agencies, including several DoD services, to the commercial sector. In its view, SBIR has proven itself to be a highly cost-effective means for maintaining and growing engineering expertise to maintain the U.S. defense and economic competitiveness.

# Defense Research Technologies

*Zoltan Acs*
*University of Baltimore*

## FIRM OVERVIEW

Defense Research Technologies, originally located in Rockville, Maryland, and currently (since 2006) in Lady Lake, Florida, specializes in sensor and control systems. The company's nonelectrical sound-amplification system was first developed by DRT president Tadeusz Drzewiecki and others at the U.S. Army's Harry Diamond Laboratories in the late 1960s and early 1970s, when they, too, faced a problem with sound and electricity causing sparking and fire hazards. Crew members on the decks of noisy aircraft carriers wanted a way of talking to each other, but they were afraid that sparks from traditional electric microphones would ignite the jet-fuel fumes that waft across the deck surfaces.

In response, the Army researchers came up with the principles for a system that Drzewiecki has patented as an "acousto-fluidic" technology. The system works on the principle that sound can travel farther if it is wind-borne and can be amplified by using it to deflect flowing jets of air. The fact that this system does not use electricity has found a special application for Orthodox Jews, who are forbidden from using electricity on the Sabbath or other holy days. Some rabbis have dubbed this kosher technology a "wind microphone."

In addition to its use on aircraft carrier decks and synagogues, DRT's sound-amplification technology is finding new uses, including listening for larvae in a granary, where electric amplification systems could generate sparks that in turn could trigger an explosion in the grain dust, which is highly combustible, but more importantly because acousto-fluidic technology is the most sensitive acoustic-sensing technology ever available. SBIR has played an important role in the company's development of this as well as other innovative sensor technologies.

## THE ROLE OF SBIR

As noted above, DRT's firm's founder, Dr. Ted Drzewiecki, had previously worked at the Army's Harry Diamond Laboratories and then was briefly employed by System Planning Corporation (SPC)—a defense consulting company—where he worked primarily on DARPA programs. In 1982, Dr. Drzewiecki and three of his co-workers founded Science and Technology Associates, a company that could provide DARPA managers the same services as SPC but at a lower cost. While at Science and Technology Associates, he developed his own client base, and transitioned into his own company Defense Research Technologies, Inc., providing consulting services to the government.

In 1984, Dr. Drzewiecki founded Defense Research Technologies (DRT), a defense-consulting firm specializing in Soviet weapons, advising the U.S. government on how to counter Soviet arms technology. He started bidding on SBIR projects in 1989 and won awards from the Army, DoE, DARPA, NASA, DOA, NIH, and SDIO.

With the end of the cold war, DRT began to lose its market niche in defense consulting and refocused its operations to pursue medical applications via NIH SBIR awards. As spin-off of DRT's first SBIR award, the company developed a pneumatic sound amplification method that does not use electricity. DRT also developed a mud pulse telemetry system, based on supported by a DoE SBIR contract, and various other control systems and sensors.

The technology being developed with DRT's recent NIH SBIR contract had a potential $3 billion–$4 billion annual market. The sensor technology for missile control and projectiles, if developed, could have been in the $10 million–$100 million range. The company's main competitors are industries in the electronics field. Microfluidics technology is analogous to electronics, only it moves molecules instead of electrons. Since the electronics technology is much faster, competitors can easily mimic microfluidic devices. Nevertheless, the kosher sound system is a niche product that competitors would have difficulty marketing in an electronic version. Applications where survival in extreme environments is required, such as the control of propellant gases, are other niche areas.

In all, the company has received 13 Phase I awards and 6 Phase II awards. The company also received funding from DoD and NIH at the Phase III level. The DoD product was an active protection system for lightly armored vehicles that generated $17,000,000 in revenue. The NIH product raised $7,500,000 in venture capital. A third product, the Kosher Sound System, resulted in self-funded commercialization.

The company's SBIR awards helped in attracting approximately $1 million-$1.5 million in VC funding for a medical gas analyzer—a project sponsored by NIH that was geared towards measuring the properties (density, viscosity, heat, etc.) of respiratory and anesthesia gasses. A spin-off from DoD and DoE has also funded research on battlefield gas sensing. Further NIH SBIR awards helped the company demonstrate the measurement of cardio-pulmonary functions using the gas analysis technology developed.

The company had 7-10 employees when it first started receiving SBIR awards. In the early 1990s, the company employed the maximum of 26 people. DRT's current sales (of about half a million dollars) keep four employees on payroll. Dr. Drzewiecki claims that the decline in the overall number of employees is not correlated with the SBIR program; however, SBIR did not facilitate retention in his company.

## COMMERCIAL OUTCOMES

As noted earlier, the company's kosher sound system has been successfully commercialized.

DRT has also developed a gas analyzer that provided consumes less than 50 ml/min side stream flow at the time when the standard was an inefficient 200–300 ml/min. The gas analysis technology—which introduced competition where there was no competition before and pioneered new techniques and approaches in the field—has been licensed to a spin-off company.

The company' s mud pulse telemetry system has been licensed to a Canadian company.

DRT's active protection system has introduced new technical developments in the area of guiding projectiles. DRT's work on the DARPA/Army SLID active protection system was entirely government funded. DRT teamed with AlliedSignal (now Honeywell) for the development phase, but was down selected due to perceived risk in further development after the optical fibers used for command guidance were found to be inadequate. Six companies originally competed. The DRT/Honeywell team survived for three years before being down selected. A note in passing, the remaining two competitors, Rockwell and Raytheon, did no better, neither being able to fly their systems successfully, and the program was terminated two years later in 1997.

## KNOWLEDGE GENERATION

Of the 41 total of patents held by Dr. Drzewiecki, 10 are SBIR related: 7 patents on the gas analyzer, 1 patent on the Mud Pulser, 1 on an oxygen sensor for molten steel, and 1 on a coal water slurry fuel injection system.

Dr. Drzewiecki also published 1 peer-reviewed paper with the Acoustical Society of America on sound systems, 3 papers (one which was peer reviewed) on the development of control systems for nuclear reactors, 10 papers on the gas analyzer (four peer reviewed), 1 peer-reviewed paper on the active protection systems, 1 peer-reviewed paper on space-based interceptors, one peer reviewed article on the USDA insect detection system, and 1 peer-reviewed paper on the Mud Pulser.

Dr. Drzewiecki's received the Instrument Society of America Gilmer Thomason Fowler Award for the best paper on the NIH-sponsored gas analyzer presented at the ISA Analysis Division 2000 Symposium.

## IMPROVING SBIR

### Surviving Funding Gaps

DRT has had to cope with funding gaps, using its own money to bridge this gap during work on the Mud Pulser, the company used. DRT has diverted staff into other projects between 1987 and 1994, when three-fourths of his funding was derived from the SBIR program. In one instance, the company had to lay off employees during a 9-month gap between Phase I and Phase II funding. Later DRT found venture capital and partnered with a company established in North Carolina.

Defense Research Technologies suggests that DoD should allow recompetes or resubmissions. Such flexibility across Phase I, II, and III is seen as essential for small companies. Big companies are more likely to be able to withstand the gap between funding.

### Award Size

In today's market, a Phase I award is just enough to write a Phase II proposal. The size of the companies receiving awards is essential in determining the size of the awards. Start-up companies thrive on half a million dollars in funding, larger companies cannot survive on one award. There are firms that receive 15–20 SBIR awards in a year, and the SBIR program is their sole source of revenue.

Additionally, companies with significant human resources have the advantage of specialized internal infrastructures in writing proposals. Small companies, where only one individual writes proposals, have difficulty competing in this environment.

### Paperwork and Bureaucracy

The paperwork involved in applying for SBIR awards is relatively little. Nevertheless, a small company invests a sizeable resource when its principal dedicates two weeks of full-time work to writing a Phase I proposal.

DRT's experience with SBIR differs widely. The Department of Agriculture is worst. They offered the smallest award, about $45,000, and sustained a heavy bureaucracy.

The DoE and NIH have great programs that promote companies and help them with commercialization. The NIH model, although a true peer-review process, is the most generous and allows for recompetes and resubmission three times each year.

DoD lacks commitment to the program and provides little technical assistance to companies. The Department of Defense has very specific guidelines and uses the program in its procurement process. Resubmissions with the DoD would have been helpful when DRT proposed its microfluidic gas analyzer. Because the current systems are electronic, fuel vapors in empty tanks may cause aircraft explosions. DRT lost that particular award because the company did not have a chance to clarify particular aspects of the system in response to the AF reviewers' comments.

## Award Selection and Funding

DRT finds that the award selection process tends to be arbitrary and relative. Companies with already developed products are more likely to win awards. Respectively, companies that lack manufacturing experience are at a disadvantage, even if they develop better processes or technologies.

DRT also perceives two distinct paradigms in SBIR funding: The program at the Department of Defense is *procurement based* and the program managers don't view it as a process of fostering new technologies. Although in the last few years DoD showed improvement in organizing technical topic areas, the agency still lacks technological focus. By contrast, the DoE and NIH SBIR topics are listed by technology areas and these awards are more like genuine *seed funds*.

For example, NIH will evaluate any technology that is related to a specific topic (i.e. gastrointestinal diseases). This approach allows companies to be more innovative and produce a product with good chances for commercialization. NIH is willing to fund three different approaches to one topic, while the DoD does not (with the exception of DARPA, a very flexible agency that uses the SBIR program to spread out its funding).

Additionally, NIH is more serious about and efficient in providing post-award technical assistance. For example, after winning a Phase I from NIH, in the first three months of receiving the award DRT was invited to present its work (even in an unfinished state) at a minimum of three colloquia. The company had a chance to network with other projects and was very satisfied with the agency's efforts to promote their work.

NIH regards the SBIR program as a chance to innovate and diversify their technology. The DoD uses it merely to foster its own technologies and validate commenced work.

## Commercialization

Current SBIR funding is enough for concept development, but insufficient to do product development for the commercial sector. The importance of commercialization within SBIR grows daily and companies are required to do market research on their own and get investors in Phase II. This is a very difficult task for most companies.

For instance, DRT had a provisional licensing agreement with a company called Vesuvius International, a huge international conglomerate that produces sensors for steel mills. DRT developed an oxygen sensor that functioned in the molten steel for 10 hours, while the competition had only disposable solutions. This technology could have revitalized the U.S. steel industry, because it was the lynch pin able to close the loop on automating the steel process, but the company could not secure enough funding in time.

While the SBIR program is very good in providing seed money and getting a concept developed, it has an inefficient and ineffective strategy for commercialization. DRT licensed its mud pulse telemetry system to a Canadian company, but slow commercialization process allowed other technologies and suppliers to break into the market first.

# FIRST RF Corporation

*Irwin Feller*
*American Association for the Advancement of Science*

## FIRM HISTORY

FIRST RF Corporation was founded by Farzin Lalezari and Theresa Boone in 2003. The firm is privately held, with ownership distributed among Farzin Lalezari, Shirley Lalezari, his wife, and Theresa Boone, a former co-worker, at Ball Aerospace.

Lalezari was born in Iran, and emigrated to the United States in 1971, while a high school student, following the imposition of a death sentence on his father, Iran's Minister of Education, by the Khomeini regime. Lalezari completed his B.S. and M.S. degrees at Brooklyn Polytechnic Institute, and then moved to Colorado for further graduate studies, before leaving to work for Ball Aerospace, a major aerospace prime contractor, also located in Boulder. At Ball, he advanced to position of chief scientist and director of research. At Ball, Lalezari produced 25 patents, all of which were assigned to the firm.

Lalezari describes the decision to form FIRST RF as reflecting disenchantment with the bureaucratization and technological stagnation of large firms, and the shift he experienced at Ball from the firm's longstanding emphasis on quality design and engineering to short-term profit measures designed to meet the requirements of stock market analysts. To meet these financial performance measures, Ball, as well as other large U.S. aerospace firms, are described as having shifted away from a growth strategy based on technological innovation to

**FIRST RF CORPORATION: COMPANY FACTS AT A GLANCE**

4865 Sterling Drive
Suite 100
Boulder, CO 80301

Phone: 303-449-5211; Fax 303-449-5188

Revenues: $25 million
Employees: 30 full-time

Number of SBIR Awards: 24
Phase I: 15
Phase II: 9

one focused on mergers and acquisition. Sacrificed in this reoriented strategy was Ball's former emphasis and reputation for being the fastest and the best in developing new technologies. The result was that the firm's capacity for technological innovativeness began to atrophy. As it grew, Ball's administrative infrastructure and overhead costs increased. Associated with this increase was increased pressure of operating units to meet specific, often short-term profit goals. During his last 4–5 years with the firm, Lalezari found himself spending approximately 50 percent of his time in weekly meetings with the firm's financial personnel debating profit quotas.

Lalezari's prior work with Ball provided him with contacts with DoD and U.S. intelligence agencies, but he left Ball with a clean slate, seeking to develop new technologies for new markets and new customers.

FIRST RF's core technology focus is advanced antennas and RF systems. The area of research represented a new departure for Lalazeri, who had not worked on the topic while at Ball. Reflecting the motivation to "think" about new problems and new solutions, which led him to leave Ball and to found FIRST RF, Lalezari used the SBIR solicitation of topics to focus on a specific problem and possible solution. Indeed, one primary benefits of the SBIR program is that it is seen as forcing firms to "think out of the box," while simultaneously providing innovators with access to users.

## EXPERIENCES WITH SBIR

In the brief period since it founding, from approximately early 2003 through mid-2005, the firm has had considerable success in competing for SBIR awards. At its inception, FIRST RF viewed the SBIR program both as a major opportunity to conduct the technologically innovative work that led to its founding and as a source of needed revenue. Lalezari reports writing about 12 SBIR proposals during the firm's first year of operation. It received awards on 7 of these proposals, a number described as a national record for a new start-up company. In the 2004 solicitation round, the firm received another 8 awards. To date, it has converted its 15 Phase I awards into 9 Phase II awards.

FIRST RF's rapid growth is directly connected to its conduct of SBIR-related research. In late 2003, it submitted a Phase I proposal for an Army-generated topic related to the detection of improvised explosive devices. The topic was directed at generating a major jump in the state of the art. FIRST RF received a Phase I award; by the time its Phase I project was finished, the firm had delivered production prototypes for use by U.S. military forces in Iraq.

In 2004, the firm entered a structured competition against 27 other firms, including major defense contractors such as Raytheon and BAE for volume production of IED countermeasure devices. It won the competition, receiving an initial $21.5 million contract from the Army, with delivery scheduled for December 2005. This Army contract has been followed by several additional contracts with DoD prime contractors in excess of $5 million. Throughout 2005, FIRST RF has been engaged in scaling up for volume production to fulfill this contract. It has selected a number of subcontractors, chosen for their reputation for quality products, for manufacturing components.

Even as it makes a rapid transition to a product-oriented, manufacturing and assembling firm, FIRST RF perceives itself to be an R&D firm. The firm describes itself as totally weaned off SBIR as a source of financial life, with its continued survival no longer dependent on the program. It still competes for SBIR funding, however, as it views the program's identification of needed technologies and associated seed funding for R&D as a key to future technological developments. Lalezari reports spending approximately 50 percent of his time on R&D.

Almost all of FIRST RF's SBIR awards have been with DoD agencies, although it has recently received a Phase II award from the National Oceanic and Atmospheric Agency. The firm also has begun to diversify its customer base. It now serves as a subcontractor to large DoD primes such as Raytheon and Northrop Grumman on large-scale systems related to electronic warfare.

The SBIR selection process is considered to be very fair. This fairness in turn is linked to the highly competitive nature of SBIR's selection process. Indeed, FIRST RF expresses fascination at the number of what it considers to have been very good proposals for which it has not received awards. The inference it draws is that the ideas and approaches embedded in the proposals of other firms must have been better.

Both winning a Phase I award and even more so being invited to compete for a Phase II award is described as entailing an essentially Darwinian process. Firms must work their tails off to win SBIR competition. For many small, start-up firms, success in SBIR competitions can be a matter of survival. This process is seen as highly beneficial to the United States. It is a needed antidote to the innovative lethargy and atrophy that has beset large defense and aerospace contractors.

The SBIR program also has provided opportunities for FIRST RF to collaborate with researchers at the University of Colorado. These collaborative research projects also have yielded considerable educational benefits. Under its SBIR awards, FIRST RF has employed four M.S. level and one Ph.D. level student. It has augmented the funds available to pay wages to these students with tuition grants. It sees this added investment as serving to attract the university's best students to its projects, and as the basis subsequently for attracting them to be interested in long-term employment.

## CAPITAL FORMATION

FIRST RF's initial capital came for personal loans obtained by Lalezari and Boone, using their homes as collateral. In the firm's first 6 months of operation, when it had 4 employees, Lalezari and Boone drew no income. The firm's recent growth has been based on revenues from its contracts and retained earnings. It has not had occasion to secure external capital.

## INTELLECTUAL PROPERTY PROTECTION

FIRST RF has filed for 1 patent. The technology covered by the patent is described by Lalezari, who, as noted above, holds 25 patents, as the toughest he's developed. The technology in effect is basically the technical objective described in his initial SBIR proposal.

## STATE GOVERNMENT INVOLVEMENT

FIRST RF has not participated in or received assistance from any state of Colorado high-tech or economic development program.

## RECOMMENDATIONS

The most important role of the SBIR program is to start new products and technologies. SBIR is one of the few remaining paths for new technologies to enter the DoD system. DoD project offices and laboratories have no discretionary R&D funds. Unless already included within the "black box" of the R&D tied to large weapon systems, there are few paths or avenues for outsider firms with novel approaches to enter into the DoD technology development system. This role must be maintained.

The SBIR program strengthens the United State's technological competitiveness. In the field of electronic warfare, FIRST RF's areas of technological expertise, Lalezari sees the United States as facing international competition from a large number of countries, including Egypt, Iran, Pakistan, Canada, and France. Lalezari expresses concern though about incipient pressures to tie the generation and funding of SBIR topics away from the DoD laboratories which generally have served as lead authors to acquisition, "mother ship," units.

The present dollar level of Phase I and Phase II awards is about right. The dollar and time ceilings do put pressure on firms, but the pressure is helpful in forcing firms to perform or get out of the way.

## SUMMARY

SBIR program has achieved the dual objectives of generating a technology of high value to DoD's mission needs and contributing to the fast start and rapid growth of a start-up firm. The DoD SBIR program has provided the U.S. Army with a new countermeasure technology to meet the threat that IED's pose to U.S. military forces in Iraq. If not for the SBIR award, FIRST RF never would have had the resources to work on the technology. SBIR awards provided the revenue that made FIRST RF a viable firm during its first year of operations, and the procurement contracts that have followed upon the firm's initial SBIR awards have led to the firm's rapid growth.

# Intelligent Automation, Inc.

*Zoltan Acs*
*University of Baltimore*

## FIRM OVERVIEW

Intelligent Automation, Inc. (IAI), is a woman-owned R&D firm founded in 1987 by Drs. Leonard and Jacqueline Haynes. IAI conducts research on distributed intelligent systems, networks, signal processing, controls, robotics, artificial intelligence, and education technology.

Key application areas include defense, transportation, forensics, space, communication, and training. In addition to research and product development, IAI has also established service capabilities built around core technology areas in the development and application of artificial intelligence-based techniques.

IAI's technologies in the marketplace today include three dimensional forensics imaging equipment used for matching bullets; platforms and tools for developing agent-based systems; high precision machine tools using a hexapod-configured device; tools for fault diagnosis and prognosis in complex systems; ad hoc mobile network protocols; and assistive learning devices for children with learning disabilities.

Since its founding, IAI has expanded to an organization of over 100 outstanding scientists and engineers (40 holding a Ph.D. degree) and anticipates an excess of $18 million in revenues in 2007. IAI is now located in a 20,000 sq. ft. facility in Rockville, Maryland, and is in the process of negotiating to add another 5,000 sq. ft. by December 2007.

The company's success in the SBIR program was reflected in its 2000 selection to receive the prestigious Tibbetts Award from the Small Business Administration (SBA) for excellence in technology research.

## ROLE OF SBIR

### Founding the Company

The company grew from a $10,000 investment by its founders. As of December of 2004, IAI was granted 218 Phase I and 69 Phase II awards. The SBIR program facilitated a growth in the company's employment from two people, at inception, to 87 employees.

## Supporting Research

Historically, SBIR contracts and Broad Agency Announcements have founded much of IAI's research. IAI's participation in major government programs—the DoD programs such as the Army Future Combat System, NASA programs to develop the next generation air traffic control system, and Homeland Security programs for controlling movement through land border crossing—is also growing in momentum.

IAI has a great track record in turning Phase I projects into Phase II awards. Additionally, the number of Phase II projects have increased over time. This continuity suggests that the company is highly involved with its projects and highly concerned with the quality of its work. Previous experience with SBIR projects provides a competitive advantage when the company is evaluated and reviewed for receiving additional awards.

In 2004, approximately 52–54 percent of the company's revenue was derived from SBIR. This percentage has been falling consistently over the years, making IAI less and less reliant on SBIR funds. Additional revenue is earned from product sales and service contracts. At this time, however, most non-SBIR revenue streams can be traced back to previous SBIR awards.

## Employment Effects

As approximately 54 percent of IAI's funding comes from SBIR, it is probable that the same percentage can be associated with the impact on employment at the company. Employees often share their time among several projects and a large part of Intelligent Automation, Inc.'s expenses are labor costs.

## Networking Effects

The SBIR program changed the structure of contemporary research, and positioned small research companies as an essential part of the process. The program connects government agencies, big corporations, and small businesses. As corporations realize that the program is a great source for innovation, they became the drivers of the acquisition program. Many big companies refrain from developing technology themselves because they lack agility required in the innovation process and because most research is highly specialized. They reduce their risk by monitoring the market place and acquiring technologies that may influence their capabilities.

In addition, IAI's management invests considerable time in trying to team with appropriate universities and win SBIR awards. The SBIR program provided IAI the opportunity to network and make connections that may ensure the company's survival if SBIR funds were withdrawn. In that case, IAI would lose an essential part of its operations and grow at a considerably slower pace.

### A Tool for Staff Development

IAI has used the SBIR program to develop an innovative model of staff development. Young people who start with the company are given the chance to make a difference within the first year of working experience. At other firms, this probability is very slim. At Intelligent Automation, Inc., employees are free to apply for any SBIR grant, under any topic, without executive approval. In 2004, several IAI staff members won awards within their first year with the company.

## COMMERCIALIZATION

Under IAI's business model, commercial activity is derived from the research conducted for various federal agencies. In recent years, the company significantly expanded that base with important new contracts related to signal processing, sensors, fault diagnosis and prognosis, autonomous agents, robotics, educational technology, and forensics among other areas.

The company uses its SBIR track record as a marketing tool to attract contracts from commercial companies. IAI lists its past results on SBIR and STTR projects on its Web site and in its marketing material.

IAI believes that it has been very effective at developing technologies starting with the concept stage through design, building, and testing of a prototype system. Their basic approach to commercializing is to team with partners who have existing products, strong marketing position and capability, and a reputation as a producer of related products.

### Partnering with Primes

In large-scale government efforts, IAI often partners with major corporations—such as AT&T, BAE Systems, CSC, Honeywell, Lockheed Martin, Motorola, Northrop Grumman, and Raytheon—and the company also evolved to become a developer of productized services and technology and an important R&D provider to such major first-tier integrators.

IAI often partners with big companies via subcontracts, consulting agreements, or teaming agreements. Details of such partnership are negotiated in terms of each party's responsibilities and share of contract funds. Partnering is constrained by the SBIR program subcontracting limit of 33 percent for Phase I contracts and 50 percent for Phase II contracts.

IAI establishes such partnerships to take advantage of a big company's specialized expertise or other resources that they can contribute to a project that was granted SBIR funding. Big corporations, like Lockheed Martin, are willing to work with small companies like IAI because of their past contracting experience. IAI staff, familiar with different divisions within particular big companies, have developed connections with their personnel, and may be subcontractors to them on other projects.

In system diagnostics and prognostics IAI has subcontracted with Honeywell and Northrop Grumman. In forensics, IAI partnered with Forensic Technologies, Inc., and obtained sales commitments on its 3D ballistic identification tool. Additional partnerships have been established with Computer Science Corporation and Time Domain Corporation on perimeter security applications. In the area of educational technology and distance learning, IAI began work with Schepp-Turner Productions, executed an agreement with Market Visions, Inc., supported the educational surveys of Westsat, Inc., and provided technology to the University of Maryland Transportation Research Institute.

The company has also negotiated a paid licensing agreement with NASA on Cybele and Cybele Pro, its intelligent agent infrastructure.

## KNOWLEDGE CREATION

The company obtained approximately nine patents and is waiting on other pending patent applications. In addition, Cybele, DIVA, GradAtions, SciClops, Rotoscan, and TalkTiles are trademarks of IAI.

## IMPROVING SBIR

### Surviving Funding Gaps

Intelligent Automation, Inc., manages the delay between Phase I and Phase II funding by managing several contracts at the same time.

### Award Size

According to Intelligent Automation, Inc.'s management, the size of SBIR awards should be reconsidered. For example, NASA has not changed the amount of its awards for at least the past eight years. Accordingly, for the same amount of money, the company is able to finance less and less work. Inflation and overhead costs are taking their toll on IAI's profitability at an increasing pace. Additionally, the timing of the SBIR awards is not conducive to commercialization because of delays in funding.

### Award Criteria

With respect to its own proposals, IAI believes that SBIR awards have always been made fairly and that debriefings have been well thought out and constructive. There is always a conflict between funding high-risk-high-payoff research and funding research where specific plans for commercialization have been made. These two goals are often conflicting. Some agencies appear to be more interested in the former and others in the latter, but generally IAI believes the SBIR program is moving more in the direction of favoring proposals that are closer to commercialization. This is not intended as a criticism—only as an observation.

An important issue for IAI relates to protection of IP. Proposals detail a company's best ideas and present the essence of the concept as clearly as possible to convince the reviewers that the concept is feasible. For those agencies where proposals are reviewed by government employees or dedicated support contractors, a proposer can at least find out who might have reviewed the proposals. For those agencies where external reviewers are used, IAI recommends that proposers should be able to obtain a list of the reviewers in the reviewer pool. Currently it is our understanding that in the case of external reviewers of SBIR proposals, if a reviewer improperly exploited IP learned from being a reviewer, it would be essentially impossible for the proposing company to ever bring any action because they could not find out any information as to who was given access to their proposal.

### Paperwork and Bureaucracy

The amount of paperwork involved in reporting on SBIR projects, relative to other federal technology and procurement programs, is average and in some cases a little more flexible. Nevertheless, some agencies require monthly reporting on six-month projects. In the company's experience, scientific staff would prefer less frequent reporting, especially for Phase I contracts.

IAI recommends that all government agencies use the same Web portal for submitting proposals, reporting, and contracting. DoD's and NASA's electronic systems are more user friendly compared to other agencies'. It is suggested that government agencies be allocated funding to standardize and compile their reporting venues.

### Commercialization

Regardless of the quality of the work performed, not all SBIRs are focused in an area that make a commercial product feasible. In addition to conventional products, IAI's model of commercialization includes "productized services," and the establishment of a working relationship with one of the big integrators leading to participation on large contracts such as the Army's FCS. IAI has been very successful in all three of these types of commercialization.

# Isothermal Systems Research

*Irwin Feller*
*American Association for the Advancement of Science*

## FIRM HISTORY

Isothermal Systems Research was founded in June, 1988 by Don Tilton. At the time, Tilton was completing his doctorate in mechanical engineering at the University of Kentucky. His graduate fellowship included participation at research related to developing Star Wars technology being conducted at Wright Patterson Air Force Base, Dayton, Ohio. Tilton's specific area of research related to spray cooling as a technique for reducing the temperatures in weapons systems, computers, and communications equipment. The research involved addressing fundamental questions about the properties of spray cooling as well as the design of equipment to transform the approach into a functional technology. The original projected end use for the technology was as part of the Star Wars system, although its prospective use in the supercomputer and power electronics industry also received early recognition. As the Star War's project ended in 1992, technology development increasingly focused on these alternative markets.

**ISOTHERMAL SYSTEMS RESEARCH:**
**COMPANY FACTS AT A GLANCE**

Address: 1300 N.E. Henely Ct.
Pullman, WA 99163
Phone: 509-366-78701

Year Started: 1988

Ownership: Private

Annual Sales: $45 million

Number of Employees: 275

SIC Code: 3679

Number of SBIR Awards—Phase I: 13
Number of SBIR Awards—Phase II: 3

Number of Patents: 28

Tilton formed Isothermal Systems to be eligible to submit and then receive an SBIR proposal to the Air Force. The Phase I award was directed at experimental work related to the characteristics of sprayed liquids. Progress on the Phase I project led to Phase II awards, which were directed at developing essential components of an operational system, such as the power electronics and the atomizer. The firm received a series of Phase I and Phase II awards, mainly from the Navy and Air Force for designing and assessing the applicability of its spray cooling technology for a diverse set of weapons systems. Success in these Phase II projects led to Phase III funding from the Air Force and Navy. This process continued throughout much of the 1990s, with SBIR funding representing approximately 60 percent of the firm's annual revenues throughout this period. In part, the dependence on SBIR funding represented the then as yet underdeveloped state of commercial sector demand for the technology.

The firm describes the gestation of its technology as requiring 15 years of technological development before it was ready to pursue commercial markets. The sequencing of this R&D development process is seen as having occurred in 3 stages: (1) from 1988 to 1992, work focused on understanding how spray cooling worked. Also requiring work in this period was development of components such as spray nozzles, since off-the-shelf nozzles then available were not suitable; (2) from 1992 to 1996, work focused on integrating all the now developed components into a functioning and reliable system; (3) from 1996 on into the present the firm began receiving contracts for testing, demonstration, and deployment. Throughout much of first and second R&D phases, the firm's work was supported by a series of Phase II work, which, in turn led to follow-on Phase III contracts.

ISR is now more actively pursuing commercial markets, and has begun to hire marketing staff. In its entry into commercial markets, it is focusing primarily on data centers. In a sense, according to the firm, new demands for its technology have emerged that have made it more commercially valuable.

ISR's relocation from Kentucky to Washington represented Tilton's personal life style choice.

## CAPITAL FORMATION

ISR is a self-financed start-up. The years during which it was working through development of its technology are described as a long and difficult bootstrap period. Its first infusion of what was described as a modest amount of venture capital occurred in 2001, after it had a commercially viable product. It received additional rounds of venture capital funding in 2005 and 2006.

## INTELLECTUAL PROPERTY

Patents are an important part of the firm's strategy for protecting its intellectual property. It presently has 28 issued patents, with another 55 pending.

## STATE GOVERNMENT INVOLVEMENT

Washington state offers several tax breaks for R&D firms, and additional financial incentives for firms to locate in the eastern part of the state. In addition, the state and the regional port authority for the Snake River have recently underwritten the construction of an R&D facility in Spokane.

## RECOMMENDATIONS FOR SBIR

ISR views it overall experience with DoD's SBIR program as positive, but sees the need for several improvement in its procedures. The award process is too slow. Of increasing concern to the firm is its view that over time DoD's SBIR program has become subject to creeping cronyism. The pre-selection of topics to favor frequent winners is increasing. Selected firms are receiving repeated awards to work on topics that at times don't represent new technological advances. ISR reports little success in its more recent efforts to get back into the DoD SBIR program after having not submitted proposals for several years.

The gap between Phase I and Phase II funding was "brutal" for the firm. The impacts of the gap were especially hurtful during the firm's first 8 years of operation, when it was very small. The firm's owners were required to go without pay during this period, and to downsize by laying off employees.

The size of Phase I and Phase II awards are seen as reasonable. The real challenge for a firm receiving an SBIR award is moving into post-Phase II activity. The outcome of a Phase II award is not even close to being a product. The award does not provide for the life cycle development of the technology: It provides no support for determining manufacturing feasibility, reliability of the product, or scaling up from bench prototypes to large-scale production. The gap between where a Phase II project ends and the beginnings of product development are enormous. DoD's technical personnel in charge of setting SBIR topics and overseeing proposal selection and project monitoring do not fully understand these differences.

Increased attention needs to be paid to how DoD selects its SBIR topics, as well as to how these topics align with DoD's overarching objectives.

## SUMMARY

ISR credits DoD's SBIR program with its founding and long-term growth. Without SBIR, it wouldn't be where it is. SBIR is described as a program that gives people a good chance to make a go of it. The firm appreciates what DoD has done for it. In return, ISR highlights that the SBIR program has produced a lot of good technology both for DoD and the commercial sector.

# JX Crystals, Inc.

*Zoltan Acs*
*University of Baltimore*

## FIRM OVERVIEW

JX Crystals, Inc., located in Issaquah, Washington, is an innovator in Infrared Cell Technology. The company's key technology is the gallium antimonide (GaSb) photovoltaic cell, which responds to longer wavelength radiation than either traditional silicon cells or newer gallium arsenide cells. This new infrared cell can be used in thermophotovoltaic generators as well as in space and terrestrial solar applications.

In thermophotovoltaic systems, photovoltaic cells respond to infrared radiation from a fuel-fired emitter, rather than the visible light energy from the sun. The company's expertise is in GaSb cell development and its plans are to continue developing these applications. JX Crystals sees tremendous potential in manufacturing its latest commercial products using its innovative technology.

## THE ROLE OF SBIR

### Founding the Company

JX Crystals was founded in 1992 by Lewis and Jany Fraas. The founders developed 30 percent efficient solar cells, a world record performance, but Boeing, where Lewis Fraas worked, decided against funding the project. While at Boeing, Dr. Fraas contacted NASA officials, who showed interest in the development of infrared cells. They founded their firm after winning two small SBIR contracts.

The company has received 12 Phase I and 10 Phase II awards over a 12 year period. These include funding from NASA for space solar cells; from DoD for quiet battery chargers using thermophotovoltaic IR sensitive cells; and from DoE for thermophotovoltaic cells for home cogenerators and hybrid solar lighting systems.

### Support for Nontraditional Ideas

According to Dr. Fraas, if a company is fairly well established with an ongoing product and still wants to do R&D, then participation in the SBIR program can be very useful. From concept idea to award, it takes anywhere from six months to one year to get an SBIR. Many companies prefer to avoid government involvement if they can finance ideas internally. Nevertheless, there are limited opportunities to fund longer-term projects. In such cases, the SBIR is a good way of feeding the next product.

### Enabling Role in Building Research Capabilities and Commercialization Opportunities

According to Dr. Fraas, the SBIR program is not enough in itself to accomplish these objectives, but it plays an enabling role that give small innovative companies a chance to suceed.

### Employment Effects

Beginning with just two employees, JX Crystals employed as many as 20 individuals before 2000. During the time of recession, it had to reduce its payroll to 5 employees. At the time of the interview the company currently employed two administrative staff, two senior engineers, and an accountant.

## COMMERCIALIZATION

While the company markets its infrared cells around the world, the market at present remains limited, consisting mostly of university research laboratories. JX Crystals' revenues, by the end of 2005, will exceed $2 million. Currently, JX Crystals is negotiating with a furnace company the establishment of a consortium that takes the new infrared technology to market.

SBIR awards funded the development of prototypes for use by DoD. The company's related civilian products are being sold mainly outside the United States. While SBIR awards did not contribute directly to the development of these civilian technologies, the program is nonetheless credited with helping the technology's development and the company's growth.

According to Dr. Fraas, existing regulatory policies inhibit the commercialization of this innovative technology in the United States. British Petroleum and Shell, he said, receive subsidies in California based on 20-year-old silicon manufacturing technology. These subsidies are available to a list of qualified vendors. Even if a company has a cheaper product with new innovations, it must pay for qualification testing and approval in order to get on the list—a significant barrier for small innovative firms.

Ms. Jany Fraas, who is from China, has brought the technology to the attention of Chinese officials, who have recognized the technology's potential. The Chinese government has since offered to pay for the qualification testing of their technology. China is interested in manufacturing the cells and selling them back to the United States.

JX Crystals' involvement with China is mainly due to their focus on production processes. China is looking 50 years ahead, while the United States is looking 50 years behind in trying to promote existing technologies. U.S. companies are disadvantaged on the international market because they are competing with foreign businesses that receive commercial and manufacturing support from their respective governments.

## ADDITIONAL FUNDING

The company received a small amount of state funding, mostly from the Washington Technology Center. The company has also collaborated with Western Washington University under this program. While this funding did not provide money for operating expenses, it has helped to move the technology forward.

Over a period of 12 years, JX Crystals partnered with big companies like ThermoElectron, ABB, and Energy Innovation, a company in California. Additional funds have been acquired from sources in Israel, China, and a Buddhist temple.

In earlier years, SBIR funding represented approximately 40 percent of total company revenue. Currently, JX Crystals derives 10 percent of its revenue from an SBIR contract with OSD.

## KNOWLEDGE EFFECTS

With approximately 20 patents, the company has secured " a good lock" on the technology. In addtion to having published about 50 scientific papers, the company also hold a trademark. It also holds four achievement awards: 1 from NASA and 3 best paper awards.

## IMPROVING SBIR

### Surviving Funding Gaps

Delay between Phase I and Phase II funding makes the survival of small businesses difficult. Even if the company survives, it can lose important capabilities as it diverts staff to other projects or even loses staff. Time has to be spent on finding bridge funding and work on other projects have to be juggled. This takes up time and money. The funding gap between Phase I and Phase II was painful and JX Crystals survived it only because the company had other contracts by that time.

### Award Size

A larger Phase I award would make it slightly easier to wait for the Phase II awards.

## Paperwork and Bureaucracy

The company finds the SBIR process to be fair but not timely. Extensive paper work and long delays before start up were found to be significant hurdles, especially given the small size of this firm.

JX Crystals has received SBIR awards from NASA, DoE, the Army, and DARPA. DoD is rated as being relatively efficient. NASA and DARPA are pretty good. The Army is financially diligent, but technologically unsophisticated.

While JX Crystals is interested in sustainable energy projects, DoE is more focsed on university research than on small business research. According to JX, DoE focuses on "long range unrealistic projects that lead nowhere," or on "large billion-dollar companies that focus on coal gassification and similar traditional technology extentions."

## Commercialization

The steps involved in bringing a new product into the marketplace start with proof of concept, which can be easily funded by Phase I awards. The second step of building the first prototype is supported by Phase II in the SBIR program. Beyond that, companies must build commercial prototypes and do beta site testing. The SBIR program locks companies into perpetual research and, at most, helps them build the very first prototype. However, it fails to support awardees in completing the award cycle with production and commercialization. While several agencies are aware of this multistep process, the SBIR program abandons small businesses at a fairly early point. Thus, there needs to be a Phase III plan. Phase III could include more extensive testing and improvement of prototye for potential investor funding.

According to JX, the heavy focus on DoD SBIR awards and associated defense systems limits the development of related commercial products for peaceful appications. There is also a lack of follow-through for commercial applications used beyond the military prototypes. This means that small businesses have to look abroad to support for technology transfer and exploitation—"a sad state of affairs." The exclusive focus on support of only military technology development is a handicap for the U.S. economy because most foreign governments help their industries in the manufacturing and commercial sector.

# JENTEK Sensors, Inc.

*Zoltan Acs*
*University of Baltimore*

## FIRM OVERVIEW

JENTEK Sensors, Inc., of Waltham, Massachussetts, produces diagnostics and prognostics sensor technologies and systems providing nondestructive testing solutions, health and process monitoring solutions, and materials characterization solutions. These sensors, for example, help monitor damage to aircraft and other high value assets.

Dr. Neil Goldfine founded JENTEK in 1992 as a company of one. Today, JENTEK has 27 employees. In 1999, Deloitte & Touche placed JENTEK 14th on its list of Fastest Growing Technology Companies in New England. In 2007, JENTEK received the FAA/Air Transport Association's Better Way award for its engine component inspection technology. The company's customers include the Navy, Air Force, Army, NASA, DoE, FAA, several foreign militaries, and Fortune 500 companies in the aerospace, materials, automotive, petrochemical, manufacturing, and consumer products industries.

## THE ROLE OF SBIR

### Fostering Rapid Company Growth

For JENTEK, the SBIR program has proved critical in providing the financial resources and infrastructure to facilitate the development of key technologies necessary for JENTEK's current and planned products. The program is credited with allowing the company to grow to critical mass in 10 years instead of 30 years. According to Dr. Goldfine, providing direct access to DoD customers and their defined needs is one of the most important contributions of the SBIR program.

### A Key Source of Funding

From 1995 to 2000, SBIR awards served as the company's principal source of funding and—complemented by personal resources and private investment—helped drive the company's technology development and expansion. Throughout its history, JENTEK has submitted 87 Phase I proposals and received 25 Phase I awards. Of these Phase I awards, three are ongoing Phase I programs. Of the 20 completed Phase I programs, 17 Phase II proposals have been submitted with 16 Phase II contract awards.

Over the last three years, SBIR represents 40–60 percent of the company's budget. It used to be 70–90 percent. While in the long term JENTEK plans to keep SBIR funding under 40 percent of the company's total revenue, the SBIR program remains a cornerstone of its technology development strategy, serving as a primary source of the company's applied R&D.

### Leverageing Private Investment

JENTEK leverages its SBIR awards to develop other sources of private investment. For example, the company's Phase III commercialization award from NAVAIR is included in JENTEK's marketing materials.

### Building Relationships

In addition to providing funds, SBIR also provides JENTEK the opportunity to build relationships with OEMs and with government customers, such as NAVAIR, WR-ALC, OO-ALC, NADEP Cherry Point, NADEP Jacksonville, Kennedy Space Flight Center, etc. These relationships are further strenghtened through subcontracting agreements.

JENTEK also advises its target customers of opportunities to team on SBIR projects. In fact, when the company selects SBIR topics, a top criterion is the potential to team with and enhance relationships with an OEM or other target customer.

## COMMERCIALIZATION

JENTEK applies physics-based models to provide reliable and robust solutions to multivariate property measurement and defect detection applications. JENTEK specifically targets those applications that are causing customers a high level of "pain" and cannot be solved using conventional methods. JENTEK can also deliver relatively low-cost turn-key solutions, such as for engine disk slot and Friction Stir Weld Inspection, that offer improved reliability and speed at a very competitive cost compared to conventional methods offered by competitors.

JENTEK estimates that the potential market in the nondestructive testing (NDT) solutions area is approximately $250 million annually; the target market in the health and process monitoring solutions area is substantially greater than $1 billion (including potential applications in electronics, life sciences, and real-time industrial process control); and the target market in the Materials Characterization solutions market is on the order of $250 million. JENTEK focused on building credible business/product lines in well-defined niches, often tied to SBIR dollars, such as (1) engine disk and blade inspection, (2) on-board sensor networks for fatigue detection and monitoring, and (3) magnetic stress gages for aircraft, rotorcraft, and bridges.

Major competitors vary depending on the market and specific application. In the NDT solutions area major competitors include GE Inspection Technologies, Olympus, Zetec, Wyle Laboratories, Boeing inspection technologies, and in-house NDT groups at OEMs and even inside government agencies. Recently, the number of JENTEK's competitors was reduced to approximately 10 or 15, as smaller companies were systematically purchased by bigger corporations.

JENTEK has developed and delivered solutions to a wide variety of specific DoD and commerical applications for (1) coating characterization, (2) weld characterization, (3) fatigue monitoring, (4) corrosion imaging, (4) gun barrel inspection, (5) engine disk slot inspection, (6) composite damage imaging, and many more. Also, new developments in through-wall temperature and stress monitoring, as well as in stress sensor networks offer breakthrough capabilities in a wide range of fields including aerospace, energy, and even life sciences.

Examples of commercial deliveries include engine slot inspection systems to NAVAIR, fatigue monitoring systems to Northrop Grumman and Lockheed Martin, Space Shuttle leading edge inspection systems to NASA, C-130/P-3 propeller inspection systems to numerous DoD and foreign military customers, coating inspection systems to Siemens, and FSW inspection systems to Eclipse Aviation and other aircraft OEMs. JENTEK sells products to the private sector, primes, as well as the agencies that funded research.

## KNOWLEDGE EFFECTS

JENTEK has had a significant impact on understanding of eddy current sensing capabilities overall. JENTEK is recognized as a leading developer of eddy current technology and is clearly setting the bar for next generation inspection technologies.

In addition to publishing numerous scientific papers, the company has been issued seven patents associated with government funding and twenty-three associated with private funding. Several other patents are pending.

When small companies have patents, they are taken more seriously. Patents (1) create an aura of success, facilitating more serious business relationships with primes; (2) are considered in the award evaluation process; and (3) make the company less susceptible to competitive pressures.

The company holds the trademarks for GridStation, MWM, IDED, and JENTEK Sensors.

Reflecting the scope of its innovations, the company has received numerous awards, including the NAVAIR Phase III commercialization award; Outstanding Phase III Transition Award, 2004, awarded by the Navy Transition Assistance Program; Outstanding Paper Award for Materials Evaluation: "Eddy Current Sensor Networks for Aircraft Fatigue Monitoring," published in the ASNT Materials Evaluation Magazine, July 2003, Aerospace Health Monitoring, Volume 61, No. 7; Technology 2007 Spin-off Achievement Award from NASA Tech Briefs;

and the 2007 FAA-ATA Better Way Award for its engine component inspection technology.

## IMPROVING SBIR

### Funding Delays

The delay between awards is not usually a problem for JENTEK, now that the company's size has increased. JENTEK typically has many other programs and can offset funding delays with other work. It is always better when there is a shorter delay; nevertheless, it is far better for decisions to be sound rather than rushed. The company generally does better (at receiving Phase II contracts) when the TPOC takes his or her time and makes decisions based on the best and most complete information. Thus, JENTEK prefers that this process is not rushed. However, once an award decision has been made, delays in Phase II start dates can be painful.

JENTEK typically keeps low levels of R&D funding to support efforts between Phase I and Phase II relevant to JENTEK overall R&D goals.

### Award Size

For Phase I, $70,000 to $100,000 is reasonable. Too large a Phase I program becomes a distraction until feasibility is demonstrated. However, larger Phase II awards would be helpful. Ideally, a $750,000 base program with a $500,000–$750,000 option would be advisable. The option should be directly supported/sponsored by the target customer.

Often there is a gap between SBIR funding and program funding (transition money is seen as R&D). However, the NAVAIR matching funds approach is a great solution to this problem and has helped JENTEK succesfully transition several systems (saving the DoD substantial funds).

### Award Selection

The award selection process is seen as "surprisingly fair": "We have blind bid on many projects where the customer had never heard of us and succeeded in winning both Phase I and Phase II awards." The delays from proposal submittal to award to program completion are not too long to prevent successful technology transfer. These delays are better than a rush to judgement.

However, there is often a conflict of interest in small businesses being evaluated by college professors who compete for the same money. Small companies can find themselves in the awkward position of competing with universities that have infiltrated the funding agencies. In applying for transition money, SBIR companies often have two or three university representatives on the committee who seemingly decide between transitioning a technology or funding themselves. Naturally, the universities choose to fund themselves. This often delays commercialization of deserving technologies. However, diligent small businesses can navigate this process by finding DoD customers with real and pressing needs.

## Commercialization Programs

JENTEK notes that NAVAIR events (commercialization forums) have been particularly useful. The Air Force, particularly WR-ALC (Robbins Air Force Base), is great at supporting transition to use as well. NASA has also been very good at helping the company transition systems to use. Other agencies are more research focused and not as interested in near-term commercialization, but this balance is not necessarily a bad thing.

Citing the NAVAIR matching funds approach, JENTEK notes that the SBIR program should mainly promote product transition into Phase III and facilitate the connection between small businesses and primes. It is a great way to encourage small businesses to communicate with the stakeholders at the bases and deliver valuable solutions. Also, SBIR offices should encourage stakeholders at the bases to communicate with SBIR firms, by informing them about the opportunity for such matching. SBIR firms should not be limited in their ability to keep technolgy proprietary; this is the lifeline of any succesfull small technology-oriented company. It is realistic to have open architecture processes with "plug-and-play" proprietary and open products/solutions selectable by customers/users.

However, with the exception of NAVAIR, access to transition support funds is a weakness at all agencies; these funds are needed to bridge the gap between research and implementation dollars. OEMs should be encouraged more to work with Phase II SBIR firms and include these firms in larger programs. This happens, but without formal guidelines and tools. Finally, standardization and centralization impede small business processes. Such standardization often artifically raises the entry fee to compete for real applications beyond the resources of small entities. Of course such standardization, if used properly, can substantially improve efficiency and reduce costs.

## Liaising with the Primes

Because one of the SBIR program's greatest advantages is the opportunity to develop business relationships with primes, a company's efforts to connect with primes should be an evaluation criteria for Phase I awards. Agencies, in making Phase I award decisions, should consider whether the SBIR company has: (1) a clear relationship with primes or (2) funding to fill out the technology matrix combined with ability to provide value to the customer (who oftentime is the funding agency itself).

## Program Cycle

For small companies, functioning in parallel with the government's twelve month cycle is difficult. Twenty-four month programs are beneficial for SBIR firms and, at the same time, hard to find. Thus, the twenty-four month Phase II opportunity is great for stabilizing small company funding. The recent push by some groups to introduce options (toll gates) after twelve months (in Phase II programs) removes this benefit and makes life much more difficult.

## Protecting Intellectual Property

The protection of proprietary information in the proposal submission process is a key issue for small innovative businesses. There are groups that push for openness and devalue companies that hold SBIR data proprietary. The SBIR program should help primes, and particularly government employees, understand that the lifeline of small businesses is proprietary technology. If disclosure of proprietary information is a criterion in funding research, inevitably, small companies will be eliminated from the competition process. This practice, by which only primes can have proprietary information, is an increasingly debilitating burden for small companies and will dramatically reduce the quality of the technology that the agencies receive for their SBIR investments, and overall.

# Marine Acoustics, Inc. VoxTec International, Inc.

*Irwin Feller*
*American Association for the Advancement of Science*

## FIRM HISTORY

This narrative recounts the genealogy of two related firms: Marine Acoustics, Inc. (MAI), Middletown, Rhode Island, and VoxTec International, Inc., Annapolis, Maryland. Until January, 2005 VoxTec, was a wholly owned division of MAI. In January, 2005 it spun off as a separate firm. MAI has been the recipient firm for SBIR awards, with VoxTec's founding directly tied to MAI's receipt of an SBIR award.

MAI was founded in 1988 by William Ellison, a graduate of the U.S. Naval Academy graduate (1963), and a holder of a Ph.D. in underwater acoustics from MIT. Upon his retirement from the Navy, Ellison served as a freelance consultant to Navy Systems Warfare. While he was working in this capacity, Navy officials suggested to him that they would find it easier to enter into contracts and also to have him undertake larger scale projects if his work was performed through a firm rather than as an individual. Accordingly, Ellison formed Marine Acoustics,

### COMPANY FACTS AT A GLANCE

| **Marine Acoustics, Inc** | **VoxTec International, Inc.** |
|---|---|
| 809 Aquidneck Avenue | 706 Giddings Avenue |
| Middletown, RI 02842 | Suite 2A |
| Phone: 401-847-7508 | Annapolis, MD 21401 |
| Fax: 401-847-7864 | Phone: 410-626-9825 |
| Email:info@marineacoustics.com | Fax: 410-626-9851 |
| | Email: *www.ace@sarich.com* |
| Revenues: $10.6 Million in 2004 | $2.5 million |
| Number of Employees: 35 | 6 |
| Number of SBIR Awards: 11 | |
| Phase I: 6 | |
| Phase II: 4 | |
| Phase III: 1 | |
| Number of Patents: 1 | Pending: 2 |

locating in the Newport, Rhode Island area, which provided close proximity to Navy facilities and a sought after life style.

MAI provides a range of engineering, technical, operational and environmental planning/compliance support services to a number of government agencies, commercial firms, and universities.

The VoxTec division of MAI was formed by Ace Sarich, a graduate of the U.S. Naval Academy, a Seal veteran with two tours in Vietnam, and a graduate of the Naval Postgraduate School. Sarich also served as a faculty member at USNA, where he taught mechanical engineering and Naval Systems Engineering. After retiring from the Navy in 1986, Sarich worked briefly for a small R&D firm and as a freelance consultant on a variety of engineering projects. Several of these projects involved designing equipment to meet the special needs of DoD's special operations commands. Sarich joined Marine Acoustics in 1987, where he worked on classified R&D projects out of the firm's offices in Arlington, VA. Upon forming VoxTec, Sarich first worked out of his Maryland home, but then relocated to an office in Annapolis as the firm grew.

## EXPERIENCES UNDER SBIR

Marine Acoustics is reported as having had 1 SBIR award prior to the award that underpins VoxTec's formation. SBIR thus represents a small portion of Marine Acoustics' core operations. SBIR's impacts are manifest though in the events that led to the development of the firm's handheld voice translator (Phraselator) and VoxTec's subsequent founding and growth.

The sequence begins with discussions between Sarich and a DARPA program manager, a former high school classmate and Naval Academy classmate, about DARPA's ongoing efforts to develop field usable voice translators. The state of the art through 2000 consisted of pc-based voice translators, but this platform was too cumbersome for use in field operations. The mission objective was to develop a handheld translator. Sarich, a Marine Acoustics employee, expressed an interest in working on the technology. Subsequently, DARPA developed an SBIR topic for the technology.

Sarich applied for a Phase I award, which he received in 2000. Using this award, he built a PDA version of the translator. This progress led to a Phase II award in January, 2001, and resulted in the development of a working prototype by September, 2001. Following 9/11, and the subsequent deployment of American military forces in Afghanistan and Iraq, Marine Acoustics received a Phase III award to tool up to begin limited production of the handheld translator. Initial field deployment of the technology began in 2002.

The successful development of the handheld translator led to changes in Sarich's relationships with Marine Acoustics. As noted, Marine Acoustics started out and has continued to specialize in marine engineering design. Much of its work is done on a cost-plus fee basis. After the firm scaled up to begin Phase III production, it began to encounter difficulties in meshing the accounting and managerial systems associated with being primarily a service-oriented firm that used cost-plus-fee accounting, with those required to grow a high-tech, product-oriented start-up firm. Growth of the latter type of firm, especially one based on internal sources of revenue, was seen as requiring a pricing structure that included overhead and profit margins different from those found in Marine Acoustics R&D contracts. Also, as it perceived broader nondefense markets for its translator technology, such as in law enforcement and medicine, VoxTec saw the need for a new, different orientation to marketing and pricing.

These considerations led to a decision to gradually separate Marine Acoustics and VoxTec. In January 2005, VoxTec International, Inc., was spun off. Under the present arrangement between the two firms, product marketing and sales are handled by VoxTec, while finance and accounting are handled by Marine Acoustics. For every Phraselator that it sells, VoxTec is committed to paying a royalty to Marine Acoustics. When a predetermined total level of payments have been made, the intellectual property underlying the Phraselator will accrue to VoxTec, and no additional royalties will be paid. At that time, the government R&D contracts will be novated from Marine Acoustics to VoxTec.

## CAPITAL FORMATION

Operating primarily as a contract R&D firm, providing engineering design services to the Navy, primarily the Navy's Space and Naval Systems Command, required little in the way of initial capital. Marine Acoustics thus began and has remained a privately held firm, with its stock distributed between its founders and an ESOP.

VoxTec also is privately held. Its activities are funded primarily by existing DARPA and Army contracts. It views going public as a desirable future outcome.

## INTELLECTUAL PROPERTY

Marine Acoustics, as a supplier of contract R&D services, has tended not to produce patentable inventions. VoxTec, as a product-oriented firm, however, sees patents as an important source of intellectual property protection. It is beginning to file patents on its inventions, but has not received any patents to date. The firm notes though that much of the technology embedded in its translator already exists in the public domain. Its "technological leap forward" in large part was based on a distinctive integration of components and an acute awareness of and sensitivity towards the operational needs of end-users, that is frontline warfighters.

## EXPERIENCES WITH STATE GOVERNMENT PROGRAMS

VoxTec reports no involvement with Maryland's technology development or economic development programs.

## RECOMMENDATIONS

According to VoxTec, the core features of the SBIR program, including the open solicitations and the sequencing of Phase I and Phase II awards are sound. SBIR's value is that it provides opportunities both for the sponsoring agencies to learn about new technological possibilities and for firms to pursue them.

The downside to these procedures is that they can make the topic generation and solicitation processes "fishing expeditions" on the part of agency program managers. Managers may use the process to see what's out there without having a serious intention of making an award. In such cases, firms may spend a lot of time putting together a proposal, with little realistic prospect of receiving an award.

The perception also exists among some firms that selected competitions are wired. Thus, firms are required to spend time deciphering the messages conveyed by program managers and other industry sources about whether or not a solicitation truly represents an open competition.

## SUMMARY

Without SBIR, according to VoxTec's founder, DARPA would not have an operational technology. In a period of approximately 2 ½ years, years, the research funded by DARPA has rapidly progressed from a concept to a prototype to a technology deployed in Afghanistan and Iraq. The award also has served as the basis for the formation of a new firm, VoxTec. As described by the firm, without SBIR, VoxTec would not be where it is today. Although in its early stage of development, and focused at present at scaling up production of its core defense-oriented product, the firm sees its technology as representing a significant advance over existing off-the-shelf products, and as having considerable potential in public and private sector markets.

# Multispectral Solutions

*Zoltan Acs*
*University of Baltimore*

## FIRM OVERVIEW

Multispectral Solutions, Inc. (MSSI) is an industry leader in ultra wideband (UWB), an emerging wireless technology for communications, precision localization, RFID & radar applications. With a core competency in RF (radio frequency) and high-speed digital design, the company has applied this technology to a wide range of military, government and commercial products including low probability of detection communications, high-precision ranging and radar, radio frequency identification (RFID) and precision real-time location systems (RTLS).

MSSI is located near Washington, DC, in Germantown, MD. Dr. Robert Fontana is the company's principal founder.[3] He believes that a key strength of the SBIR program is its ability to fill definite technology gaps in DoD's arsenal, and its ability to grant small companies an opportunity to develop their businesses.

## THE ROLE OF SBIR

The founder recognizes the significant positive role of SBIR, among a number of interrelated factors, in helping the company establish strategic partnerships, acquire external funding, and establish itself in the market, among other outcomes.

### A Source of Early-stage R&D Funding

The company realized, very early in its history—winning two SBIRs in its first year of operation—that the SBIR program was a valuable source of nonequity diluting R&D funding. Even when venture capital funds are available for early-stage R&D, the innovative small business is often required to give up as much as 70 percent of the company, often compromising its ability to react to new opportunities.

[3]Dr. Fontana earned a masters degree in from MIT and a doctorate from Stanford University, all in electrical engineering. Prior to forming MSSI in 1988, Dr. Fontana worked in the defense industry for Raytheon, Hughes Aircraft and Litton Industries, and as a professor at Carnegie Mellon University.

### A Signal of Quality to Private Investors

Multispectral Solutions used the validation provided by successful performance on a wide variety of SBIR programs as an important leverage for both non-SBIR and angel funding. Winning numerous Phase I and II, and two Phase III, SBIR awards allowed the company to attract additional funds and to advance the technology to a more mature state. Over the course of its history, MSSI has received in excess of $30 million in government R&D funding, largely as a consequence of techniques and products which it developed under SBIR grants. As a consequence, the company has not required any outside funding, although it successfully completed a round of angel funding for $1.75 million which it used to accelerate its commercial applications of the technology.

### A Path to Government Contracting

An important contribution of SBIR is that it allows small companies to form relationships inside of the government arena. This task is very difficult without programs like SBIR. For example, one SBIR award, which transitioned into an acquisition contract (Phase III SBIR) for an aircraft wireless intercom system with the Naval Air Systems Command, was the basis for partnering with the Raytheon Technical Services Company. At the delivery point on the Phase II award, Raytheon paid for a license from MSSI to integrate the Multispectral Solutions' UWB RF card stack into an existing Raytheon product (AIC-14) for use aboard aircraft.

### Addressing Agency Missions

Multispectral Solutions has always strived to understand the technology needs of the defense sector, and the SBIR process has been an invaluable mechanism for obtaining such insight directly from the organizations and commands. Furthermore, the SBIR process strongly encouraged the commercialization of the company's technology, enabling it to produce commercially viable products which can capitalize on the private sector markets.

### Employment Growth

It is hard to quantify the effect of SBIR on staffing, because while the program is a prime mover, it is not the only thing that drives employment requirements. Multispectral Solution has grown from employing two individuals in 1989 to 29 individuals as of this interview, of which approximately 25 are full-time R&D staff. In addition, the firm hired some high-level executives as a direct consequence of staffing requirements generated by Phase II and III SBIRs.

### Establishing Technological Leadership

The SBIR program helped establish Multispectral Solutions as one of the leading national authorities on UWB, a field that has experienced dramatic technological changes. During the docket proceeding for the FCC rulemaking on unlicensed use of UWB technology, Multispectral Solutions was instrumental in helping the FCC craft a workable draft. The company was a leading responder under the docket and was heavily referenced throughout the proceeding by other respondents.

Multispectral Solutions continues to apply for further SBIR awards but has been selective in responding to solicitations. It invests its resources into writing proposals only if the solicitation seems to advance the company's core technology.

## COMMERCIALIZATION

Within the past few years, the FCC changed its policies regarding ultra wide band technology allowing it to emerge into the commercial sector under its rules for unlicensed (Part 15) equipment. As a result, a number of small UWB companies were formed; however, due to its seniority in the field, Multispectral Solutions considers itself an industry leader. As a consequence of its extensive SBIR experience in designing and building operation systems, the company has specialized in UWB hardware and complete systems. On the other hand, MSSI sees most UWB contenders as having specialized in chipset development for short range, consumer-oriented, wireless communications applications.

MSSI's commercial products include its *Sapphire* DART UWB-based Real Time Location System (which received the Frost & Sullivan 2005 *Product Innovation of the Year* Award), a wireless audio distribution product *SpectraPulse* sold and marketed by Audio-Technica under exclusive license to MSSI, and a commercial version of its UWB radar.

Interestingly, *SpectraPulse* is a commercial version of a wireless intercom system that was developed under Phase I, II and III SBIR programs with the U.S. Navy. The Navy Phase III award, for which MSSI was named in 2004 as a Navy SBIR Success Story, is an excellent case story for both the good and bad sides of SBIR contracting. The Navy set aside $24.5 million in acquisition funds for this project. An initial Phase III SBIR award, of roughly $4 million, was used to produce the first article system test for the Navy. Multispectral Solutions was the prime contractor for the award and was responsible for designing mobile units, a wireless card stack, antenna designs, etc. Raytheon was selected as a production partner in charge of integrating the technology within existing military equipment onboard fixed and rotary wing aircraft. Multispectral Solutions, although a relatively small company, planned to complete 40 percent of the work involved in the project, with Raytheon handling the production deliveries to the customer.

Unfortunately, this contract did not come to fruition. Since then, MSSI has significantly downsized its participation in the SBIR program.

## KNOWLEDGE EFFECTS

The company's revenue approached $6 million last year. In this income range, the company is big enough to establish competitive advantage over small start-ups. The company is in the process of doing a first article build for an aircraft ICS system in Navy helicopters.

The company has been issued 10 patents and has filed for several more. In addition to publishing several papers, the company also coordinated the first IEEE conference on UWB in Baltimore in 2002. There were 400–500 attendees from around the world and Multispectral Solutions produced five of the papers presented at the conference.

Dr. Robert Fontana, MSSI founder, president and CEO, was recently elected (November 2006) Fellow of the IEEE for "contributions to short pulse electromagnetics as applied to ultra wideband systems."

## IMPROVING SBIR

### Surviving Funding Gaps

Factors in surviving funding gaps include good management (and sometimes good fortune) and, importantly, the SBIR program managers' timeliness in administering awards. By and large, MSSI believes that the Navy has a good administration process and their program managers are more conscientious than others.

The company does not build funding delays into the standard budget and work force process. Depending on the job, the Multispectral Solutions may stop work until funding arrives. It does not actively seek bridge funding.

### Commercialization Assistance

Multispectral Solutions attended commercialization conferences and was recently highlighted as a model company. At these conferences, companies can present their technologies and gain information on the government agencies' procurement needs. The company finds such support activities very useful.

### Award Size

The size of Phase I awards could be increased to $100,000 or $125,000, with the condition that awardees develop an actual product. This strategy would also allow SBIR program managers to make a more educated determination on the viability of Phase II awards. The size of Phase II awards, typically $750,000, seems to be adequate.

Various government agencies offer different sized SBIR awards. In the company's perspective, SBIR awards should be standardized, using the Navy as a model. The program should also establish procedures that encourage independence from SBIR funding.

### Paperwork and Bureaucracy

SBIR paperwork is minimal in comparison to the formalities related to the company's production contract.

Multispectral Solutions had contracts with several DoD agencies: Army, Air Force, Navy, etc. Navy seems to have a well-organized program with more oversight. They give careful consideration to their expected outcomes and their program managers have a good work ethic.

### Award Selection

The selection process may be biased in certain cases. Multispectral Solutions successfully challenged a solicitation that seemed custom-tailored to a particular company's proposal.

### Multiple-award Winners

The company noted that the SBIR program is very beneficial when utilized for its originally intended goal. However, some companies develop business models that live off the SBIR program. A small business that is writing dozens of Phase I proposals each year may not be focused on technology and true commercialization. It may just be trying to turn out volume and pump up revenue rather than leveraging on some specific aspect of their technology and capitalize on the private sector market. Firm representatives believe that some multiple-award firms develop human resource infrastructures for the sole purpose of writing proposals that knock out commercially promising proposals from firms like Multispectral Solutions.

## CONCLUSION

In sum, SBIR awards seem to have played a significant role in the firm's development of its technologies and subsequent growth. At the same time, its experience demonstrates that even successful completion of SBIR awards does not automatically lead to success in production contracts. Nonetheless, MSSI's experience does underscore the positive role of small firms in meeting DoD technology needs and SBIR's role in facilitating this contribution.

# Next Century Corporation

*Irwin Feller*
*American Association for the Advancement of Science*

## FIRM HISTORY

Next Century Corporation's founding reflects a set of personal and business decisions made by its founders following 9/11. One of them, John McBeth, had formerly been employed with a sequence of computer software firms that had gone through a series of acquisitions. He was first with Century Computing, a company formed in 1979, and a recipient of SBIR awards for several agencies. Century Computing was acquired by AppNet in 1998, as part of that firm's acquisition of 12 other firms and a subsequent public offering. AppNet, in turn, was soon acquired by Commerce One.

Seeking a more creative and entrepreneurial firm environment, in 2001, McBeth, along with 2 other former senior officers from AppNet, left to start a new company. In September 2004, they founded DigitalNet, raising $100 million in venture capital to buy computer software firms.

The 9/11 bombing of the World Trade Center found McBeth at a software conference in San Diego. Forced to drive cross-country for part of his trip back to DC because of the freeze on domestic air flights, and realizing that were it not for a last minute change in flight plans, he would have been on one of the planes that

**NEXT CENTURY CORPORATION:**
**COMPANY FACTS AT A GLANCE**

8101 Sandy Spring Road
Laurel, Maryland 20707

Phone: 301.939.2600
Fax: 301.939.2606

SIC: 7173

Revenues: $5 million
Employment: 23

Number of SBIR Awards
- Phase I: 3
- Phase II: 2

crashed, McBeth came to the decision that he no longer wanted a career based on buying and selling firms but instead wanted to form his own company.

This company was to function according to a specific set of rules; these rules were a reflection of his personal values and prior corporate experiences. The rules were that the firm would never be sold and would not be taken public (thus ruling out recourse to venture capital or outside investment); its goal would be to help protect the United States, and its products would be designed to save lives.

McBeth was joined by three partners, each of whom invested $500,000, in founding Next Century. Each of the four had experience as senior managers of software firms, and each was well recognized in the industry for their technical expertise and business experience. The $2 million initial investment was intended to permit the firm to function without a need either to secure bank loans for working capital or outside investments. The new firm's goal was to have a positive cash flow by the time their capital dropped to $1 million, a goal they effectively met. Next Century remains an employee-owned firm.

Next Century located in Laurel, Maryland, in part because of the very favorable rent it received from a local developer and in part because of its convenient location midway between Baltimore and DC, near its major customers.

Next Century's customer base is similar to the one that McBeth and his partners had previously worked with. Its software has been used by the Navy on the submarine fleet, first responders, and U.S. Special Forces. Next Century also has begun to work with nongovernment customers, providing, for example, software used by a firm that repairs electrical transmission lines and a firm that supplies software services to police departments. It is now expanding its business with U.S. intelligence agencies.

Next Century sees itself as remaining a firm devoted to providing R&D services, but not becoming a producer of the technologies it develops. In his previous employment settings, McBeth had seen 3 efforts to reorient a firm from being a service provider into providing both services and equipment; each effort failed. In his view, a firm must decide between being one or the other. This conclusion is not the same though as saying a firm cannot provide services to both the government and commercial sectors. In his view, and in Next Century's experiences, it is possible to do both. Moreover, Next Century does foresee the possibilities of spinning off 1 or more firms if its R&D leads to viable products requiring volume production.

## EXPERIENCES UNDER SBIR

A central thrust of New Century business strategy has been to be proactive, that is to decide who they want to target, in contrast to what is seen as the reactive—simply respond to customer requests—strategy of the firms with whom its founders were formerly involved. Additionally, in keeping with its founding mission of helping protect the United States and saving lives, the firm decided to focus on the U.S. military, specifically the warfighter, the ultimate end user. Here too, reflecting the immediacy of the tragic outcome of U.S. special military experiences in Mogadishu, Somalia, the firm decided to concentrate on the needs of the U.S. Special Forces. It asked itself: what type of detection and sensor systems were needed by the frontline warfighter?

Its answer initially was a concept, Strategic Insight™ systems, that provide end users with the right information at the right time no matter where they are. The technological manifestation of this concept was a personal threat warning system that could be used by Special Operations forces for reconnaissance and detection. Next Century prepared a concept paper on the technology. It then sought the advice of Colonel van Ardsdale, a former Delta Force officer who had commanded U.S. forces in Somalia. Its next step was to submit an unsolicited white paper to the U.S. Special Operations Command in Tampa, Florida.

Upon receiving the concept paper, Special Operations invited Next Century to present a briefing on its proposed technology. At the time, Special Operations had been pursuing a related R&D program conducted at a DoD research laboratory, but budget cuts had led it to scale back work on the project. The SBIR program however provided Special Operations with an alternative route to fund research on threat warning systems, and it had listed such as need as one of its SBIR topics. More generally, Special Operations is seen as using SBIR as a parallel R&D system, funding multiple Phase I awards that are then winnowed down (down selected) for further development. (Use of this approach appears to be increasing across several services.)

Next Century's concept for a threat warning system consisted of software to alert the war fighter to the presence and location of friendly and hostile forces. SOCOM encouraged Next Century to submit a Phase I proposal, which it received. In an unprecedented fashion, SOCOM issued several other SBIR awards to companies developing other parts of the overall solution. These included multiple awards to develop an antenna vest, RF receivers, and a wrist-worn display device. The award to Next Century was 1 of 4 made by Special Operations on the topic.

Within this competitive environment, Next Century's strategy is to push Phase I R&D into the development of a prototype that can be demonstrated to a sponsor, rather than stopping work at the "paper" design stage. In fact, as part of its Phase I effort, Next Century reached out to the other Phase I contractors and arranged for integration of the overall solution, which it demonstrated at the conclusion of Phase I. This "extra" step, which required that it invest R&D funds above the size of its award, is seen by the firm as having contributed to its success in receiving a follow-on Phase II award of $750,000. This award resulted in the successful completion of a demonstrable technology, leading in turn to Phase III funding (of $660,000) and the transition of support for the technology from SBIR to the DoD's Advanced Concept Technological Development stage. The threat warning system underwent field trials in May, 2005, with Next Century now serving as both the primary software developer and the systems integrator for the system.

The SBIR program has contributed to Next Century's founding objectives, and its technological advances under the program are also seen as a great marketing tool. As a source of the firm's revenue, however, even after its Phase II and Phase III awards, SBIR contracts have seldom exceeded 10 percent of the firm's total annual revenues.

Next Century continues to see SBIR topics as a fruitful source of R&D funding, and as a means of developing new products and markets, both in the federal government and commercial sectors. It routinely reads the SBIR solicitations from several agencies. When it sees a topic of possible interest, it convenes a meeting of its technical staff to discuss the level of interest of individual firm members and the feasibility of forming a research team. This consultative process typically leads to a severe winnowing of interest, reducing in one case an initial list of 12 topics of possible interest of which only one reached the proposal stage.

Next Century's approach both to SBIR solicitations and awards for which it might successfully compete also entails investing time and effort in meeting with the topic author to refine the definition of the agency need. This approach is seen as helping clarify both to themselves and at times also to the topic author the precise character of the problem that the agency is seeking to solve. In one case for example, such a conversation led the firm to reconfigure and broaden its approach to its Phase I award from how to deal with weather uncertainty to the mission-planning question of what equipment to take given weather uncertainty.

## PARTICIPATION IN STATE GOVERNMENT PROGRAMS

Next Century has had no direct involvement with the state of Maryland's economic development programs. Previously, while at Century Computing, McBeth had experience with the state's Department of Business and Economic Development, which provided professional training to the firm's employees.

## RECOMMENDATIONS

Next Century views SBIR as a productive program both for the agencies that sponsor its awards and for the firms that participate in it. In McBeth's view, the view that SBIR takes too long, makes awards in amounts too small to generate technological advances, and overall is not worth it is dead wrong.

The funding gap between Phase I and Phase II awards is a recurrent source of concern for small firms, which has been aggravated by the increasing practice of DoD and other agencies of making multiple Phase I awards. Whereas formerly, in McBeth's experiences with his earlier firms, Phase I awards lasted 6 months with provisions for finishing early, thereby allowing awardees to quickly commence work on Phase II activities, under the evolving practice of multiple awards, now down selection for Phase II awards may not occur until all firms have completed their Phase I work. This arrangement can lengthen the gap for a Phase I awardee who finishes sooner. Relatedly, the firm sees DoD and other agency practices of issuing multiple Phase I awards as having merit, especially on "blue-sky" topics as having some merit, but the absolute number of such awards issued on specific topics can be excessive.

The sizes of Phase I and Phase II awards should be increased to $100,000 and $1 million, respectively, even if these changes resulted in fewer awards.

## SUMMARY

The SBIR program has permitted Next Century to realize its founders' objectives of helping protect the United States and saving lives. SBIR is seen as of considerable value to DoD. In Next Century's experience, SBIR provided U.S. Special Operations with a contractual device to pursue a parallel R&D strategy at a time when programmatic funding had been cut. Next Century's outputs under its Phase I and Phase II SBIR awards have already been field tested and advanced to Phase III funding. Although a relatively small percentage of the firm's total revenues, SBIR serves as a repeated signal to the firm of new R&D needs; its successful performance of SBIR contracts, in turn, serves the firm well as a marketing tool in being able to document its performance.

# Pearson Knowledge Technologies

*Nicholas Vonortas*
*George Washington University*

## COMPANY BACKGROUND

Pearson Knowledge Technologies is an educational software development firm based in Boulder, Colorado. In late 2004, the firm, formerly Knowledge Analysis Technologies, joined with Pearson Education, a multiscope education unit of the Pearson parent company, an international media concern that also owns Penguin Books, the *Financial Times*, as well as major text books in the United States, such as Prentice Hall and Addison Wesley. Knowledge Analysis Technologies was founded in 1998 in the city of Boulder, Colorado by Drs. Thomas Landauer, Darrell Laham, and Peter Foltz. Between 1999 and the 2004 merger with Pearson Education, Knowledge Analysis Technologies earned twelve Phase I, five Phase II, and one Phase 3 SBIR awards. Additionally, four STTR Phase I and three STTR Phase II grants were awarded. SBIR and STTR grants were principally from the Department of Defense (DoD), while the Department of Education and National Science Foundation also contributed. As the parent company, Pearson Education, exceeds the size limit that defines a small business, Pearson Knowledge Technologies, as the company is now known, is no longer eligible for the SBIR program.

Pearson Knowledge Technologies focuses on the application of advanced software tools to products for the education, publishing, government, and defense/intelligence markets. Specific products allow a computer to analyze free-form text, up to gigabytes in length, for content in order to provide feedback regarding essay grading, indexing, information retrieval, and job or mission requirements. The firm is also able to adapt these technologies to software platforms in any language.

Most of the company's products are based on the Intelligent Essay Assessor, the product around which Pearson Knowledge Technologies was built. That innovation will be discussed more extensively later in this report. Underpinning the Essay Assessor is a technology termed Latent Semantic Analysis, a software application that automatically understands text in a manner similar to that of a human.

Product examples centered on the Intelligent Essay Assessor are described here. Summary Street, currently being used by middle and high school students through the Colorado public school system, is designed to automatically grade and review online summaries of readings. Feedback can include an evaluation of content, essay coherence, spelling, grammar, redundancy, and irrelevant sentences. Standard Seeker is a semi-automated, online tool that correlates standards, objectives, and skills lists to publication manuals, including textbooks, lesson plans, and test items. It also allows for more rapid and accurate automatic indexing of these items. SuperManual is an electronic manual designed to allow for fast information searches, especially when compared to hard-copy text versions of the same information. Applications include military training and maintenance manuals and other traditionally large volumes. Career Map matches mission and job requirements with an individual's personal work and training history. Currently, this product is being used to allow military commanders to semi-automatically assign tasks to specific soldiers and units.

## SBIR AND THE FIRM

The SBIR played an important role in the formation of Pearson Knowledge Technologies and its subsequent growth. The founders of the company first came in contact with the SBIR program while serving in the capacity as subcontractors to other SBIR award grantees. Dr. Landauer was one of the original developer of Latent Semantic Analysis, the software code that underpins the Intelligent Essay Analyzer. At first, the innovation was developed and applied in the capacity of subcontractor functions for other SBIR awardees. Then, a Phase I grant was received from the Air Force for an intelligent data-mining agent for rapid and optimal deployment of war-fighting knowledge. Pearson Knowledge Technologies was then formed to handle the increased demand for latent semantic analysis-related projects.

More or less continuous SBIR funding maintained the new company in its early phases and allowed it to survive the late 1990s dotcom bust. Specifically, the Phase II award resulting from development of the first Phase I product kept the company from going under. Automated essay scoring was initially viewed with skepticism by the general public. Pearson Knowledge Technologies (its predecessor to be exact) decided to self-finance the company realizing that market acceptance might be a long process. Venture capital was not considered an option as the founders desired to keep full control of the innovation's application in order to branch out beyond the armed forces into the educational market.[4] More importantly, the technology required more advancement, thus more research while, according to the interviewees, a venture capital investor would have wanted rapid results. The SBIR awards allowed the technology to mature before the company relied on market forces for long-term success.

[4]The founders had educational experience. Several of them, including Dr. Landauer, had started from Bell Labs in New Jersey and then moved to Colorado where Dr. Landauer taught for the University at Boulder. Dr. Laham was his student.

## COMMERCIALIZATION PROCESSES

The most important technological advance for Pearson Knowledge Technologies has been the Intelligent Essay Analyzer (IEA). Based on latent semantic analysis, IEA forms the basis of most of the firm's commercial products. IEA's main function is to examine a piece of text and extract the overall meaning of that text. The technology can be used to automatically grade essays, retrieve information, and to sort information. One application is in interactive, electronic technical manuals being used by the Navy. Another application is used by some public school systems in Colorado. Students submit essays to an online assessor, which provides instantaneous feedback regarding content, grammar, and spelling, all without human intervention. Students are then challenged to refine and resubmit the essays in order to build writing skills.

Pearson Knowledge Technologies maintains all of the information processing in-house. The firm's business model is to sell a service, rather than sell licenses to their product, or sell their product as a physical package. Proprietary computer code is thus not compromised and the company can maintain control over who has access to this technology. In order to make the IEA products attractive and affordable for school systems, the firm has, in the past, partnered with various textbook publishers. When a school system purchases a suite of textbooks, it is also given access to the services on the Pearson Knowledge Technologies' Web site. Teachers are then able to tie in lesson plans from the textbooks directly with online essay review. This is also presumably the major complementarity with Pearson Education, its current parent company, which owns several large publishing houses.

The company does not use patents for intellectual property protection. It considers the enforcement of software patents to be fraught with difficulties. Rather than reveal the necessary information in a patent, Pearson Knowledge Technologies treats its knowledge base as a trade secret and works to stay ahead of the competition through research and development.

The interviewees see writing skills as a fundamental aspect of the learning experience and are concerned that those skill sets are degrading in the public school system. IEA-related products provide access to cost-effective writing aids that do not require additional teacher input. Pearson sees a potential market of around $50 million in the next ten years for these products. One current impediment to a wider use of IEA technology is the lack of access to computer hardware in many schools, as an Internet connection is required to use the essay review product. However, the firm believes that this barrier is being reduced and forecasts wider commercialization in the near future.

## COMPANY IMPRESSIONS OF THE SBIR PROCESS

As indicated earlier, the company founders first became aware of the SBIR process before establishing the company while working as subcontractors on a

DARPA SBIR award. Most of the firm's awards have come from DoD agencies that have also been some of the earlier users of the resulting data mining and educational products.

Overall, Pearson Knowledge Technologies has been quite pleased with the SBIR structure. The interviewees stressed that granting funds to small companies in order to perform risky, cutting-edge research is one of the best ways to spend R&D monies. Venture capitalists and big firms, according to Pearson, do not foster the same level of risk taking. Academic grants, while helpful to academics, do not solve the "real" technology problems facing the country, and are not nearly as effective in generating knowledge as are small private companies.

The firm has experienced some minor differences among the granting agencies. For example, the NSF Fastlane process was very cumbersome at first, though Pearson now appreciates the fact that it tells the applying firm exactly what steps need to be taken and in what order they should be performed. Pearson sees the SBIR programs of the National Institutes of Health, the Department of Education, and the National Science Foundation as more academically oriented, while the Department of Defense has a much more specific, mission-oriented application process in place.

Regarding suggestions for improvement, the interviewees made mention of the amount of money involved. In line with many other award recipients, Phase I funding is no longer considered adequate. It was characteristically observed that the money may be enough for very tiny companies with no overhead, but anyone with a formal office and more than one or two employees will find the amount of $100,000 to be insufficient. An award closer to $150,000 would be more appropriate. Phase II awards should be increased as well, up to around $2 million.

The interviewees indicated their warm support for the program: They indicated that federal agencies should be excited to give out these grants for the good of the country. They suspect, however—and, in fact, had heard from at least one employee of a large federal agency—that if the program was not obligatory, the agency would have abolished it.[5]

Being a small company does impose some logistical hurdles when seeking SBIR grants. Pearson mentioned the large amount of regulations, and changes to those regulations, that accompany the SBIR program. Those regulations can be time-consuming and difficult to understand, and may lead to some firms being intimidated and not applying for the grants. A good solution from the firm's point of view would be for the government to provide some sort of assistance with navigating through the application, award, and follow-up processes.

For Pearson Knowledge Technologies, a major lacuna with the SBIR program is that the individuals within agencies who approve the grants are not the same or are not connected to those individuals who handle procurement (provid-

[5]This corroborates anecdotal information from other government agencies that the authors of this case have come across.

ing contracts to purchase final goods or services). In order to sell a product or process that has been developed through SBIR awards to a government vendor therefore requires getting to know a whole new set of people. This comment implies the understanding that the chances of winning a government contract increase with networking between the vendor and the procurement end of a federal agency. The outcome, according to the interviewees, is increased barriers for new or small businesses to get a first contract.[6]

The interviewees also perceived inefficiencies in the structure of federal agencies from the point of view of a small firm. The structure of the armed forces was, for example, said to be highly fragmented, making it very difficult for a new/small unconnected firm to reach various divisions and alert them to their product offerings. One possible solution may be to have some kind of trade show for everyone wishing to do business with a specific federal agency. However, such a trade show would most likely be dominated by federal prime contractors (large diversified companies) leaving little room, in Pearson's view, for smaller, more innovative companies. The firm recommends that a method be put in place by which those who award the grants can mention specific projects to those who work in the contracting areas of their respective federal agencies.

Finally, the interviewees indicated their perception of a funding gap between federal R&D awards and the market. Such a gap, sometimes known as the Valley of Death or the Darwinian Sea, means that not every commercializable innovation will make it to the marketplace.[7] Pearson would like to see more commercialization assistance from the SBIR program than is currently available.

[6]This comment becomes more interesting when it comes from a company with exposure to the DoD that, among all federal agencies, is viewed as supporting primarily research closest to final products and processes and most effective in making the link from research to application.

[7]Lewis M. Branscomb and Philip E. Auerswald, *Between Invention and Innovation: An Analysis of Funding for Early-Stage Technology Development*, Gaithersburg, MD: National Institute of Standards and Technology, 2002.

# Physical Sciences, Inc.

*Irwin Feller*
*American Association for the Advancement of Science*

## THE COMPANY

Physical Sciences, Inc., (PSI) was established in 1973 by Robert Weiss, Kurt Wray, Michael Finson, George Caledonia, and other colleagues at the Avco-Everett Research Laboratory, a Massachusetts-based, DoD-oriented research firm. The founders left Avco-Everett to start their own firm in part because they sought a smaller firm research and working environment than was possible at Avco-Everett, which at the time of their leaving had grown to about 900 employees.

PSI is located in Massachusetts, and has retained its major laboratories and corporate headquarters there because it is where its founders live. It has additional operations in Bedford, MA; Princeton, NJ; Lanham, MD; and San Ramon, CA.

The firm's growth was modest at first. Its initial contracts were with the Air Force and the Department of Energy. By the early 1980s, it had approximately $10 million in revenues and a staff of 35–50. Sizeable reductions in DoE's R&D budget in the early 1980s caused its contract revenues to fall by approximately one-third. The firm recovered after that period, in part by diversifying the range of its federal customers, such as participation in NSF's Research Applied to National Needs (RANN) program. As the breadth of its technical competencies kept pace with rapidly changing advances in laser and optics technology, and as it become more actively involved in the SBIR program, it was able to expand its range of technological expertise as well as of federal governmental and private sector customers. For FY2005–2006, its revenues are estimated at $35 million, and its employment level at 175. (These figures include sales and employment levels at its wholly owned subsidiaries Q-Peak and Research Support Instruments, Inc, but exclude employment at Confluent Photonics Corp., a commercial spinout.) Approximately 40 percent of estimated FY2005 revenues is derived from SBIR awards. SBIR awards have contributed a diminishing portion of firm revenues, falling from a peak of about 60 percent in the late 1990s to a projected 35 percent in FY2006.

The founding vision for the firm was to do worldclass basic and applied research and prototype product development under contracts from government and private sector sponsors. It has grown primarily by self-financing and employee stock ownership. This strategy rather, than one based on pursuit of external angel or venture capital, has been adopted in order to avoid dilution of owner/employee direction of the firm. Related to this vision of being a premier research organization was the expectation that the firm's staff would publish research findings in the open literature. These foundational principles have continued in force to the

**PHYSICAL SCIENCES, INC.:**
**COMPANY FACTS AT A GLANCE**

Address: 20 New England Business Center
Andover, MA 01810

Phone: 978-689-0003

Year Started: 1973

Ownership: Employee Stock Ownership Trust

Revenue: $35m (estimated FY2005)

Total Number of Employees: 175 (Physical Sciences Inc and its two subsidiaries, Q-Peak, Inc. and Research Support Instruments, Inc.)

Number of Patent disclosures: 100 since 1992; approximately 12 per year since 2000

Number of Patents issued: 39 U.S., 54 foreign patents (24 issued to PSI, or pending, and 5 issued to Q-Peak, a PSI subsidiary, are directly related to SBIR/STTR programs)

SIC: Primary (8731)

Secondary (none)

Technology Focus: Optical sensors, contaminant monitors, aerospace materials, weapons of mass destruction detectors, power sources, signal processing, system modeling, weapons testing

Funding Sources: Federal government R&D contracts and services (80 percent); sales to the private sector, domestic and international (20 percent).

Number of SBIR Awards: Phase I: 435
Phase II: 176

DoD SBIR Awards: Phase I: 337
Phase II: 98

Number of STTR Awards: (included in above)

Publications: Total: Over 1,200 to date, probably 50 percent of which are SBIR/STTR-related (an accurate number has not been determined)

present, accounting for the firm's emphasis on R&D and prototype development rather than manufacturing, which would require additional external capital.

The firm's initial financing came from the assets of its founders, including mortgages on their homes, and funds from family and friends. The firm has drawn little support and seen few benefits in the various technology development programs operated by the state of Massachusetts. Massachusetts is seen as lagging behind other states in the scale and flexibility of programs targeted at fostering the establishment and growth of small, high-technology firms.

In keeping with its pursuit of autonomy and a concentration on contract R&D, PSI has limited its involvement with venture capital firms. Its engagement with them has generally involved the launching of spin-off firms to commercialize products derived from PSI's R&D technological developments, all of which flowed from SBIR funding. To date, this involvement has been infrequent, and the economic record has been mixed. One such firm in the area of medical instrumentation failed when it couldn't raise sufficient (third stage) venture capital funding to complete clinical trials and obtain FDA approval. Another spin-off firm underwritten by venture capital funds did succeed, and was acquired by a strategic partner. A more recent venture in the area of optical communications is currently manufacturing components for the telecommunications and cable television industries.

From its inception, the firm's business strategy has been to specialize in the performance of contract research and development and prototype development. In terms of DoD's categorization of R&D, the firm sees itself as oriented towards 6.2 and 6.3 projects. In the terminology derived from Donald Stoke's classic work, *Pasteur's Quadrant*, it has strategically chosen to position itself in Pasteur's Quadrant, that is as a performer of R&D characterized by the pursuit of both fundamental understanding and utility.

PSI's initial research expertise was based upon and has continued to center on the development and application of laser and optics technology. Reflecting the experience of its founders, the firm initially targeted the aerospace industry as its primary customer. As optical technology has evolved to an ever-wider set of applications, the firm's technological and market bases have widened to encompass applied R&D, production operations, and bundling of "hands-on" service delivery with the application of newly developed products, especially in the areas of instrument development, diagnostics, and monitoring.

Over time, PSI has applied its core research expertise to a widening, more diversified set of technological applications and for an increasingly diversified set of government and private sector clients, both in the United States and internationally. It has strategically positioned itself in an R&D market niche defined by multidisciplinary expertise and research infrastructure in specialized high-tech areas too small to attract major investments by large DoD prime contractors, while at the same time too mission-driven to elicit competition from universities. Its interdisciplinary orientation reflects its origin: Its founding partners represented expertise in aeronautical and mechanical engineering, physical chemistry, and physics. Its current R&D projects encompass optical sensors, laser systems, space hardware, contaminant monitors, aerospace materials, weapons of mass destruction detectors, power sources, signal processing, system modeling, and weapons testing. Reflecting the breadth and interdisciplinary nature of this R&D portfolio, its research staff has R&D expertise in fields extending from astrophysics to zoology.

Given this breadth of activity, the firm operates on a matrix model; it has multiple divisions, arrayed across general topical areas. Its research staff, however, operate across divisions, employing their specific expertise to multiple projects It also employs cross-division review procedures to set priorities, sift prospective responses to DoD solicitations for Phase I proposals, and provide critical technical evaluation of work in process.

The firm's primary customer is the Department of Defense. DoD R&D contracts drawn from across several Services account for an estimated 70 percent of firm revenues. In recounting the impacts of its R&D endeavors, PSI emphasize the application of its technologies and the beneficial impacts these applications have had on the ability of DoD sponsors to achieve mission objectives. Given this emphasis, it sees concepts and related measures of technology transfer, applications, contributions to mission needs, and impact as more important indicators of the quality of the work it performs under Phase I and Phase II SBIR awards than more commonly used measures of commercial impact.

Some of the firm's contracts with DoD involve development of specialized, one-of-a-kind, technologies, that are seen as significantly contributing to the service's mission, but for which the total market, public or private sector as measured by sales volume, is quite small or nonexistent. Other DoD contracts lead to the development of technologies, mainly in the area of instruments, that the firm seeks to market to the private sector. For example, PSI's development under SBIR awards of sensor technology to detect methane gas leaks has been sold to gas companies. In general, sales to the private sector are largely based on technologies developed for DoD under SBIR awards.

PSI's strategy of emphasizing contractual R&D has led it to purposefully limit the degree to which it seeks to move beyond bench and field prototypes, especially in the scale of manufacturing activities. Thus, it engages in limited production for specific instruments for DoD and other federal agencies. When its technological developments lead to commercially viable products, PSI follows a mixed strategy. One strategy is to form new firms, with new, independent management, that operate as partially owned spin-outs. Shaping this business decision is the firm's view that the "cultures" and operational needs of contract R&D and manufacturing firms differ sufficiently that it is more efficient to operate them as separate entities rather than attempt to combine them into one larger firm. Conversely, wholly owned subsidiaries, which are focused on R&D activities, have become eligible for SBIR competition on their own. Since 1990, PSI has formed four new product companies employing technologies developed by their R&D activities. In general, though, PSI sees itself as operating in technologically sophisticated areas whose commercial markets are too small to attract the interest of venture capital firms.

Another strategy is to sell directly to a customer. This strategy is followed for products where the scale of production is low and does not require extensive capital investment in new plant and equipment. In cases where the product has larger market potential but primarily as a subcomponent in a larger complex technological product, PSI has licensed the (patented) technology to an industry leader.

PSI also notes that the gestation period of the technological advances contained in several of its DoD R&D-funded projects is often quite lengthy, with the implication that attempts to measure the commercial import within short periods of time, say 3 years, can be misleading. Its experiences with DoD also have demonstrated the multiple but at times different uses to which a technology has been applied rather than that projected in an initial proposal. PSI also conducts classified research, one effect of which is to limit public disclosure of the technological or economic impact of some of its activities. Its 30-year history also highlights cases in which a different service than the one that supported the initial Phase I award has made beneficial use of the resulting research findings or technology.

## EXPERIENCES WITH SBIR

PSI received its first SBIR award in 1983, 10 years after its founding. The SBIR program however is credited by the firm for contributing significantly to growth and diversification since then. As stated in its corporate material, "The Small Business Innovation (SBIR) program has played a pivotal role in PSI's technical and commercial success, and has been responsible for a family of intelligent instrumentation products based on proprietary electro-optical, and electromechanical technologies."

Since its first award, PSI has been a highly successful competitor for Phase I and Phase II SBIR awards. As of 2005, summed across all federal agencies, it had received an 435 Phase I and 176 Phase II awards, placing it among the top 5 recipients of SBIR awardees. PSI has received SBIR awards from multiple agencies, including several DoD services, NIH, NSF, NASA, DoE, NIST, and EPA.

Acknowledging its distinctive performance in SBIR competitions, PSI, however, rejects the label that it an "SBIR mill." Rather, it sees itself as winning SBIR awards because it provides valuable R&D services to its (repeat) federal agency customers, who have limited discretionary resources other than SBIR.

SBIR awards are seen as an especially flexible mechanism by which DoD can contract for the development and application of advanced instrumentation for monitoring and testing. SBIR Phase I and Phase II awards are seen as an especially effective and appropriate mechanism to further DoD's 6.2 R&D objectives, especially in advancing technologies to the stage of a bench prototype. It notes that the Phase II award frequently culminates at that stage; additional R&D support is seen as needed to move the technology through the stages of field prototypes, engineering prototypes, and eventually to manufacturing prototypes, with each of these stages being necessarily preludes to the commercial introduction of a new product.

Addressing the delays between Phase I and Phase II awards, PSI considers it prudent to avoid spending money on Phase II awards until it receives formal notice that its proposal has been successful. However, at times, it will allocate company funds to bridge the gap between awards in order to keep an R&D project going. Since this support invariably involves closing down or deferring other R&D projects, at times those being conducted by other divisions, decisions about the use of internal funds involve extensive consultation with R&D staff. PSI's current size and matrix organization are seen as enabling it to somewhat buffer these delays, an advantage it now sees itself as having as compared with smaller firms or those with limited SBIR award portfolios. It will shift staff among projects, as needed, to minimize interruptions in the course of work on projects deemed likely to win Phase II competitions. (The Navy is singled out for commendation on its ability to compress the time between Phase I and Phase II awards).

## RECOMMENDATIONS FOR SBIR

PSI believes that SBIR needs to maintain and indeed increase its emphasis on breakthrough technologies. It is concerned that the increasing emphasis being placed on and within SBIR towards commercialization will cause it to "die by incrementalism." Commercialization is conventionally measured by sales, at times with the implication that only those to the private sector "count." In the view of the firm, this narrowing of the objectives of the SBIR program omits or obscures the contributions that the "application" of the outputs of specific SBIR projects can make to the mission requirements of DoD. As stated by PSI representatives, a root cause of this problem is the failure at times to recognize that the legislative intent of SBIR is both to meet the mission-oriented needs of the sponsoring agencies *and* to produce commercial spin-off, wherever possible. Over time, the two objectives have incorrectly been interpreted as one, with the latter one becoming the exclusive criterion for evaluating the aggregate performance of SBIR awardees, and the program itself.

PSI also sees an increase in the dollar size of Phase I and Phase II awards as needed, even at the trade-off of DoD and other federal agencies thus being able to make fewer awards.

Administrative expenses chargeable to the SBIR program also are seen as needed to reduce the unduly lengthy review processes for both types of award and to shorten the time between Phase I and Phase II awards. PSI recognizes that its views on administrative costs differ from those of most other participants in the SBIR program, who see such charges as subtracting from the amount available for awards to firms.

SBIR's award processes are described as fair, but not necessarily competent. Acknowledging that agencies may encounter difficulties in recruiting competent reviewers who do not have conflicts of interest, PSI's reaction to some reviews of its proposals is that some reviewers are flat-out incompetent. Among federal agencies with SBIR programs, DoD is viewed to have the most efficiently run program, with the Navy being deemed the best of all services. One reason is the DoD culture that encourages one-on-one conversations with program managers and cutting edge technology. Similarly, NIH is held to have a highly effective SBIR program. It is seen as truly viewing small firms as contributing to technological innovation, and as understanding that multiple Phase II awards are frequently necessary to convert findings generated from Phase I awards into marketable products and processes. NIH also is commended for the breadth of its outreach activities; these include meetings between the firm and NIH program managers, and opportunities at larger forums for small firms to interact with university researchers and other, larger firms. NIH review procedures though are criticized for the propensity of some reviewers to confuse SBIR proposals with R01 submissions. While the proposals are arguably of equal quality, the scoring system used for SBIRS is different from that used to evaluate an R01.

At the other end of the distribution, NSF's SBIR program is said to be the worst among federal agencies, both because of its protracted review and award processes and the confusing commercial emphasis of its (mostly academic) reviewers. It is also the only agency that restricts companies to four proposals per year. DoE is seen as having very smart personnel, but lacking respect for the R&D capabilities of small businesses. Instead, in its operation of the SBIR program, DoE sees small businesses mainly as vendors of new products, particularly instruments, that are to be used in national laboratories. NASA has mission objectives similar to the DoD, but needs to improve on communicating its goals and requirements through program manager-to-company interactions.

# Procedyne Corporation

*Irwin Feller*
*American Association for the Advancement of Science*

Procedyne was established in 1961 by three professors from Stevens Institute of Technology: Dr. H. Kenneth Staffin, Dr. Ernest J. Henley, and Dr. Robert Staffin. The firm started as a consulting business, drawing upon H. K. Staffin's and Henley's expertise in chemical engineering and R. Staffin's expertise in electrical engineering. Soon after, in about 1964, the firm took on its current configuration of being a technology-based firm that provided services to large manufacturers. The firm's core competencies were in the optimization of manufacturing processes, specifically those found in the automotive and chemical processing industries.

The firm located in New Brunswick, New Jersey, because this was the area in which the founders lived. Also, over time, Procedyne has found it valuable to be located near Rutgers University, which has strong research and educational programs in science and engineering, especially those related to material science and process engineering.

Earnings from consulting activities generated by the three founders were a major source of the firm's initial capital—both for operations and for expansion of facilities. Another source of early capital, circa 1970, was an investment of $300,000, by the new enterprise department of Hercules Incorporated. This investment stemmed in part from H. K. Staffin's prior involvement with Hercules, where he had served as the operations manager of its polyolefin plastics division.

**PROCEDYNE CORPORATION:**
**COMPANY FACTS AT A GLANCE**

Address: 11 Industrial Drive
New Brunswick, NJ 08901
Phone: 732-249-8347

Ownership: Privately held; equity
SIC: 3567
Technology Focus: Fluid-bed furnaces; industrial R&D; engineering systems design

Number of SBIR Awards—Phase I: 7
Number of SBIR Awards—Phase II: 2

The investment also reflected the then short-lived practice of large firms in many established manufacturing sectors to create in-house venture capital units to invest in embryonic new technologies and start-up firms. Hercules's investment gave it a 10 percent equity stake in Procedyne, coupled with an option to invest an additional $350,000 for a larger ownership stake, which it subsequently exercised. Hercules subsequently held their equity for an extended period, but chose the stance of a passive investor. The company's founders subsequently bought back these shares. Similarly, in the early 1980s, the firm attracted investment from an external investor, who invested $1 million for a 30 percent stake. Again, soon after, Procedyne bought back these shares.

Drawing on their expertise in chemical engineering, Procedyne moved into engineering systems, specializing in the application of fluid-bed processing, a technique long in use in petroleum refining but little used in manufacturing industries. Over time, Procedyne's growth has been based on a combination of downstream and upstream integration—first moving downstream to build fluid-bed furnaces that would embody its engineering systems design, and subsequently in an upstream manner, becoming a source of technical expertise in testing and applied, industrially oriented research and development.

This technological expansion began about 1980, when Procedyne developed new techniques for maintaining constant temperature baths in fluid-bed processing that in turn permitted a more reliable and less costly method for applying coatings to various surfaces. The core technology had broad applicability, being useful to the Navy in coating surfaces as well as to the plastics industry in forming molds. At about this time too, Procedyne purchased a nearby failing metal working firm, that extended its capabilities into the production of equipment that embodied its technical designs.

The firm describes its vertical expansion and movement into new applications as being heavily shaped by its customer's orders. Procedyne would accept a contract that involved a stretch in both in technological and manufacturing capabilities, and then would have to deliver. This "bootstrapping" strategy was described as risky, as it required the firm to solve technical problems while meeting contract requirements for operational pieces of industrial equipment. The strategies however served to both enrich Procedyne's in-house technical expertise, while broadening its end-user markets. The firm credits the breadth of scientific expertise of its founding managers along with their combination of a research and an industrial applications orientation for successfully carrying off this strategy.

Procedyne experienced cyclical fluctuations in business that rocked the company. In the 1980s, it had begun a major expansion tied to large orders from the Abex Corporation, a major manufacturer of cast steel railroad car wheels. To meet these orders, Procedyne built large furnaces, requiring heavy capital investments. By the late 1980s, the railroad industry was contracting sharply, as did orders from Abex, which fell from $4–5 million annually to about $1 million.

This fall-off in sales and Abex's subsequent bankruptcy led to major reductions in Procedyne's revenues and employment. In turn, these setbacks have led the firm to seek to diversify in its activities and markets.

The SBIR program has been a small portion of Procedyne's revenues, but is seen as having contributing significantly to the expansion of the firm's technical capabilities and product lines. Procedyne's participation with the SBIR program began in the mid-1980s. The firm had performed prior work with DoD and was familiar with its contracting system. Once it became aware of the SBIR program from general sources, it saw it as a useful means of extending the applications of its core technology. The potential for commercialization served as the lens through which the firm evaluated its technical capabilities against DoD and other agency solicitations.

The firm received a few Phase I awards from the Department of Defense and the Department of Energy related to extending its fluid-bed technology to the application of coatings on metal parts to improve their wear-resistance. They also were successful in part in winning Phase II awards on some of these Phase I projects. The Phase II projects related to fluid-bed technology produced good technical results, and led to a patent, but proved not to be economical for industrial use. From the firm's perspective, however, the Phase II projects produced a significant advance in technology as well as adding to its knowledge about chemical reactions in fluid beds. It was then able to use this knowledge to apply the technology in other industrial settings.

Procedyne's SBIR projects had two other distinctive impacts on the firm. First, the firm's renewed activities in applied R&D led it to move systematically into becoming a supplier of R&D to other industries. It established an R&D division that provided technical consulting services related to chemical processing and heat treating to other industries interested in determining whether or not fluid-bed processing technology could be applied to their components. The output of this division was typically either a technical report, which was the final product in cases where fluid-bed processing proved not to be applicable, or a technical report that at times led to follow-on business in cases where the technology was found to be applicable. Industrial R&D has been a staple part of Procedyne's business for approximately 15 years, serving about 30 firms per year and generating about $1 million in revenues annually. Its combination of a general-purpose R&D division with core engineering design and manufacturing capabilities is seen by Procedyne as one of its hallmark competitive advantages.

Second, SBIR played a major role in catalyzing the firm's interaction with a faculty researcher at Rutgers University that led to the development of a new firm built about technical advances in nanoscale microstructure materials. In 1986, following attendance at a seminar at Rutgers on nanoscale microstructure research on tungsten carbide, Procedyne saw the potential of applying its fluid-bed technology to substantially scaling up production of this material, which at the time was being produced at the milligram level for laboratory purposes. At the time, Rutgers already had a patent pending on the new material. Procedyne and Rutgers jointly submitted a Phase I proposal to the Navy to make nano-phased tungsten carbide and cermets for high wear parts performances. Both the Phase I and a subsequent Phase II proposal were technically successful.

Following these technical successes, in about 1990. Rutgers and Procedyne shared equally in the formation of a new firm, Nanodyne, which designed and fabricated facilities for a demonstration plant in New Jersey, but which was incorporated in Delaware. Formation of the company had ripple effects on university-state government relationships. Rutgers, as a state institution, was bound by state legislation, which at the time limited the equity share that a faculty member could receive in a start-up firm based on his/her research. The negotiations leading to the formation of Nanodyne bumped up against this ceiling, leading in turn to changes in state legislation that removed this ceiling.

Nanodyne received several infusions of venture capital and investments from established firms, such as Ampersand and De Beers. After several years of operation in New Jersey, in about 2000, the firm was bought by Union-Manjere, a Belgium based firm, and moved to North Carolina.

SBIR is seen as an excellent program. In the firm's view, Nanodyne would not have been established if not for the SBIR program. The size of SBIR awards is seen as reasonable; $75,000 is seen as a realistic amount to assess the feasibility of a technology. Procedyne experienced no major problems with maintaining research activity caused by delays in decision making and funding between Phase I and Phase II awards, in part because it was able to secure two $10,000 bridging awards from the New Jersey Commission for Science and Technology.

# RLW, Inc.

*Irwin Feller*
*American Association for the Advancement of Science*

RLW was founded in 2000 by Lewis Watt and William Nickerson, former researchers at The Pennsylvania State University's Applied Research Laboratory (ARL). (ARL is a university-based research facility, specializing in underwater acoustics; historically, its primary sponsor has been the U.S. Navy.) Both Watt and Nickerson had prior experience with DoD, when they formed RLW. Watt is a retired Marine Corps officer; Nickerson had worked as a civilian program manager in Navy depot management. In 1985–1986, while in this role, he had written topic statements for SBIR solicitation.

Watt and Nickerson were recruited from ARL by Oceana Sensor Technologies, a Virginia-based firm that sought to combine its expertise in hardware with their expertise in software development and familiarity with management of ship maintenance and related logistics. At first, in 1999, the relationship was based on consulting agreements between Oceana Sensor Technologies and Nickerson.

**RLW, INC.: COMPANY FACTS AT A GLANCE**

Address: 2029 Cato Avenue, State College, PA 16803
Phone: 814-867-5122

Year Started: 2000

Ownership: Private

Annual Sales (FY2004): $3.8 million; FY2005 (estimated) $7.6 million

Number of Employees: 39 full-time; 11 part-time

3-Year Sales Growth Rate: Compound annual doubling

SIC Code: 7371

Technology Focus: Software development for monitoring smart machines

Number of SBIR Awards—Phase I: 12
Number of SBIR Awards—Phase II: 9

Number of Patents: 1 pending

When first formed, RLW's primary revenues were subcontracts from this firm, which also took equity in RLW. The relationship between the 2 firms however soon proved unsatisfactory, and ended after about 18 months.

During this period, Watt and Nickerson wrote three SBIR proposals, two as subcontractors for the Virginia firm, one to the Navy through RLW. Only the latter proposal was funded. Coming at the time of the separation from the Virginia firm, the SBIR award is seen as having been critical to RLW's establishment. Without it, the firm likely would not have survived for long.

During its first 3 years of operation. RLW also received a $240,000 award from Pennsylvania's Ben Franklin Partnership Program to support development and marketing of one of its first products. Under terms of the BFP program, RLW must pay back twice this amount as the project generates revenues. Repayment is made in the form of a percentage (3 percent) of gross revenues on the product line. An appealing feature of the BFP award is that unlike bank loans, only the firm is liable for the amount of the loan, not the firm's principals.

Although the BFP program is credited with assisting RLW's survive its early years, Pennsylvania's programs to support high-technology small and medium-sized firms are seen as lagging behind those of other states. Virginia was singled out as a state that actively facilitates the efforts of such firms to compete for SBIR awards. Relatedly, Pennsylvania lacks tax credit or loan programs for small firms whose assets are primarily "intangibles," as is common for firms based on software development. The general lack of such capital and the onerous terms under which state funds are made available made life "miserable" for RLW in its initial efforts to secure working capital. Because the state wouldn't take a secondary position on bank loans, RLW founders were required to pledge their personal assets to borrow working capital. The firm continues to locate most of its operations in Pennsylvania because that is where the founders prefer to live; overall though, it reports "no business reason to stay in Pennsylvania."

RLW expresses wariness about involvement with the venture capital community. To receive venture capital, a fledging firm must commit itself to an expansion plan that is likely to be infeasible. The result is that a firm is forced to turn again to venture capital for additional rounds of funding. This process however dilutes the founders' ownership of the company. The SBIR program fills an early-stage funding gap when a technology is too risky for industry to invest in and when the perceived market is too small or too uncertain. It also helps protect the founder's equity. The SBIR program is the best way to fund start-up companies.

RLW's initial core technology was software that provided for the internal monitoring of the "health" of machines, including vibration, pressure, temperature, electrical current, and voltage. About this core, the firm is engaged in expanding its product lines and markets in several directions. In terms of product lines, given the tight integration of hardware and software, it has sought to reduce it reliance on outside suppliers of components, which now represent about one-third the value added of a product, and to expand its manufacturing capabilities. Its goals are to reduce the size and cost of its product, while simultaneously increasing its reliability. It also is seeking to expand its market, both within the Navy and to the commercial sector, adding new capabilities for Web-based transmission of data.

RLW has transitioned from being a firm that sold its man hours, that is its consulting expertise, to one that is selling products. An additional transition is its shift from being a supplier of software to one that increasingly provides an integrated software-hardware product. It is selling its product to equipment manufacturers of pumps and motors. One potentially attractive market is the power generation industry.

RLW has received 12 Phase I awards and 9 Phase II awards. SBIR Phase I and Phase II awards presently represent about 49 percent of the firm's revenues. Its plans are to outgrow its dependence on SBIR awards by increasing sales both to the DoD and commercial sectors. The SBIR program however has distinctive features that make it attractive to small firms, and RLW plans to continue to submit proposals. In particular, SBIR awards provide good protection for intellectual property, that enhances the position of small firms in negotiations with larger firms, especially DoD prime contractors. Also, SBIR awards in effect serve as a form of early venture capital, akin to the allowances for independent research and development available to larger DoD contractors. This source of funds typically is not available to small firms. Also of importance to the firm, SBIR funding permits it to venture into new R&D areas and expand market potential without having to yield ownership to external investors.

The firm has one patent pending, but more generally, given its emphasis on software development, has relied primarily on copyrights to protect its intellectual property. Research publications, presentations at professional meetings and state government workshops on the SBIR program, and frequent briefings to DoD officials, have followed from its SBIR-funded research. Overall though, reflecting its departure from a university setting, the firm sees itself as being finished by publications, having been there and done that.

Drawing on Watt's and Nickerson's prior experiences with DoD contracts, RLW has experienced few audit or management problems working under federal government or DoD reporting requirements. Its accounting system has been designed to meet DCAA requirements.

Given that it's a new firm with limited capital, delays between Phase I and Phase II awards have "nearly sunk" the company. Cited was one case with an 8 month delay that required the firm to incur considerable additional debt in order to keep operating. No explanation was provided to RLW for the delay, although it is attributed to the slow pace at times with which DoD contract officers process SBIR awards (relative to larger program awards). DoD's processing of SBIR awards though is seen as having improved over time.

The current sizes for Phase I and Phase II awards are seen as reasonable. The Phase I level approximately supports 1 person for 6 months. The firm's experience though is that each of its Phase I awards has incurred costs equivalent to 1.5 times the award.

# Savi Technology

*Irwin Feller*
*American Association for the Advancement of Science*

## FIRM HISTORY

Savi Technology was established in 1989 by Robert Reis, a Stanford University engineering graduate and serial entrepreneur, about the core technological concept of installing radio frequency emitters, or tags, in products as a means of identifying their location. Based on an experience in which Reis had difficulty locating his young son in a store, the original market concept was to install the technology in children's shoes as a way for parents to monitor their whereabouts. This concept quickly proved technically and commercially unworkable.

The value of integrating radio frequency identification devices (RIFD) with the Internet for purposes of supply chain management soon became evident, and it is along these lines that Savi has developed, becoming an international leader. From the late 1980s through the early 1990s, the firm experienced modest growth. Its growth has increased rapidly since then, especially after adoption of its tech-

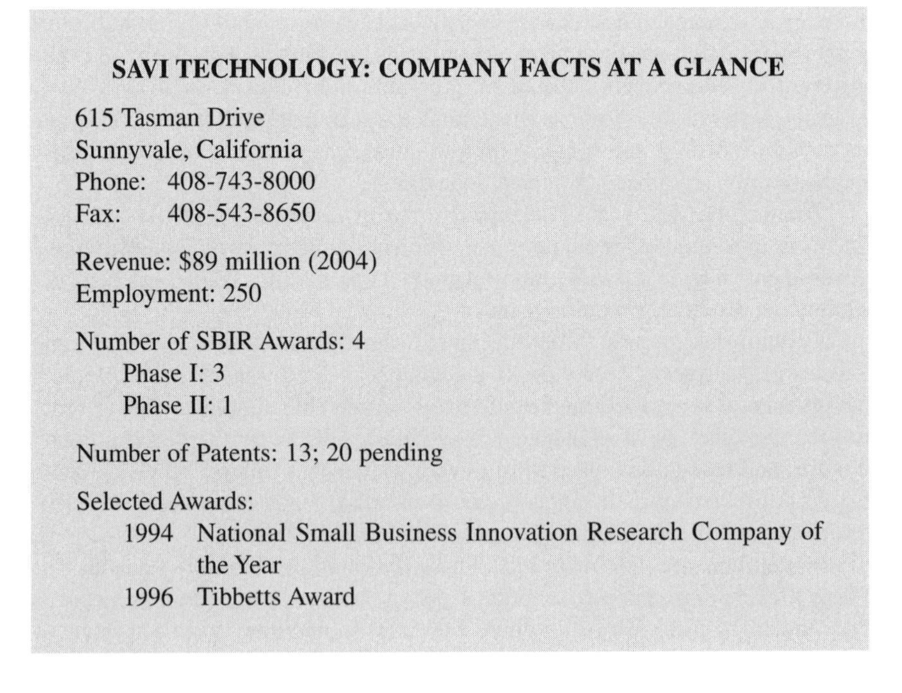

**SAVI TECHNOLOGY: COMPANY FACTS AT A GLANCE**

615 Tasman Drive
Sunnyvale, California
Phone: 408-743-8000
Fax: 408-543-8650

Revenue: $89 million (2004)
Employment: 250

Number of SBIR Awards: 4
Phase I: 3
Phase II: 1

Number of Patents: 13; 20 pending

Selected Awards:
1994 National Small Business Innovation Research Company of the Year
1996 Tibbetts Award

nology by the Army, where it is credited with greatly improving the efficiency and effectiveness of DoD's logistic management capabilities.

By the mid-1990s, Savi made a major business decision to systematically focus on the defense market. This strategy has led to a sequential extension of its technology to a widening set of DoD requirements, international defense customers, homeland security, and asset security management.

Savi is now the major supplier of RFID technology to the Department of Defense, and one of the key technologies provides to its Global Total Asset Visibility Network. It has developed a strong international presence, being a major supplier of RIFD and related technologies to the United Kingdom's Ministry of Defense, NATO, NATO member nations, and Australia. It also is increasingly engaged in the development of globally interoperable logistics monitoring systems with major international ports. It has also built a steadily increasingly commercial business, especially among multinational firms.

If its technological path has been relatively straightforward, consisting of a continuing stream of improvements and widened applications of its radio frequent identification technology, Savi's history as a firm has been circuitous. As a relatively small firm, with about 40 employees and $10 million in sales but with limited markets, Savi was sold in 1990 to Texas Instruments for $40 million, which at that time was following a diversification strategy. Soon after however, following the death of TI's chief executive office, Savi's place within TI became unclear. In 1997, TI sold Savi to Raytheon, which was in the process of acquiring several firms as part of a diversification strategy. Raytheon's business strategy soon gave way to one of concentrating on core businesses, with Savi at the margins of Raytheon's operations. In 1999, Savi's management entered into a buy-out agreement, purchasing the firm from Raytheon for $10 million.

## CAPITAL FORMATION

Begun as a start-up operation, augmented with an infusion of angel capital in 1992, Savi remains a privately held firm, albeit with several rounds of venture capital since its management buy-out. Being sold twice and then regaining its autonomy via a management buy-out, although at one level detracting from Savi's ability to articulate and operate a focused technology development and business strategy, has over time proven beneficial to the firm. TI and Raytheon are estimated to have invested $50 million in Savi's R&D. As a consequence, when Savi regained its independence, it was on a stronger technological and production basis than when it was first sold. As described by Vikram Verman, Savi's CEO, the firm's history thus resembles the story of Jonah, albeit with a positive outcome: as a fledging firm, it was swallowed up by a whale—actually 2 of them, nurtured inside their bellies, and then disgorged as a stronger unit, better able to fend for itself.

Savi has had 4 rounds of venture capital funding since 1999, raising a total of $150 million. Among these investors are Accel, UPS Strategic Investment Fund, Mohr Davidow, Temasek, an investment holding company for the Port of Singapore and Neptune Orient Lines, Hutchison Whampoa and Mitusi, among others.

## EXPERIENCES UNDER SBIR

The SBIR program provided two key inputs into Savi's long-term growth. First, a combination of DARPA and Navy SBIR awards to the firm in 1989 and 1990 provided it with the seed capital that enabled it to refine the initial technological concept of radio frequency identification tags, such that it performed as needed when its initial market opportunity surfaced during the First Gulf War. Second, it provided the funding that led to the employment of Vikram Verman, then a Stanford University Ph.D. student in engineering. Verma was born in India, moving to the United States at age 18 to study electrical engineering. Upon completing his undergraduate degree at Florida Institute of Technology, he moved first to the University of Michigan and then to Stanford University for graduate work. He joined Savi in 1990, advancing steadily from staff engineer, to vice-president for engineering, to his current position as CEO.

In total, Savi received 38 SBIR awards, with the last of these awards being received in 1992. Savi credits its subsequent technological and business success to these early SBIR awards, even though it did not commercialize the technology identified in the initial SBIR projects. Rather, in its formative period, essentially between 1989 and 1991, SBIR awards were critical in helping Savi build the organizational infrastructure, engineering teams, and knowledge base that undergirded its subsequent growth. Furthermore, although successful in competing for SBIR R&D awards, Savi never saw itself as a government contract R&D firm. Rather, its founding and continuing objective has been to be a commercially oriented, product-based firm. R&D is a means to obtain a competitive advantage for its products and related services.

Savi sees its greatest asset to be the know-how and organizational infrastructure gained from multiple R&D projects. These internal, often intangible assets, permit it to deal from a position of strength when it negotiates with external investors, such as venture capital funds. Verma estimates that Savi received a total of $3 million in SBIR awards for the development of RFID. For its part, the firm invested about $150 million in development, and that development took approximately 10 years.

## INTELLECTUAL PROPERTY

Intellectual property protection in the form of patents is seen as of modest importance to Savi. It files patents primarily to defend it technology and market position, not as a means though of securing license revenues or of entering into cross-licensing agreements. As viewed by the firm, its know-how and organizational capacity to assemble high-performing engineering teams are the main sources of its continuing technological innovativeness.

## INVOLVEMENT IN STATE GOVERNMENT PROGRAMS

Savi has not participated in any state of California high-tech or economic development programs.

## RECOMMENDATIONS

Savi's participation in the SBIR program ended by the mid-1990s. Accordingly, its assessments of the SBIR program and recommendations for its improvement relate more to the general place of the program within the U.S. national innovation system than to its specific programmatic details. In its view, the SBIR program accords with the federal government's role of financing R&D before a technology is commercially viable, and before a fledging firm can attract external capital. SBIR should be viewed and used as a source of seed capital. It is a means to an end, the end being the development and production of a usable and competitively marketable product. Firms that make a habit of living off the SBIR program are misusing the program. To guard against this practice, selection criteria should include requirements that firms detail how they plan to commercialize a product. Also, the importance and contributions of the SBIR program need to be more fully and effectively communicated.

## SUMMARY

The SBIR program served both as a source of seed capital for Savi's early R&D on RFID devices that have been the base of its employment and revenue growth and as the foundation for the knowledge and organizational infrastructure that have made it an internationally prominent firm in supply-chain management technology. Although it no longer participates in the program, Savi views SBIR as an essential element—a national treasure in America's long-term capacity to compete internationally on the basis of technological innovation.

# Scientific Research Corporation

*Zoltan Acs*
*University of Baltimore*

## FIRM OVERVIEW

Scientific Research Corporation (SRC) is a privately held company specializing in defense applications. With fifteen locations throughout the United States, SRC offers products and services in Communications, Signal Intelligence, Radar Systems and Test & Evaluation. The company's Atlanta operations, which account for approximately 15 percent of company revenue, participates in the SBIR Program and focuses on research, product development, technology insertion and engineering services to enable Assured Communications for Deployed Warfighters.

SRC is a winner of the Small Business Administration 2000 Tibbetts Award.

The company was founded in 1988 by Dr. Charles Watt, whose prior responsibilities includes service at Clemson University, the Georgia Institute of Technology, the Office of the Secretary of Defense (OSD), and the U.S. Navy, as well as with the Bell system and Bendix Cororation. While working as lab director at Georgia Tech Research Institute (GTRI), Watt focused radar testing and evaluation programs. He founded SRC to pursue a program that captured his interests but GTRI would not pursue. In a period of 5–10 years he bought out the initial investors, and SRC became privately owned by the Watt family.

In the 1990s, the company landed a contract on a satellite communication program with the SPAWAR Systems Charleston. That program attracted experts in the communications area like Alan Harris and David Chapman. The primary customer, for a long time, was SPAWAR Charleston. As the company expanded, it started employing people on-site at the government location.

SRC now now employs approximately 850 management and staff. SRC passed SBIR bid eligibility in late 2001.

As SRC transitioned engineering work to offices on-site at the customer location, it turned to the SBIR program to fund engineers at the Atlanta office. The company's entry into SBIR is intended to bring work to a part of the labor force and attract some previously unavailable R&D funding.

## ROLE OF SBIR

### Unique Features

SBIR provides basic research funding to small businesses, without requiring them to compete with large organizations. SBIR is the only source of R&D

available to small business today, given the consolidated acquisition practices seen across the government.

The program also provides intellectual property protection in the form of SBIR Data Rights.

### Developing Innovative Products

SRC has utilized the SBIR awards to develop technology, products, and engineering product support in three areas:

1. "Assured C4ISR Wireless Networks for the Deployed Warfighter,"
2. "Rapid Replacement technology insertion for replacing failing, obsolete mission critical electronic systems deployed in the field and the associated support equipment," and
3. "Unattended Ground Sensor (UGS) Systems."

The "Assured C4ISR Wireless Networks for the Deployed Warfighter" area incorporated and integrated several specific technologies as acquired under seven SBIR topics to produce an integrated product family of mobile wireless network software and hardware modules covering multiple layers of the Open Standards Interconnect model. This approach allows SRC to deliver specific network management, wireless network operations, covert communications and intelligence collection capabilities in combinations and packages specifically tailored to meet customer requirements while maintaining interoperability and economies of scale.

SRC's "Rapid Replacement" technology provides form, fit, and function replacements for failing, obsolete mission critical electronic systems deployed in the field (N01-188 & N97-060) and the associated support equipment (N01-013 & AF97-234) for life cycle sustainment. SRC has developed this technology over the last 15 years to provide reusable hardware and software modules for integrating form, fit, and function replacement solutions that meet U.S. Military requirements. "Rapid Replacement" technology developed from previous efforts was integrated with commercial-off-the-shelf and newly developed hardware and software modules to provide solutions for SBIR Topics N01-188, N01-013, AF97-234, and N97-060. New hardware and software modules from these SBIR topics have been integrated with SRC's "Rapid Replacement" technology to provide form, fit, and function technology replacement solutions for the United States Air Force, Army, and Navy.

The Unattended Ground Sensor (UGS) Systems Technologies incorporated and integrated several specific technologies as acquired under six SBIR topics including Intersensor Information Assurance, Netted Full Spectrum Sensors, Energy Efficient Routing, Compressed Voice Over Variable Bit Rate Links, Common Object Request Broker Architecture (CORBA) Security Services, and Adaptable Packet Switched, Battle Command Information.

In total, SRC has received 30 Phase I awards and 20 Phase II awards between 1995 and 2002. It has also received over $3 million in funding from non-SBIR DoD programs, over $1 million from private industry, and over $1million has been invested by SRC.

### A Marketing Tool

SBIR has also been used as a marketing tool, both during Phase I research for proving the feasibility and during Phase II prototyping. SBIR has help attract over $4 million in investments—much of this by customers outside of the SBIR programs who needed specific capabilities engineered into SRC's product, and were willing to pay to have those changes incorporated. The company also invested in itself.

### Employment Effects

The SBIR program helped SRC establish a product development group in Atlanta when the market for engineering services was declining in the late 1990s. This office would be drastically different if it had not participated in the SBIR program. If there was no SBIR, the company would have probably sustained a 7 percent employment increase over a period of about 10 years. SBIR also changed the company's focus more to product development and led to the development of several products especially in the wireless networking area.

## COMMERCIALIZATION

SRC operates in a multibillion dollar market, with major competitors that includes Boeing, Lockheed Martin, BAE Systems, Northrup Grumman, Raytheon, SAIC, Telcordia, BBN, AT&T, General Dynamics, Cisco. The firm believes that it maintains that its technical superiority and depth of technical capability, responsiveness, flexibility, and cost provide its competitive edge.

### Sales

SRC estimates over $25 million in sales to date for products related to TurboLink, Covert Communications and "Rapid Replacement" technologies. SRC was recently awarded a five year $25 million Phase III IDIQ contract (a type of contract that provides for an indefinite quantity of supplies or services during a fixed period of time) for "Assured C4ISR Wireless Networks for the Deployed Warfighter" from AFRL/IF (Hanscom Air Force Base). SRI is also expecting a five-year $25 million Phase III IDIQ contract from NAVAIR for replacing the T-45 CRT Multi-Function Display with the Touch Thru Metal (TTM) Multi-Function Color Display. In addition, SRC is working with the Army to establish a five-year $25 million Phase III IDIQ contract for "Unattended Ground Sensor (UGS) Systems."

The largest product that came out was a turbo link product that served as a gateway between ATM and other networks including IP. An Asynchronous Transfer Mode (ATM) is a cell relay, packet switching network and data link layer protocol which encodes data traffic into small fixes sized cells. This differs from other technologies based on packet-switched networks (such as the Internet Protocol [IP] or Ethernet) in which variable sized packets are used. SRC ended up selling turbo link to several programs within the UK that supported UK army and navy applications. Drawing from this expericne, SRC developed gateway technologies that have been successful within the United States. It also developed through SBIR several ad hoc routing protocols that are predominantly IP based.

### Licensing Agreements

SRC holds 37 licensing agreements for evaluating and testing MobileRoute® and Wireles Adhoc Routing Protocols, 2 licesning agreements for evaluating and testing Wavelet Packet Modulation (WPM), and 1 licensing agreement each for TurboLink®, NetTempo, Covert Wireless Network Sensor, and Computer Network Attack.

### Procurement

SRC estimates that about 50 percent of products developed with SBIR were sold back to the agencies funding it. The other 50 percent is nonrecurring engineering that adapted the product to a specific application.

Is procurement commercialization? Yes, SRC belives that selling the product back to the agency is commercialization. However, they belive that a lot of good technology with good applications and good uses are not taken advantage of by the agencies. "The SBIR program could improve its procurement process by enabling procurement from small businesses without going through some of the primes. The Navy has been pretty good about that."

## KNOWLEDGE EFFECTS

The company filed for three patents. 3 pending patents for TurboLink® (U.S. & EU), MobileRoute®, and Wavelet Packet Modulation (WPM).

It holds the following trademarks: TurboLink® (U.S. & Canada), MobileRoute®, BandShare®, IPOverdrive®, and EasyConfig®.

The company has published a number of technical papers:

- SDR Forum 2003 Technical Conference, November 2003 "Hardware-in-the-Loop Simulation Techniques for Validating SDR Software"

- MILCOM 2003, October 2003 "Energy-Efficient Networking Techniques for Wireless Sensor Networks"
- MILCOM 2003, October 2003 "Exploiting the Synergies of Circular Simplex Turbo Block Coding and Wavelet Packet Modulation"
- SPIE Wavelets X Conference, August 2003 "Enabling Time-Frequency Agility: Wavelet Packet Modulation in Practice"
- SPIE 2003 "Solving Bezel Reliability and CRT Obsolescence"
- Proceedings of the 36th Asilomar Conference on Signals, Systems & Computers, Pacific Grove, CA, November 2002 "Receiver Timing Recovery for Adaptive Wavelet Packet Modulated Signals"
- SDR Forum's 2002 Software Defined Radio Technical Conference, November 2002 "A Portable Software Implementation of a Hybrid MANET Routing Protocol"
- MILCOM 2002, October 2002 "Intersensor Information Assurance for DoD Tactical Networks"
- MILCOM 2002 "Experimental Comparison of Hybrid and Proactive MANET Routing Protocols"
- SPIE Aerosense 2002 "Wavelet Packet Modulation: solving the synchronization problem"
- Invited Presentation at Second Annual Communications 21 Conference, Washington, DC, March 2001 "IP Quality of Service in Mobile Ad Hoc Networks"
- MILCOM 2000 "Alternate Path Routing in Mobile Ad Hoc Networks"
- MILCOM 2000 "Adapting The DOCSIS Protocols for Military Point-to-Point Wireless Links"
- IEEE Computer Magazine, September 1998, "Algorithm-Agile Encryption in ATM Networks"
- IEEE MASCOTS '98, July 1998 "IPB: An Internet Protocol Benchmark Using Simulated Traffic"

## IMPROVING SBIR

### Surviving Funding Gaps

SRC was large enough and well established in other business opportunities prior to the SBIRs so that personnel could be assigned to other efforts during the delay between Phase I and Phase II. A key component for this was a strategy of selecting new Phase I contracts that leveraged prior technologically related Phase I/II efforts so that a common work force could be shared as needed.

Within the R&D group the firm made sure that both the first and second solicitations in a given year funded related projects. In this way, the people that might be displaced by the transition gap for the first project could work on the second project. SRC acknowledges that "along the way we did lose a couple of principle investigators because of gap in funding, and when we got the Phase II, we had to provide for acceptable substitution."

## Award Size and Duration

SRC believes that the size and duration of the awards should be commensurate with the challenge to be addressed. "Some efforts can be easily done for the $100,000, and some can't event scratch the surface for that amount. There should be some variation maybe depending on the nature of the problem that needs to be addressed." A similar problem is seen with Phase II awards. "There is more money there, but if you're developing a product, especially if you want to have a true product at the end vs. a prototype/proof concept; there's a pretty big difference there."

"You should be able to propose what it will take to do the work versus what you can do for the budget. Phase I efforts of 6 months and $100,000 are barely adequate for Principal Investigator to produce a paper feasibility study. However, customers are increasingly demanding to see some physical demonstration of feasibility in Phase I. Additional funding of $50–100,000 is needed in Phase I. The current Phase II funding profile of $750,000 spread over 24 months is not adequate for development of product prototypes suitable for transition directly to military programs."

SRC has found that 3–4 sequential Phase IIs are required to get a prototype to sufficient maturity for transition. Hence, their practice of bidding new SBIR's that leverage prior work. Consequently, the first or second customer for the SBIR technology frequently doesn't see mature prototypes until after his program is over; the third or fourth customers are the beneficiaries. This is frustrating to customers.

Spreading the current $750,000 over two budget years also results in SBIR technology being made outdated by commercial developments. Funding levels of at least $750,000 per year for 2 years or $1.5 million in a single allocation are needed to deliver reasonable prototypes. Much additional work is needed beyond that to produce ready for market products.

"The Phase I & II funding levels should be increased, just for reasons of cost growth over the last 15 years if nothing else."

## Award Selection

Phase I award selections are seen as being fair and are timely. SRC considers the most critical aspect in the bid selection process to be the direct contact discussion with the topic point of contact. Failure to have this dialogue, preferrably face-to-face, usually results in a no-bid decision because SRC has learned through many failed bids that the topic write up does not provide all of the necessary information to prepare a successful bid. Phase II awards vary by agency, internal agency priorities and availability of funding.

The Navy now has the superior process in SRC's opinion, because it couples Phase I and II with the end customers. For a Phase I topic to be advertised, there must be a program willing to advance the effort to production. NAVAIR in particular has significantly advanced this with the use of Phase III IDIQ contracts and direct program office participation. Funding levels are still too low and slow.

The Air Force comes in second, in SRC's estimation, because they assign an R&D engineer to each topic so that a firm can at least identify the R&D customer and tailor the proposal and program to that customer. Historically, the Air Force did not tightly couple SBIRs to end customers. This is changing now with the inception of IDIQ Phase III contracts to transition technology to program office products.

The Army's review process is multitiered and a small businesss has no insight into the process outside the Army's technical point of contact, who frequently had no involvement in the editing of the topic orginally. The Army also does not couple their R&D efforts to program offices making Phase III transitions difficult.

### Bridging the Small Business Definition

SRC recommends that firms retain their small business status if they grow over the course of their involvement in the SBIR program. "One of the objectives of the SBIR Program is to grow businesses. If you were less than 500 employees and eligible to bid for Phase I then grew to over 500 employees while executing Phase I you should be eligible for Phase II. This opinion is common in the community, but it is subject to individual office policies."

### Transition to Commercialization

SRC believes that the disconnect between the SBIR acquisition and an end-user/product customer is a major weakness of the SBIR program. This disconnect can be seen in Phase I where the technical point of contact is not the topic author and has no vested interest in the topic with no time, money, or interest in the topic's success. Even with a successful Phase II, there is often no identified Phase III customer—with the prototype, in some cases, too immature for an end customer. Significant program additions are needed to improve the trasitions to Phase II and Phase III.

SRC believes that the Phase II Enhancement/Plus program needs to be expanded and simplified to encourage product customers to risk a little money to evaluate a prototype and begin the process of transitioning to a Phase III effort.

SRC further notes that the Phase III IDIQ contract should be made a universal final step in all successful Phase II programs. This removes the contracting barrier to transitioning to any number of program customers and allows the company to build a business on those contracts. Program offices should receive encouragement to use SBIR products, either through direct funding of the effort or a reward scheme.

# Specialty Devices

*Irwin Feller*
*American Association for the Advancement of Science*

## FIRM HISTORY

Specialty Devices was founded in 1986 by Paul Higley. Higley's background is in electrical engineering and ocean engineering. Prior to founding the firm, he had worked for Raytheon and Science Applications in Rhode Island, but then left for a position with Atlantic Richfield, Texas. The attraction of the Atlantic Richfield position was that it was to work R&D projects directed at identifying emerging technologies. Higley found this prospect more exciting than location financial considerations. The sharp decline in the price of crude oil in the mid-1980s and associated cutbacks in oil industry R&D led Atlantic Richfield to curtail this R&D program, which led Higley to move out on his own. Life style considerations led him to remain in Texas, even though only approximately 20 percent of the firm's business is with the oil industry.

Specialty Devices started out as a one person firm, operating out of Higley's spare bedroom. He drew on his previous business contacts with oil firms and Department of Defense personnel to launch a business that combined the conduct of marine surveys for the oil industry with engineering design of new marine

**SPECIALTY DEVICES: COMPANY FACTS AT A GLANCE**

1104 Summit Avenue
Suite 104
Plano, TX 75074

Phone: 972-578-7501
Fax: 972-423-8480
<*http://www.SpecialDevices.com*>

Number of Employees: 15
Revenues: $1 million (FY2005)
Number of SBIR Awards: 5
  Phase I: 3
  Phase II: 2

Number of Patents: 1

technologies. Since its founding, the firm has grown to 15, a number of whom were hired from former customers. The added employees have added to the firm's capabilities in physical oceanography. Specialty Products also engaged in joint funding of R&D projects with several universities.

As suggested by its name, Specialty Devices' strategy is to specialize in a small, niche market, offering high-tech approaches to a variety of engineering design assignments. This specialty is the design and construction of prototypes of bathymetric, reservoir management, sediment mapping, sediment coring, and hydrographical survey systems. Its 3 main business areas are display systems, marine survey equipment, and deep ocean observatory design. At the technological core of these services and products is the firm's expertise in optics.

The firm has performed contract work for a diversified set of defense, aerospace, and oil industry clients. Its primary markets have shifted back and forth over time, depending on the annual flow of contracts. At present, its sales are distributed approximately between 60 percent military and 40 percent private sector; in earlier years, this ratio was reversed. The absolute size and relative importance of military sales is likely to increase in the immediate future as it is currently engaged on a contract to upgrade the radar system displays on Navy destroyers, and in line to possibly perform such work on larger Navy ships.

Although it continues to see itself as a contract R&D firm, Specialty Devices also sees larger opportunities in commercializing its technological innovations. However, it sees its ability to move its new designs from R&D prototypes and limited production to larger scale, commercial introduction as being constrained by several factors. As described by the firm, at times it encounters internal competition between its engineers' R&D interests and final product manufacturing efforts. To alleviate this pressure, the firm is outsource manufacturing of some components and separating production from engineering.

Another obstacle to the firm's objective is the cost of scaling up for volume manufacturing. It estimates the medical market for 3-D dimensional displays at $1 billion, but notes that it would take sums in excess of $10 million to scale up to reach a competitive level of output. Given these market realities, its sees its most viable business strategy as one of licensing its technology to the larger firms that dominate either the medical equipment industry, such as GE, or the flat display industry, such as Sharp, Samsung, or NEC.

Specialty Devices also has considered spinning off its display technology into a new firm, using venture capital as a source of funding. Its plans along these lines remain in a formative stage, though. In part, the firm's hesitation about adopting this strategy is that it would require a considerable portion of Higley's time and effort to raise the necessary capital, which, in turn, might detract from the firm's current performance. Not thinking the time as yet right for a major transition, it has chosen not to participate in SBIR sponsored activities, in which consultants have been employed to tutor Phase II awardees on how to write business plans to be included in Phase III proposals.

## CAPITAL FORMATION

Specialty Devices' establishment was financed by Higley's personal funds, an initial stake that required borrowing from funds set aside for his children's education. The firm remains privately held; its growth has been financed by retained earnings.

## INTELLECTUAL PROPERTY

Specialty Devices views intellectual property protection as important for some of its products, and has patented its 3-D display technology. For other of its products however, it sees the combination of a small potential market and the costs of securing a patent too high to make filing a profitable approach to securing intellectual property protection. Instead, it believes that the technical sophistication of its products serves to make reverse engineering by its competitors too expensive for them to undertake, thereby providing the needed degree of proprietary protection.

## INVOLVEMENT IN STATE GOVERNMENT PROGRAMS

Specialty Devices has collaborated with Baylor University on a collaborative R&D project, funded in part by a Texas state government program directed as fostering small business-university cooperation. The state grant was used primarily to fund students.

## EXPERIENCES WITH SBIR

Specialty Devices received its first SBIR award in 1997 on a project submitted jointly with a faculty member at the University of Texas-Dallas. The faculty member knew about the SBIR program and approached Higley requesting his collaboration. The project in turn yielded a patentable technology, with the patent held by the university.

In total, the firm has received 3 Phase I and 2 Phase II SBIR projects. These awards have come from the Air Force, DARPA, and the Navy. Highlighting the range of uses to which its core technology can be put, Specialty Devices' initial Phase I and Phase II awards from the Air Force were for R&D on an early version of a 3-D display system. The Air Force is described as interested in the technology emerging from the awards, but has not provided funds to enable the firm to scale up to volume production. Further developments of the display technology, as noted above, have occurred under SBIR awards from the Navy. Overall, Specialty Devices' emerging display technology is seen as having great market potential, but is described as not yet having reached the point where it is ready for commercialization.

The firm has experienced several, lengthy funding gaps, extending as long as 1 and 2 years, between its Phase I and Phase II awards. In part, the longer gap occurred because it extended its effort on a Phase I award to enhance the quality of its work, and thus missed the deadline for a funding cycle. In general, it has responded to the gaps in funding by cutting back on the level of effort directed towards the projects, but continuing to pursue them, using internal funds. It also notes though that its Navy SBIR contact officer fought hard to accelerate funding of its Phase II award.

## RECOMMENDATIONS

Specialty Devices views the SBIR competitive selection process as fair, the size of the Phase I and Phase II awards as reasonable and, overall, the structure of the SBIR program as working effectively.

A concern though to the firm about the program is the presence of firms that survive on SBIR awards. Such firms are seen as being good at writing proposals, but not at innovation. They are seen as only being interested in Phase I and II awards and then work on the project becomes minimal. This attitude is antithetical to Specialty Devices' perspective on the role that SBIR is intended to play in the U.S. national innovation system. It also harms broader public perceptions and assessments of the SBIR program. SDI has been impressed with the capability and knowledge of the DoD SBIR program technical representatives on its projects, and believes the quality of the DoD people involved is a major contribution to the success of the SBIR program and to minimizing the abuse of the SBIR program.

## SUMMARY

SBIR is a fabulous program, in the firm's view. Without SBIR, Specialty Devices would not be what or where it is today. The SBIR program has served as a critical source of funds for the firm to undertake new R&D. SBIR has contributed to Specialty Devices' increasing role as a provider of goods and services to DoD and to its core technological and product base, which is seen as having considerable private sector potential. In seeking to grow further and to move its technologies into the private sector, however, the firm finds itself at a transition point where it needs to resolve business strategy issues related to transitioning from a contract R&D performer, specializing in niche markets, to one engaged in larger scale production. This effort is underway.

# Starsys Research Corporation

*Irwin Feller*
*American Association for the Advancement of Science*

## FIRM HISTORY

Starsys was formed in 1986 by Scott Tibbitts and Daryl Maus, initially as Maus Technologies. Mr. Tibbitts and Mr. Maus were colleagues who had worked for Rockwell International, in the Boulder, Colorado, area and who believed that by pooling their expertise they could make improvements on existing technologies and direct them to underserved markets. The specific technology related to nonelectronic thermal control systems; the specific intended market was commercial water heaters.

The projected market however proved not to be a profitable one. The technological concepts underlying the firm's initial products were quickly found to have value in spacecraft. In 1988, the firm reorganized as Starsys Research, and reoriented itself towards the development and fabrication of innovative aerospace actuators and mechanisms. Starsys' initial sale of spacecraft hardware was as a subcontract supplier of components to JPL, which had mission responsibility for major NASA space launches. JPL's acceptance of Starsys technology led to follow-on sales for NASA space missions and to the opening of new markets

**STARSYS RESEARCH CORPORATION: COMPANY FACTS AT A GLANCE**

4909 Nautilus Court North
Boulder, CO 80301
Phone: 303-583-1400
Fax: 303-530-2401

Employment: 130
Revenues: $21 million

Number of SBIR Awards:
Phase I: 7
Phase II: 7

Number of Patents:
3 active; 1 pending

with commercial and government customers (e.g., Air Force and MDA). Further contributing to Starsys' growth was the incorporation of its technology in the launching of commercial telecommunications satellites, such as Iridium. In total, its mechanical and electromechanical subsystems technologies have been successfully flown on over 200 spacecraft.

Over time, the firm's technological emphasis shifted to the design and development of components (e.g., deployment mechanisms, latches, locks, antennas) for launch release systems and satellite capture systems. Starsys now produces a range of products, ranging from off-the-shelf catalogue products to unique products developed to meet specific customer requirements. Approximately two-thirds of its current revenues are derived from sales to DoD services, with the balance being distributed among sales to NASA and commercial customers, including international sales. Reflecting its founding orientation, the firm has a strong market, end-user orientation. The company's R&D and product development is geared towards customer needs, and it seeks to develop technologies that are applicable across a broad spectrum of uses, markets, and customers.

In 2000, Starsys acquired the American Technology Consortium, ATC. The expertise and product history brought by ATC increased the product offerings in electromechanical subsystems. Sales growth within this market sector was significant from 2001 through 2004.

Starsys' location in Boulder reflects a combination of life style preferences of its founders and its close proximity to a number of larger aerospace firms, such as Ball Aerospace and Lockheed Martin Space Systems. Locating in Boulder also has facilitated interactions with the University of Colorado, with which the firm has teamed on several proposals and works closely on programs.

## CAPITAL FORMATION

Starsys started as a two-person, garage shop, bootstrapped business. Its initial capital came from the personal savings of its founders. Its growth has been financed largely by retained earnings. Throughout the company's first 17 years, it made no efforts to secure venture capital or angel capital.

## EXPERIENCES WITH SBIR

SBIR awards have represented only about 5 percent of the firm's revenue, but are viewed as having made important contributions to its market position. If work on its current SBIR projects takes hold, it could lead to strong sales potential.

Starsys expresses acute concern about the fairness of the SBIR topic generation and selection process. Rather than being "straight," these processes are seen as dependent on networking and negotiations, in which some firms devote time to building relationships with laboratory personnel to insert selected topics into DoD solicitations. The topic specifications in effect are geared to meet the qualifications of a single firm. Even though they may write "killer" proposals, "outsiders" responding to the solicitation are at a competitive disadvantage.

Starsys has written many proposals, only to receive a "thank you for participation," but essentially "good-bye" response from the sponsoring agency. Its own successes in SBIR competitions indeed are held to be higher when it has had prior contacts with a sponsoring laboratory.

DoD's feedback to firms in cases of unsuccessful proposals also needs improvement. Current feedback is negligible. All the firm learns is that is has not been selected. DoD laboratories are unwilling to provide more specific information about the shortcomings of the technical merits of the proposal. This information would be of considerable value to the firm in submitting proposals under future competitions.

Even as it recognizes the needs for careful technical reviews of Phase II proposals, Starsys considers the gap between Phase I and Phase II proposals, even under the Fast Track system, as unacceptable from a schedule and execution perspective. Under a typical sequence, if a Phase I project ends in December, with concurrent submission of a Phase II proposal, the proposal would not be reviewed until February, and funding not received until May. This funding gap detracts from a firm's ability to maintain work on a project, sustain the technical team, and prevent long delays in the achievement of the technical goals sought from the onset of the project. Typically, it would need to reassign individuals who were working on the SBIR project to other firm projects. At the same time though it experiences direct or indirect pressure from DoD to continue work on the project. To do so however requires the use of internal firm R&D and/or marketing funds. Further detracting from the attractiveness of the SBIR program in such situations is that its pre-contract expenses are not reimbursable. The investment of firm resources on Phase I and subsequent Phase II awards can readily exceed the available profit derived from the awards. Thus, even as it sees topics that fit within its technical competence and hold potential value to the firm, it must still weigh whether or not submitting a proposal is a worthwhile investment.

## INTELLECTUAL PROPERTY

Despite its emphasis on products, intellectual property protection in the form of patents is seen as having only modest importance for the firm. Among the reasons for the limited efforts to secure patents is the small size of the market for much of the firm's innovative technology and the small number of direct competitors. Many of the products are specific to a mission application and thus unique to the customer. Products that can be commercialized are often produced for niche applications and thus have high barriers to entry for competitors. Additionally, the cost and time to secure patents are often prohibitive. More generally, the firm sees its intellectual property as residing inside the heads of its personnel, rather than being embodied in specific mechanisms, susceptible to reverse engineering.

## INVOLVEMENT IN STATE GOVERNMENT PROGRAMS

Starsys reports no direct involvement with state of Colorado technology development or economic development programs.

## RECOMMENDATIONS

To improve the proposal submission and selection process, it would be helpful if the initial briefings provided by DoD to firms would be more specific. The SBIR workshops do not provide the true inner workings of the process and thus are of minimal value. Direct one-on-one time with principal investigators would provide better insight as to the intent of the topic and the ability of a firm to provide a credible and winnable proposal.

The amount of funding for Phase I and Phase II awards are generally reasonable, but would benefit from some upward adjustment. The Phase I award of $100,000 is a reasonable amount to attract a firm's interest, and is adequate as seed funds to produce what was described as PowerPoint engineering, essentially advancing a technological concept to the stage where design work could be completed. A preferable approach to Phase I funding, according to the firm, would be to increase Phase I funding to a level that permitted a firm to develop prototype hardware as well as complete the initial trade studies and conceptual design. The ability to generate a working model of the design concept helps add credibility to the program and reduces risk as the program advances towards the Phase II efforts. Advancing to this stage would make it easier for DoD to determine which Phase I awards should be awarded Phase II contracts. Such a modification would be especially helpful to Starsys, which views itself more as a hardware company focused on generating design engineering solutions that are put into flight applications for spacecraft mechanical and electromechanical subsystems than as a pure design engineering company.

# Systems & Process Engineering Corporation (SPEC)

*Irwin Feller*
*American Association for the Advancement of Science*

## FIRM HISTORY

Systems & Process Engineering Corporation (SPEC) was established by Randy Noster in 1986. Noster had been working at BAE Systems, Austin, Texas on the design of chips for sensor technology, but found that BAE's business strategy and areas of research emphasis were shifting away from his vision of designing chips capable of on-board signal processing for detection purposes. Noster was joined by a few former colleagues to launch what was described as a "garage start-up." The firm started with "sweat equity" by Noster and his colleagues; they wrote 14 proposals during the course of the firm's first 6 months, submitting them to DARPA, Army, and MIT.

Systems & Process Engineering's core technology has centered about the design of multipurpose integrated chips for detection, signal processing, and

**SYSTEMS & PROCESS ENGINEERING CORPORATION: COMPANY FACTS AT A GLANCE**

101 West 6th
Suite 200
Austin, TX 78701-2932

SIC: 8731
Number of Employees—2005: 50
Revenues Estimated between $10 million and $19 million
Number of Patents: 6
Number of SBIR Awards
Phase I: 133
Number of SBIR Awards
Phase II: 38

Awards: 1998 Tibbetts Award
1999 STRICOM "Star Award
2002 Tibbetts Award
2004 Heavy Hitter Award

communications. It defines itself as a product development firm. It employs an array of R&D contracts, including SBIR awards or subcontracts with other SBIR awardees, to develop new products, primarily for the defense and aerospace industries, but also, over time, for the commercial sector, such as the oil and gas industry. Systems & Process Engineering's initial contract was with MIT's Lincoln Laboratory, which had a DARPA contract on detection processing. Subsequent contracts came from firms with which Noster and his colleagues had prior contacts.

Using its core expertise in the design of sensor technology and systems integration as a base, the firm's business strategy remains flexible. It has designed chips for a wide variety of sensor technologies, specializing in on-board integration of sensor technologies, resulting in more compact, single module equipment. If and as a product is seen to have commercial potential, SPEC transitions the technology to a spin-off firm. To date, it has successfully done this twice, launching Extreme Devices in 1998 and Coherent Logix in 2005.

Systems & Process Engineering describes itself as being able to quickly response to changes in technologies and changes in markets. It describes itself as going where opportunities emerge and dropping those for which limited market potential is seen. This strategy has led to some ups and downs in its revenues and employment; employment peaked at 80, but fell after the dot-com bubble. It has now stabilized at approximately 50.

Systems & Process Engineering now has a diverse array of R&D and service contracts from several federal agencies, as well as a major subcontract with an aerospace firm. These contracts range from R&D, systems integration, and systems support, including servicing operations of one of its product in Iraq.

At present, the firm's business is derived approximately as follows: 40 percent as a prime contractor on U.S. government contracts; 30 percent on commercial contracts; 20 percent SBIR awards; 10 percent others.

## CAPITAL FORMATION

As noted, Systems & Process Engineering started as a small operation, drawing on investments by its founders. Although it had 1 aerospace firm and Motorola and IBM plants, Austin in 1986 was characterized as lacking in high-tech industries, and either a banking or venture capital industry interested in supporting small, high-tech start-up firms. When Noster approached local banks and the regional SBA office to borrow money for equipment, the response he received was that these lenders understood real estate, cattle, and oil and gas, but that they didn't do high-tech.

Austin is seen as having come a long way since then. Local sources of venture capital now exist for high-tech firms, with Systems & Process Engineering's growth constituting part of the evolution of the growth of the city's high-tech sector. For example, in spinning off Extreme Devices and Coherent Logix the company was able to secure venture capital from a local VC firm, Austin Ventures, as well as national venture capital firms, such as KLM Capital and ARCH ventures. Systems & Process Engineering though still remains a privately held firm.

## STATE GOVERNMENT INVOLVEMENT

Systems & Process Engineering has received no specific support from the state of Texas. It is cited though by state officials as an exemplar of the development of the high-tech sector in Austin.

## INTELLECTUAL PROPERTY

Systems & Process Engineering actively seeks to guard its intellectual property. It employs a full-time corporate counsel, who devotes approximately one-half his time to intellectual property matters. The firm files provisional patents on most inventions, although it does not always pursue a permanent patent.

## SBIR EXPERIENCES

Systems & Process Engineering learned about the SBIR program in 1986 during a visit to the SBA office in San Antonio, which they had visited to obtain information about SBA programs. They immediately began writing proposals. Of the 11 they submitted, 2 were funded in early 1987, 1 from the then Missile Defense Command, the other from the Army.

Since 1987, the firm has received SBIR awards from several federal agencies, including NASA (fiber optics), Air Force (connected wireless), and the Army (austere supply chain management).

Systems & Process Engineering views the SBIR program as a co-funding for product development, not as a funding source for research, per se. It filters SBIR topic selection first by assessing whether the topic would like to a specific product that met U.S. government needs, then by whether the topic would led to a product marketable to the commercial sector. It also estimates that it invests an amount equal to an SBIR award for Phase I projects and in supplying the financial glue that maintains a project during the gap between Phase I and Phase II funding.

## RECOMMENDATIONS FOR SBIR

Its experiences with the DoD-SBIR program lead Systems & Process Engineering to offer extensive recommendations for the program's future operations. In general, its assessments highlight the positive contributions that SBIR has made to the development of small high-tech businesses, to technological innovation, and to mission needs. SBIR is seen as providing good value to the government in terms of value per dollar for leading edge technology development. Recent modifications to the program, such as combining Phase I and Phase I Option programs into one proposal, help reduce the schedule and funding delays between programs. SBIR's use of Web technology for downloading and submitting proposals also are seen as working well.

Several aspects of the SBIR program are, however, seen as requiring improvement. These include the following:

1. Incentives to Program Executive Offices (PEOs) and prime contractors to transition Phase II SBIR technologies and products more efficiently to Phase III applications in major "programs of record" are needed. SBIR topics tend to represent the priorities of the various service laboratories. Less apparent, at least from the perspective of an SBIR awardee is whether there is a procurement "socket" into which the technologies they have developed under Phase I and Phase II awards can be fit. Tremendous institutional inertia within DoD's procurement practices exists; there is an institutional bias towards large programs and thus large prime contractors, whereas smaller firms are more innovative, more on target to end-user needs, and faster in responding to changing needs.
2. The timeline between and among topic generation, release of a proposal, and selection of proposals is too long, causing several problems. The information in the SBIR Topic is dated by the time the topic gets released for proposal. The delay between the time when a topic was originally accepted by an agency and when it actually gets released as a proposal leads to situations in which when the firm's first discussion with the topic author reveals new requirements and information not readily apparent from the initial posting of the topic. A process in which the topic author had an opportunity to update the requirements and information in the topic immediately before release of the topic for proposal would provide more accurate information to prospective bidders.
3. The SBIR debriefing process could be improved. The typical debrief for a proposal that was not accepted points out many positive attributes and then states the proposal was not selected for funding by a "rigorous selection process." This provides little feedback to the SBIR contractor to make decisions on judging if the particular technology is of further interest to the U.S. government. A more useful SBIR debrief process would be to provide the same package of feedback information to all vendors that proposed, the SBIR debrief information would contain (a) number of proposals submitted for this topic; (b) summary of different technology solutions proposed for this topic, (c) agency review of benefits and costs of the different technology solutions proposed.

Also, for the benefit of firms that submitted proposals but did not receive funding, a statement about which technology solutions were funded, allowing for deletion of company names other than the names of award winners with abstracts, would be helpful, if only to guide them in determining the future course of their R&D efforts.

4. The level of Phase I funding should be increased. The basic funding levels for Phase I of $100,000 have basically remained unchanged for over 10 years. The Phase I funding levels should be raised to keep pace with inflation and to provide the SBIR contractor the ability to produce critical component prototypes for demonstration rather then just paper studies.

5. Substantial addition optional funding should be permitted for Phase II awards. The Air Force has started to make an additional $500,000 available beyond the base $750,000 Phase II funding to fill the gap between Phase II R&D and commercial development of the embryonic technology. This type of funding option should become standard for all Phase II SBIRs.

6. Reporting requirements should be standardized to reduce overhead for both the government and the SBIR awardee.

The standard Phase I SBIR program requires monthly reports. This requires the SBIR contractor to write a report, the vendor's contracts officer to submit a DD250, the government contract officer to process the DD250, the COTR to review and approve the monthly report, and the government contractors officer to submit payment. For each four weeks of technical work effort, there is considerable overhead to process monthly reports.

7. A third-party review board, independent of agency control, is needed to resolve conflicts that arise in the transition of SBIR development programs to "programs of record" status, including production and field operations by prime contractors, especially to insure protection of the an SBIR awardee's intellectual property rights.

## SUMMARY

Although it began with and continues to have diversified customers and revenue sources, Systems & Process Engineering credits SBIR as underlying everything that it is. Systems & Process Engineering core technologies have all had SBIR support. Technological innovation, as demonstrated by Systems & Process Engineering experiences, is a multiyear, multistage process. The firm estimates that its major innovations have taken from 6–8 years to develop, and have frequently been based on, and required, multiple SBIR awards. These technologies have increased DoD's ability to fulfill mission needs, and have also found use in the commercial sector. The main issue confronting the SBIR program, in its view, relates to the dynamics of the Phase III/acquisition and procurement processes. Systems & Process Engineering is right at the cutting edge of emerging technologies, but it has no ready path to introduce its technologies into larger systems.

# Technology Systems

*Irwin Feller*
*American Association for the Advancement of Science*

## FIRM HISTORY

Technology Systems was founded in 1985 by Charles (Chuck) Benton, a software specialist, who during the early 1980s had authored a series of video games for a number of California-based firms. As the demand for new video games slowed, Benton relocated to Maine, where he served as a software development consultant for a number of firms. While in Maine, he learned about the SBIR program via some of the program's brochures as well as an outreach seminar sponsored by the state of Maine to encourage the state's small businesses to apply for SBIR awards.

Benton submitted a number of Phase I proposals, succeeding with his third submission. The successful proposal was to DARPA to examine the applicability of arcade level technology to military training. This Phase I proposal led to a Phase II proposal. Overlapping with these awards, Benton also successfully competed for an Air Force Phase I award to develop a flight simulation training program. (This project served dual purposes: At the same time that Benton was developing the program, it was being treated as a component of a larger Air Force research project, in which research psychologists were seeking to study how

**TECHNOLOGY SYSTEMS: COMPANY FACTS AT A GLANCE**

35 Water Street
PO Box 717
Wiscasset, ME 04578

Phone: 207-882-7589
Fax: 207-882-4062

Number of Employees: 11
Annual Revenues: $1-5 million range

Number of Patents: 1

Number of SBIR Awards
- Phase I: 10
- Phase II: 7

people learn.) The DARPA and Air Force Phase II projects served as the basis for Technology Systems' formation, and permitted it to add employees.

Since its inception, Technology Systems has reinvented itself several times to remain viable in the face of competition, changes in DoD's organization of its R&D and procurement systems, and the complexities of transitioning from a contract R&D to product development and manufacturing firm. The willingness to engage in this reinvention in large part reflects Benton's preference to remain an independent entrepreneur who lives in Maine. Had he been required to make the frequent adjustments to his start-up firm's business strategy while living in Massachusetts, he likely would have gone to work for a large firm.

From 1987–1991, Technology Systems essentially was a contract R&D firm, heavily dependent on SBIR awards for its revenues. In August, 1991, as its Phase II awards ended and with no opportunity to secure Phase III funding, the firm had zero income, and was forced to furlough all its employees.

Following this experience, Technology Systems sought to diversify its markets. From 1992 to 2000, it began a dual process of diversification: first, in expanding the number of clients for whom it conducted contract R&D; second, in beginning limited production of products that embodied its new software. In this period, it reached a stage at which its revenues were apportioned approximately equally between contract R&D and product sales. The 50 percent of its revenues derived from contract R&D were distributed about equally between the federal government, primarily DoD, and industrial firms. Among the latter customers were major defense and consumer product firms, such as Computer Sciences Corp, Loral, and SAIC. In effect, Technology Systems sought to position itself further up the learning curve in simulation technology, selling this technology as an input in larger technological systems. In its emphasis on product sales, the firm's competitive strategy was to offer favorable price-performance bundles; thus, in one sector, offering what it described as 70 percent of the functionality of a rival's product at 10 percent of the prices.

The firm's strategy changed again about 1999, as a series of events among the services that each resulted in reduced revenues. At about that date, DARPA and Army funding of R&D began to decline and the Air Force laboratory with which the firm had been dealing was closed as a result of a wave of DoD base closings.

In its search for new customers and markets, Technology Systems has had to extend itself to the edges of its core competencies. It has done this by moving into the design of software and optimization models for industrial controls, at the same time finding its new primary customer in the needs of the U.S. Navy. Contemporary shipbuilding technology in effect requires the customization of single I-beams. These new production requirements place new demands for software systems that can optimize production. This is the niche that Technology Systems has sought to fill. Its success in developing this software has reached the point where the firm has just spun off a new firm to specialize in further development and applications of industrial control software. In this new venture, Technology Systems is teamed with another firm.

The firm has also become involved in developing communications and mission planning software for undersea warfare systems. This technology derives from an integration of research support it received from the National Science Foundation for work on distributed networking with SBIR grants from the Navy.

Technology Systems also has begun to develop a product line in geo-registered visualizations that are added to video to support tactical and navigational operations. This technology is of interest to the Navy, both in mine clearing operations and maritime navigation. Technology Systems' efforts to commercialize this technology via establishment of a spin-off firm, Looksea, however have not been successful to date. While a product launch management team was assembled, Looksea failed to produce commercial sales or attract outside investment. Technology Systems now sees itself as having to liquidate Looksea because it is draining the firm's resources. This experience has led Technology Systems to rethink its commercialization strategy, pointing to increased reliance in the future on licensing its technological advances rather than attempting to create spin-off firms.

## CAPITAL FORMATION

Technology Systems was a bootstrap operation. It was underwritten by the SBIR awards and Benton's own finances. One attractive feature in the forming of software firms, according to Benton, is that they typically do not require as much initial capital as prototype development and/or manufacturing firms. Technology Systems remains a privately held firm.

According to the firm, constraining its ability to grow, especially, as noted above, in transitioning from a contract R&D firm to one engaged in even modest scale production, is the meager supply of external capital available to start-up, high-tech firms in Maine. The venture capital/angel capital market in Maine is described as "functionally nonexistent." At best, to the extent that outside investors do exist, they are seeking to acquire ownership of high-tech firms on terms that offer little return to the founding entrepreneurs.

## INTELLECTUAL PROPERTY

Technology Systems holds a somewhat dualistic view about the importance of intellectual property protection. It does not hold patents as generally needed to protect the firm's intellectual property, which typically are shielded by know-how and tacit experiences. However, it recognizes that patents may generate several external benefits: Both to customers and potential investors, they may be a sign of technological innovativeness and prowess. They may also constitute an intangible asset that would serve to increase the market value of the firm, were it to be bought by other investors or go public. Accordingly, the firm does seek patents: It currently holds 1 patent, and has several pending applications.

## INVOLVEMENT IN STATE PROGRAMS

Technology Systems has benefited from several State of Maine R&D and economic development programs. As noted above, its early awareness of and information about SBIR was fostered by outreach workshops sponsored by state agencies. Technology Systems also has received both seed grants ($5,000–$10,000) and development grants (which range from $100,000 to $250,000) from the Maine Technology Institute, a state program designed to foster the growth and development of selected industrial sectors. Technology Systems also received technical assistance from the Maine Patent Program in filing for its first patent.

In all, the firm gives the state of Maine high scores for the assistance it has provided start-up, high-tech firms.

## EXPERIENCES UNDER SBIR

Technology Systems describes the fairness and timeliness of SBIR's review procedures as generally being "pretty good." It recognizes, but accepts as a fact of life, that some agencies/laboratories have "favorite sons" that at times give specific firms a lead in competitive races. It considers the size of Phase I and Phase II awards to be reasonable. It has not found the funding gap between Phase I and Phase II awards to be a major problem, as it has generally been able to maintain the momentum of a project by using internal resources to redeploy its staff. It also accepts as a fact of life the ebb and flow of funding for specific topic areas; indeed, it sees this as a positive feature of the SBIR program, as this ebb and flow leads over time to the entry and exit of firms from successive competitions, thus providing for a process of natural selection in who wins. However, it does not look with favor on the proposal to set aside a portion of DoD SBIR funds for the biotech sector.

Of especial concern, and harm, to the firm has been the reorganization of the Army's R&D programs. This reorganization entailed the establishment of designated battle laboratories to address the specific R&D needs of various branches, e.g., artillery, tanks. The support contracts for the laboratories were then awarded to Lockheed Martin. Accordingly, Technology Systems soon found itself in a position that whenever it proposed a technological idea to a battle laboratory, Lockheed Martin, as the support contractor, was able to interject its claim to conduct the needed R&D. From the perspective of Technology Systems, it was dealing with a stacked deck, which made pursuit of Army R&D contracts not worthwhile. It has thus exited the simulation industry.

## RECOMMENDATIONS

Increasing the number of annual solicitation rounds in DoD competitions from 2 to 4 would be of great benefit to small firms, such as Technology Systems. In the last round, the firm prepared 6.5 proposals; this was a grueling experience.

DoD's Fast Track process also is seen as unintentionally distorting good business practice. In one case, the firm undertook to do a Fast Track submission to extend its R&D on lasers. It entered into discussions with several high-tech firms to secure the required matching funds. The firms were interested in Technology Systems' technology, but unwilling to advance funds on the grounds that they typically did not do so for R&D projects. Technology Systems thus found it necessary to submit its proposal as a standard Phase II submission. Its proposal was not funded, in part, according to the firm because the DoD program manager had expected it be submitted as a Fast Track submission, and thus downgraded it when it became part of the regular competition. (The proposal subsequently was funded through an alternate channel, in part because of the interest of Bath Iron Works in the technology.)

## SUMMARY

The SBIR program was a key contributor to Technology Systems' founding, and has served as a valuable, if variable source of R&D funds for the firm as it has reinvented itself with respect to technologies, markets, and customers over its almost 20 year history. From the firm's perspective, the SBIR program also has yielded good value to the American taxpayer. By enabling Technology Systems to work at the leading edge of technological innovation, the program has generated substantial cost savings in the ways in which DoD has been able to conduct its operations. Research supported by SBIR also has had a broader national impact on industrial practice. Technology Systems' initial R&D project for DARPA on network simulation has been incorporated into ISO and IEEE standards. Estimation of the savings generated by the firm's SBIR awards is conceptually feasible although difficult, because the firm's technological advances are typically incorporated into larger training and weapons systems provided by DoD's prime contractors, such as Lockheed Martin.

# Thermacore International, Inc.

*Irwin Feller*
*American Association for the Advancement of Science*

## FIRM HISTORY

Thermacore was founded by Yale Eastman in 1970. Eastman was an employee of RCA, assigned to working on the development of heat pipe technologies related to the conversion of heat to electricity, with specific application to the nuclear power industry. When RCA ceased work on the technology, Eastman left to form his own company to continue his former work, staying in Lancaster, Pennsylvania, where RCA's plant was located.

Thermacore started as a "garage" start-up. At the same time, the shift in national energy and environmental policy from nuclear to solar power led the new firm to focus its attention on developing heat pipe technologies for solar applications. It remained a small firm with no more than 10 employees throughout most of the 1970s, working on specific industry and U.S. government R&D contracts. It began to grow but along this same trajectory, conducting contract R&D firm, primarily for NASA, DoD, and DoE under a series of non-SBIR and SBIR contracts.

Thermacore now is a globally oriented firm, providing products for the computer, telecommunications, power electronics, medical, and test equipment industries. It continues to invest in R&D to widen the range of uses of heat pipe technology. Reflecting its transition from an R&D to a production-oriented firm, contract R&D projects from NASA and DoD and original equipment manufac-

**THERMACORE INTERNATIONAL, INC.: COMPANY FACTS AT A GLANCE**

780 Eden Road
Lancaster, PA 17604-3243
Phone: 717-569-6551, Fax: 717-569-4797
SIC 3443
Number of Patents: 61
Number of SBIR Awards: 82
 Phase I
 Phase II

turer presently represent approximately only 6 percent of the firm's revenues. The balance, 94 percent, comes from sale of commercial products.

As seen by Thermacore, the marketplace caught up to its technology. As the market for personal computers grew, so too did the importance of dissipating the PC's internal heat. In the early 1990s, Thermacore was approached by Intel to discuss the prospects of transitioning from low volume to mass production of its heat pipe technology. With financial support from a venture capital firm, Thermacore then took the risk of setting up a production line before receiving orders. Subsequently, it received large orders from several major PC manufacturers, such as HP, Dell, IBM, and Sun.

This expansion into the PC market provided a quantum jump in the firm's scale of operations, as well as its transition from an R&D firm primarily oriented to product development to a volume manufacturer of a commercial product. The transition was described as requiring at times painful transitions in the firm's outlook and staffing patterns. The firm's personnel had to adapt to new sets of customers and a new orientation towards marketing. Experiences under the SBIR program are seen as having helped in making these transitions, as earlier work on SBIR projects had provided valuable training experiences for Thermacore's engineering and technical staffs, providing it with a "brain pool" of "know-how" related to manufacturing reliability. These tacit skills have contributed to the firm's ongoing competitive position even as patents on its initial core technologies have expired.

Thermacore remained a privately held firm during its first 20 years, eschewing a public offering lest it dilute ownership control of the firm. In 2001, as part of his retirement, Eastman sold the firm to Modine Manufacturing, Wisconsin, an international leader in thermal energy management, with estimated sales in FY2005 of $1.5 billion, which operates it as a wholly owned subsidiary. This new ownership arrangement has meant that Thermacore is no longer eligible to participate in the SBIR program. It does however continue to do some SBIR-funded research as a subcontractor to firms conducting Phase I and II research.

The firm has remained in Lancaster for historic reasons. As the production of PC's increasingly has shifted to southeast Asia, the firm has been experiencing increased pressure from its major customers to locate its production facilities close to them. It now has a high volume manufacturing branch in Taiwan (Thermacore Taiwan).

## EXPERIENCES WITH SBIR PROGRAM

Thermacore began active pursuit of SBIR awards soon after SBIR's program establishment, detailing one of its engineers to monitor SBIR topics. In 1984, it submitted two Phase I proposals to NASA and DoD on topics related to high performance heat pipes; it received awards on each proposal. Thermacore's pursuit of SBIR awards was tied to an emphasis on product development and commercialization. Their repeated involvement in the SBIR process is seen as highlighting the length of time, amount of financial support, and number of incremental technological advances needed to move an R&D concept into a useful product, whether for government agencies or the commercial sector. Thus, for Thermacore, SBIR awards served as the basis for sequential improvements in its core technologies in heat pipes. Heat pipe technology was seen as a dual-purpose technology: it was supported by government agencies for its usefulness in defense and aerospace, but over time, with customized modifications, has become an innovation employed in a diverse set of private sector industries.

Reflecting the resulting lesson that technological innovation is a complex process, the firm notes that no single SBIR award led to a specific innovation. Nor is it possible to attribute any single use within a larger military or aerospace technical system to any single award or contract. Instead, it was the accumulation of technical advances under these awards coupled with the firm's in-house expertise that resulted in its innovations. New products are viewed as involving the meshing of multiple ideas.

During its period of eligibility, Thermacore received (82) SBIR awards from several government agencies, including DoD, NASA, and DoE. It considers the SBIR proposal selection process(es) to have been fair and timely. Having a diversified portfolio of awards from different agencies meant that it was able to handle the gaps it experienced between Phase I and Phase II awards by redeploying technical personnel. Even so, especially in its early years, when it had fewer awards, the gap made for a roller coaster existence in meeting payroll.

## INTELLECTUAL PROPERTY

Thermacore, historically, has followed a policy of seeking patent protection for its inventions. It presently has 62 patents. Patents are viewed both as a device for protecting intellectual property and as a means of symbolizing the firm's technological leadership. Recent experiences in short-lived, unsatisfactory foreign joint venture however have also highlighted to Thermacore how fragile patent protection can be with partners who learn the technology and then seek to become competitors and in countries where intellectual property rights are loosely enforced. Increasingly, the firm has placed value on its tacit knowledge, embodied in large part in the technical expertise of its engineers responding to customer needs.

## INVOLVEMENT IN STATE GOVERNMENT PROGRAMS

Thermacore's involvement with the state of Pennsylvania technology development programs has been limited and unsatisfactory. It participates in one project under the state's Ben Franklin Partnership Program, a state program that funds private sector R&D, typically on a cost-sharing basis, as well as typically in partnership with a university researcher. The collaboration with researchers at Penn State University did not prove satisfactory. It was seen as benefiting the researcher's laboratory while producing little value to Thermacore.

## RECOMMENDATIONS

Even though it is no longer eligible to participate in the SBIR program, Thermacore sees the program as having considerable value, and recommends that it be continued.

## SUMMARY

Thermacore's history highlights how DoD and other government agency SBIR awards can contribute to development of a technology whose end uses extend well beyond the mission objectives or commercial uses pursued at the time of initial project support. Thermacore views SBIRs as having helped it to push the envelope in technical performance and to establish its credibility in the marketplace. SBIR awards played a critical role both in terms of supporting R&D on its core technologies and in enabling it to shift from an R&D, product development firm tied primarily to government contracts to a world leader in selected commercial markets. The combination of technical expertise coupled with its increasing knowledge of marketplace needs and opportunities has made Thermacore the international leader in the field of heat pipe technology. Also evident in Thermacore's establishment and growth is the lengthy process at times required to bring a technology to practical uses.

# ThermoAnalytics, Inc.

*Zoltan Acs*
*University of Baltimore*

## FIRM OVERVIEW

ThermoAnalytics, located in northern Michigan, develops Computer Aided Engineering (CAE) software for commercial and military thermal analysis and infrared signature prediction. The company also develops custom software for thermal and infrared signature management and provides consulting services for design and analysis. ThermoAnalytics is now a $5.5 million company.

ThermoAnalytics was founded by Keith Johnson and Allen Curran, both mechanical engineers and specialists in the thermal sciences. The company spun off from a Michigan Technological University contract research group in 1996. The original universty group was incorporated as a nonprofit and provided defense consulting services. It specialized in applied physics projects for the Army. After the Army cut funding for R&D, the group strengthened their ties with the big three automotive companies, developing dual-use tools that determine the infrared signature, thermal-heat transfer of vehicles.

While Ford Motor Company was interested in the product, it had difficulty dealing with a university nonprofit entity, encountering issues with licensing, training, and support. The automotive company needed the services of a for-profit software engineering firm. The principals therefore determined that it was time to spin off and diversify their markets in the government and commercial sectors. Immediately after ThermoAnalytics was established, the founders started looking into opportunities provided by SBIR.

ThermoAnalytics remains an applied R&D company, even though it markets products. The company started out with 8 employees, all with R&D background. As the company grew, Mr. Johnson's role transitioned from being an engineer to being an operations manager. While ThermoAnalytics added more administrative and marketing staff to payroll, 75 percent of its 35 employees continue to work as engineers and physicists in applied R&D.

## ROLE OF SBIR

The SBIR funding has helped ThermoAnalytics grow from being a service company to a company with commercial and military product sales. ThermoAnalytics continues to grow by taking advantage of SBIR topics that are in its core areas.

Keith Johnson became aware of SBIR while working with other firms providing services for DoD. He helped co-author a proposal while at Michigan Tech.

The Army liked the proposal. However, due to delays in the funding process, ThemoAnalytics had already established itself as a small business by the time the first Phase I was awarded. The project was very successful and earned a Phase I, II, and III.

The project's success, facilitated by the fact that ThermoAnalytics already had a CRADA with Ford at the time of the award, helped prove the technology's dual use. This agreement led to further contracts with Ford. Without SBIR, the company believes that it would have still established its relationship with Ford, but would have remained a small work-for-hire contractor without ability to sell on the commercial market.

In 1996, ThermoAnalytics had the fastest radiation solver. Lacking a patent, however, the company knew that it must continue to innovate. That is where SBIR awards were crucial. SBIR awards helped ThemoAnalytics to finance product development, freeing revenue to meet sales and marketing expenses. According to ThermoAnalytics, this strategy has been commercially successful.

The company has received 13 Phase I awards between 1997 and 2005, and 10 Phase II awards between 1998 and 2005. By taking advantage of the Phase II Plus matching funds ThermoAnalytics has attracted over $1 million in third-party funding and about $1 million in SBIR matching funds.

ThermoAnalytics believes that a key feature of SBIR—not available through other government programs or sources of private funding—is the ability to create a proprietary product with the company retaining the rights to commercialization. SBIR awards also allow small business to have an R&D group within their organization that is paid through outside funds rather than IR&D.

## COMMERCIALIZATION

The firm's competitive advantage relies on the speed and simplicity of CAE software that supports rapid prototype design. The product has niche applications on the military side. In the commercial sector, the product is differentiated because it is very different from solutions povided by competitors. ThermoAnalytics emphasizes that one of its strenghts is its "low innovation inertia," that is quick product development and quick turnaround. Competitors include CFD companies such as FLUENT and several others.

In the defense industry, the software can be utilized in infrared munitions that detect and destroy an object by sensing heat. The application on the demand side is signature management: keeping objects cool so they are invisible to sensors. On the private commercial side, particularly the automotive industry, the software has applications in thermal management related to exhaust, converters, and electrical components. Innovations in the automotive field have also resulted in more fuel-efficient designs by reducing aerodynamic drag. These reductions in drag also produce problems related to overheating.

There is also demand for the company's products in Germany, which has both a strong military and a strong automotive industry. The company's product is sold in Germany through small distributor companies (employing usually two people) that specialize in export sales. The distributor finds the customers for the product, but ThermoAnalytics' Web site is also a strong referral source. The Web site contains demonstration software and a demonstration product that can be accessed by entering individual contact information. This information is forwarded to the distributor for follow-up and potential leads.

In all, the firm estimates, conservatively, that its market size is $10 million for the product and $10 million for consulting services. The split in revenue over the last 3–5 years has been 75 percent military and 25 percent commercial. This is because the military has more funds to invest into development. According to ThermoAnalytics, the automotive industry, when faced with economic downturns, is known to cut from the R&D budget first. "While this keeps the shareholders happy, it hurts the company in the long run."

Thermoanalytics develops and sells directly all its products. International sales are done through independent distributors. The software is licensed and sold to a variety of agencies: prime contractors, government agencies, and commercial companies.

## KNOWLEDGE EFFECTS

Since ThermoAnalytics innovates in the area of applied physics, rater than basic physics, the majority of algorithms exist in textbooks somewhere. While patent protection is elusive, the company notes that there are, nevertheless, unique ways during the software development process that provide intellectual property protection.

The company lists 15 scientific papers published and four trademarks registered for the company and its products. It has also won six awards for its innovative research.

## IMPROVING SBIR

### Surviving the Funding Gap

The company notes that the funding gap was a problem in the earlier years when holdup in the congressional budget delayed the Phase II award. These gaps have, on occasion, put the company's work at risk. This gap is managed by diverting staff into other projects, when necessary. In addition, the company notes that the Phase I option provides bridge funding.

The company also notes that the Army has a more structured schedule than the Navy and Air Force that makes planning easier. The Army also provides a much better Phase II Plus match.

## Award Size

ThermoAnalytics believes that the Army's model of Phase I 70,000—6 months, Option 50,000—4 months, Phase II 730,000—2 years is best. It notes that the only timing problem emerged when the Phase II proposal needed to be submitted when only about 4 months of the Phase I had passed.

## Award Selection

The award selection process is seen as fair and, timely. However, the company believes that the award selection processes should be standardized across agencies. Solicitation schedules for all agencies should be posted yearly. The agencies should ensure that adequate time has been allowed for the Phase II request for proposal and the completion of the Phase I research.

## Commercialization Assistance Programs

ThermoAnalytics has participated in a limited way with commercialization programs. Given that it is already well versed on SBIR and has previous direct commercialization experience, these events are no longer seen as very helpful.

## Commercialization Transition

ThermoAnalytics believes that the SBIR program should provide incentives for award recipients to "buy in" early into the process, work with the prime contractor, actually develop a product, and secure adequate funding for Phase III.

# Trident Systems

*Irwin Feller*
*American Association for the Advancement of Science*

## FIRM HISTORY

Trident Systems was established by Nicholas Karangelen in 1985. Karangelen is a U.S. Navy Academy graduate and was selected and trained by Admiral Rickover for service on nuclear submarines during the cold war period. Upon being commissioned in 1976, he served on fast attack submarines. Karangelen left

**TRIDENT SYSTEMS: COMPANY AT A GLANCE**

Address: 10201 Lee Highway, Suite 300
Fairfax, VA 22030
Phone: 703-691-7794

Year Started: 1985
Ownership: Privately held, equity based
Annual Sales: FY2004 $20 million
Number of Employees: 115
3-Year Sales Growth Rate: 44 percent per year on average
SIC Code: 8711
Technology Focus: Systems Engineering Research and Development; Systems and Software Engineering: C4I Systems, Touch Screen Technologies; Enterprise Collaboration Centers

Number of SBIR Awards—Phase I: 56 (DoD Phase I): 56
Number of SBIR Awards—Phase II: 32
Number of Patents: 4
Number of Publications: Numerous technical papers, articles, books, including EIA Systems Engineering Data Standard EIA-927: Common Data Schema for Complex Systems

**Awards**
17 Virginia Small Business Technology Achievement Awards
NAVSEA Value Engineering Award
Washington Navy Yard Campus Renovation Award

the Navy in 1981. He then worked for TRW, in Mclean, Virginia as a systems engineer until 1983, when he was recruited by IBM to work on an advanced submarine combat system. Work experiences with TRW and IBM were described as excellent opportunities for on-the-job training, as they added new skills to Karangelen's earlier training in physics and engineering. However, they also proved frustrating. Big firms were described as overly wedded to existing ways of doing things—they didn't like to get out of their boxes.

A series of family events, coupled with what Karangelen, a second generation Greek-American, termed his cultural heritage of entrepreneurship, led him in 1985 to decide to start his own firm. Trident hired its first employee in 1986. It operated primarily as a "services" company, selling hours of consulting services to major DoD contractors, such as GE and Westinghouse, but had no prime contracts. This work served both to build Trident's reputation as a knowledgeable, reliable performer and to give it new insights into DoD requirements.

In about 1988, one of the firms for which Trident was working was becoming too large to qualify for a DoD-Navy set-aside program in which a 5 year, cost plus fixed fee contract was to be bid on systems development for antisubmarine warfare. Trident, which then had 4 employees, submitted a proposal as the prime contractor and in a highly competitive contest, won the contract. The stability of the contract, the quality of the work it performed under it, and the business relationships developed during performance of the contract are described as having launched Trident on the growth trajectory it has experienced since the late 1980s.

Trident has grown primarily by expanding its business about its core competencies, adding to them as requested to by its customers. These core competencies began with requirements analysis for weapons systems) then advanced into systems engineering, and more recently into systems design. In this latter capacity, it had previously worked principally as a subcontractor for larger DoD prime contractors, augmenting their in-house staffs, but now serves as prime contractor on many programs.

By 2004, Trident's revenues had reached $20 million, and its employment level had risen to 115. The larger portion of its systems engineering and design work is performed at its headquarters in Vienna, Virginia. The firm has recently moved into further downstream integration, and now is beginning to build products for niche markets based on its systems design work. Toward that end, it has recently built a light manufacturing assembly plant in Pennsylvania. It also bootstrapped much of this expansion, relying on retained earnings and high levels of reinvestment in product development.

## CAPITAL FORMATION

Having little capital at the time of Trident's founding, Karangelen started the firm in his home, essentially providing consulting services. From the very beginning though, his goal was to own and operate a company. This led him to incorporate as a privately held stock company, rather than to operate as an S corporation.

Although originally envisioning a business strategy of distributing stock to employees and then going public after a period of 10 years or so, Karangelen chose instead to defer this approach lest it detract from managerial autonomy. To overcome an inherent flaw in employee stock distribution plans that builds up pressure over time to go public, namely that the firm must go public in order for employees to realize the economic benefits of their share holdings, Trident arranged for a formal appraisal of the company's value, and then used the firm's retained earning to buy back stock from employees at this price. Trident's employee stock purchase plan was then replaced with a profit sharing plan that distributes a significant portion of the companies profit to employees on a merit basis. One result of this shift to a profit sharing strategy was to increase employee attention to the firm's profit margins. Employees are described as having become more aggressive in seeking out new businesses and generating additional contracts.

As a self-financed firm, Trident experienced huge cash flow problems in the early years of operation. When Karangelen first sought a bank line of credit (of $11,000) for working capital, the bank initially required a comparable deposit before it would make the loan (at the then prevailing interest rate of 17 percent). A bank loan to provide working capital was eventually negotiated.

## EXPERIENCES UNDER SBIR

Trident began to submit proposals to the SBIR program about 1986, submitting 4 or 5 unsuccessful proposals before it won one. Reflecting in part its own experiences but presented as a more general observation on the value of the SBIR program, Trident notes that SBIR is one of the few contract mechanisms currently in place to provide "size appropriate competition," that is opportunities for small firms to compete for DoD's R&D and procurement contracts.

Not all the benefits to the firm from the SBIR program have been manifested in Phase II or Phase III awards. In one example cited by the Trident, it was unsuccessful in its proposal for a Phase II award following work it had done under a Phase I award. The software it had written as part of its Phase I work was unique, and permitted touch screen rather than mouse control of computer screens. Development of the software spawned a general purpose technology, which has permitted the firm to branch into working of specific touch screen solutions for several different computers and end user markets.

Trident's success in developing DoD related technologies under the SBIR program has not led to proportionate successes in landing procurement contracts, however. According to the firm, in the overwhelming majority of cases, small businesses that have successful relevant capabilities and technologies do not achieve major positions in DoD acquisition programs. In some cases, the small business technologies may be seen as competing with established program interests or as a distraction from the program's plan. Some program managers may be unwilling to invest program funds in alternative technology candidates when they believe (as most are seen to do, according to the firm) that their programs are on track. Prime contractors are described as often polite but generally unwilling to bring in a promising externally developed (and potentially disruptive) technology when they have an internally developed alternative or believe (as most do) they can reasonably develop an alternative internally. In most cases, even well intentioned attempts to include small business in major DoD programs fall short because of factors unrelated to the high technical quality, reduced costs, and shorter development times offered by small business and their technology solutions. These missed opportunities represent what the firm believes to be to be the largest single impediment to current initiatives to transform DoD's weapon systems acquisition processes.

One such "missed opportunity," as reported by the firm, is its development of a handheld situation awareness technology, named DISM (Dismounted Intelligence Situation Mapboard). DISM development was initiated in FY1996 as part of an SBIR project to determine if it was possible to provide standard digital military maps (supplied by NIMA) with standard military symbology and standard military digital messaging on, what was then, the early generations of commercially available hand held computers. The goal was to provide map-based situation awareness to dismounted troops on small light handheld computers at affordable cost.

DISM capabilities were successfully demonstrated in FY1999 by Trident and subsequently integrated and tested with the Army's FBCB2 program and briefed to the Land Warrior program. Using DISM, any unit can have an instant tactical digital network for situation awareness (SA) and command and control (C2) data by connecting the DISM palmtop to the unused digital channel of their fielded SINCGARS radios. However, DISM has remained outside of the traditional Army acquisition channels even after receiving a very favorable evaluation as the dismounted extension to FBCB2 and being recognized by several operational commanders (82nd Airborne and 101st Airborne) as an opportunity to field a near-term, low-cost dismounted digitization capability. In the face of strong support for DISM by the operational forces, in the wake of failures by two large prime contractors to deliver an acceptable solution (at a cost of hundreds of millions of dollars), and instead of evaluating DISM, which the Army laboratory at CECOM had supported, the Army's PEO for Soldier systems initiated development of a new system called Commander's Digital Assistant (CDA) in FY2002 which essentially copied the DISM functionality (including using DISM graphics in program briefings). CDA was recently heralded by the Army as an SBIR success story; however, there has been no widespread deployment of CDA or head-to-head test against DISM, which is now a mature demonstrated and tested technology.

The difficulties that Trident has encountered in securing procurement contracts for the products it has developed, some of them under SBIR awards, is viewed as symptomatic of systematic shortcomings in DoD's acquisition practices and cultures. The major barrier to successful transition of DoD SBIR awards is held to rest in the mindset of acquisition program managers. Whereas Karangelen sees small firms as representing the nation's most powerful transformation force, DoD acquisition offices are described as reluctant to recognize the value of small firms and their abilities to produce discrete advances in technology. Representing large programs, they are seen as motivated to maintain the status quo rather than to adapt to new possibilities. They are also described as fearful of allowing small firms to show the possibilities of technological success lest these successes call into question prevailing policies and practices that favor big, billion dollar, Star Wars-type of weapons acquisition programs. Further adding to this reluctance of program managers to responds to the technological opportunities offered by small firms, whether as a result of SBIR or other firm initiatives, are the blocking efforts of some prime contractors who erect barriers to entry for small firms through closed system architectures.

According to Karagelen, considerable resistance exists to change both in the program offices and prime contractors that are now engaged in development and upgrade of the current generation of Navy ship combat systems. This resistance, from an historical perspective, may be understandable: Current procurement and acquisitions policies and practices did lead to the development of the ships and weapons systems that won the cold war and which are arguably without peer in the world today. The above shortcomings are seen as stemming from a reinforcing combination of DoD contract practices, procurement policies, and consolidation in the U.S. defense industry. According to the firm, throughout the 1980s, DoD R&D solicitations covered a wide size range of contracts, from $100,000 to multimillion dollar awards. Relatedly, this size distribution encompassed a range of topics, that allowed small firms to identify their niches and thus submit bids. Over time, however, as a result of the bundling of DoD programs and the consolidation of contracts into fewer, larger contracts opportunities for mid-size firms to bid on research contracts began to diminish. This consolidation, in part, has been rationalized on the grounds that it is more efficient to award fewer, larger contracts than a larger number of smaller ones. The result, according to the firm, has been a steady reduction in the number of mid- and large-size DoD contractors, as a result of mergers. Karangelen also noted that as the old firms die out or disappear few new mid-sized firms capable of supplying DoD with goods and services are being born. The result is that the DoD contracting environment is becoming steadily less competitive. The major firms—Boeing, Raytheon, Lockheed—take turns winning major contracts, in the name of maintaining the defense indus base. At lower contract levels, say in the $20 million–$80 million range it likely to find a dozen competitors for an award. These contracts tho be for "services," not weapons acquisition.

Delays between Phase I and Phase II awards also created problems for Trident during the early years of its SBIR awards. It found it necessary to shift staff among projects, to deal with tight financial squeezes at the end of fiscal periods, and at times to shut down or tie off projects. Over time, it learned to hire staff who had the ability to "shift gears" among projects; its growth also provided with additional internal funds to cope with delays.

## INVOLVEMENT IN STATE GOVERNMENT PROGRAMS

Although favorably noting that Virginia has had an active state-level program that both provides technical assistance to firms on how to compete for SBIR awards and awards some small grants, Trident itself has not participated in any of these programs.

## RECOMMENDATIONS

SBIR has been a highly successful program. Because it provides one of the few niches in DoD's research program that is compatible with the capabilities of small firms, size should be expanded. Making the program larger would obviate a need to make stringent trade-offs between the number and size of SBIR awards. Moreover, DoD's topic selection process likewise has improved over time. The increase since 1999 in the number of Phase III awards also is a desirable trend, with the qualifier that the larger percentage of these awards—an estimated 80 percent—have been concentrated in the Navy. Another desirable trend has been that some agencies are starting to award larger Phase II awards.

DoD's SBIR program could still be improved. However, the focus of the really needed changes are in the Defense Acquisition System. Among the recommended changes are the following:

1. Establish an education initiative for prospective program managers at the Defense Acquisition University. The Defense Acquisition University should provide clear guidance on the advantages of using the Phase I and Phase II SBIR and STTR contracts to identify and qualify capable small businesses and their innovative technologies for transition to DoD Acquisition programs and in use of the Phase III contracting mechanism for transitioning SBIR-developed technologies into the mainstream of acquisition programs. Appropriate SBIR program employment guidance should be included in each of the online and in-class courses taught by the Defense Acquisition University in the program management career track.
2. Require ACAT 1 and 2 program managers to include program-plan specific milestones for the transition of SBIR developed technology and utilization of other small business developed commercial-off-the-shelf (COTS) technology in their program plans and budgets. Program managers and their staffs should be directly involved in generation of SBIR topics, the selection of Phase I and

Phase II SBIR awards, the evaluation of the Phase II contract products, and the transition of successful Phase II efforts into their program. The programs' long-range multiyear budget should include funds designated for Phase III contracts and other proven commercially available technologies from small businesses in the same manner these out-year budgets are established for other program activities. Program managers should be required to report on SBIR utilization at each major program milestone and specifically on Phase III SBIR contracts awarded.

3. Require all contracts awarded over $100 Million in ACAT 1 and 2 programs to include SBIR Phase III subcontracting goals for the prime contractor with attendant fee incentives for exceeding and penalties for not achieving those goals. When a prime contractor bids a small business subcontract as part of a proposal, the prime contractor should be required to execute the subcontract on award of the prime contract unless the prime can show due cause. In situations where subcontracts are not awarded, a letter report stating the reasons should be provided by the prime contractor to the program office and the SBA, and a rebuttal to that letter should be solicited from the small business. From these inputs the SBA and the program office could make a determination to either release the subcontracting requirement or not. In addition, the DoD AT&L Office should provide a plan for requiring and incentivizing prime and subprime defense contractors to subcontract to DoD SBIR firms, as they currently do with minority, woman-owned and veteran-owned small businesses. Such a plan would include recording SBIR Phase III contract award metrics just as other small business metrics are recorded and yearly report to Congress.

4. Establish a SBIR Phase III Acceleration program in the DoD that would require each service to identify at least 25 topics each year that have completed Phase II for accelerated transition to development and production in acquisition programs of record. Each of these topics would also be approved by the respective Requirements and Budget directorates of the service chiefs to ensure that they address high priority military requirements and that sufficient funds have been budgeted to complete development and production of the selected topics. This program is intended to expand the very small cadre of DoD Program Executive Officers and program managers who have successfully embraced the SBIR program and taped the wealth of affordable and innovative technology resources for their programs.

## SUMMARY

Trident ascribes a significant portion of its success and growth to the SBIR program. Without SBIR, it wouldn't have survived, grown, or flourished. The ability to compete for SBIR awards, and the technical and economic successes it achieved because of these awards, permitted the firm to follow a totally different business model than would have been possible had it been forced to secure external capital.

SBIR solicitations provide a range of opportunities for small firms to identify, bid for, and perform DoD-related R&D; the list of topics is described as akin to a college catalogue. Trident revenues and employment levels have grown steadily, although not continuously since the late 1980s, and considers its experiences under the SBIR to have been at the heart of this growth.

The contribution of the SBIR program to the firm's growth has taken several different forms. One project, described as a highly successful SBIR, led to the development of a handheld situational awareness system. Trident contrasted its success in designing and developing this product with the experiences of larger DoD contractors, which has received far larger awards for comparable technologies but which were unable to produce a useful product.

The technological and mission benefits generated by DoD's SBIR program are dissipated in the transition between R&D and acquisitions. SBIR produces fruit, which is not picked up by the acquisition system. DoD's acquisition system is overly resistant to change, especially in allowing more open competition. Program offices and prime contractors have a strong investment in the existing monolithic approach (i.e., one large prime contractor who is responsible for the program). Prime contractors are seen as firmly entrenched and skilled at constructing the case for their continuing role as monolithic system provider and gate keeper for innovative, competitive (and potentially disruptive) technologies. DoD program offices have been open to discussing the merits of the open architecture (OA) approach and quick to identify how they are currently implementing OA elements into their programs, they also are not often successful forcing significant change on their prime contractors who largely determine the fate of the program.

# ViaSat, Inc.[8]

*Peter Cahill*
*BRTRC, Inc.*

## BACKGROUND

ViaSat, Inc., was formed in 1986. Three outstanding engineers, who were in their early thirties, founded the firm. The founders were fellow employees at M/A-COM Linkabit, a San Diego based satellite telecommunication firm. Linkabit had been founded by Andrew Viterbi and Irwin Jacobs, who later founded QUALCOMM. Linkabit was very high tech, extremely innovative, a magnet for the very best in digital communications. It has spun off about 40–50 firms in southern California.

Following the classic path of newborn technology firms, the three, who were unaware of SBIR at the time, began business in a garage with under $25,000 in capital. Initially, ViaSat consulted with defense firms, which were preparing proposals for satellite programs, with agreements that a winning proposal would result in an engineering subcontract to ViaSat. After two such "proposal" contracts, ViaSat obtained venture financing of $300,000 from Southern California Ventures.

Venture funding had little impact on company growth compared to the impact of the SBIR program. The venture funding was used as a financial safety net, while SBIR fueled growth, providing research and development (R&D) dollars, and providing entry to contract dollars without the extensive red tape of competition. ViaSat won its first Phase I award ($49,955.00) from the Navy in the summer of 1987. This led to a Phase II in 1988. Subsequent modifications to the Phase II contract made its ultimate value $1.2 million.

From the beginning, every contract, whether consulting with defense firms, conducting SBIR, or doing follow on R&D, was aimed at developing products to manufacture. The first breakthrough was the initial SBIR for a Communications Environment Simulator, for use in air combat test and evaluation. That SBIR created a specialized test equipment product basis, and demonstrated ViaSat's ability to design and manufacture. ViaSat credits that product as producing $42 million in sales to the Department of Defense (DoD) and $17 million in sales to private industry. Subsequent to the Phase II, DoD contributed an additional $5 million to developing the technology.

ViaSat's initial successes in defense and government related products continues today in its Government Systems division. Products include terminals,

[8]Case study is based on an interview in July 2004 with James P. Collins, the vice president for Business Development, and on information in ViaSat Annual Reports and on the ViaSat Web site.

control systems, and training terminals for UHF and wideband military satcom; MIDS/Link-16 tactical communication terminals; data messaging processors and software for clear communication over noisy tactical channels; RF communication simulation systems; and secure networking products enabling encrypted communication over nonsecure networks.

ViaSat had its initial public offering (NASDAQ: VSAT) in 1996, an IPO that the firm attributes to the impact of their SBIR awards. Unlike management in many emerging technology firms, which change their upper level management as they obtain venture funding, grow and go public, ViaSat founders, Mark Dankberg, Mark Miller, and Steve Hart, continue to provide strategic vision and control of the company. The three had complementary skills and remain in key roles at the company today. Dankberg is Chairman and CEO, Miller is Chief Technical Officer, and Hart is Vice President of Engineering.

ViaSat "graduated" from the SBIR program in late 2001 with the acquisition of Comsat Laboratories. This acquisition of the satellite products group of Lockheed Martin Global Telecommunications brought the company size to over 500 employees. By the end of their fiscal 2005, annual revenue had grown to $346 million (18th consecutive profitable year) and employment to 1,029 employees.

Prior to going public, ViaSat was listed on the *INC.500* list of fastest growing private companies three times. The company is on the *Forbes* list of "200 Best Small Companies," and the *Business Week* list of "100 Best Small Corporations." ViaSat is an ISO9001 certified company.

## ROLE OF SBIR

The R&D of SBIR has been a huge determinant in company growth. SBIR developed products, and particularly in the early years provided credibility with prime contractors. The company would have succeeded without SBIR due to the strength of the ownership team, but they would have been unlikely to have achieved their current level of success, and it would have been a much different company. SBIR spurred the growth of technical capabilities at a much faster pace and provided opportunities to develop technical strengths in new areas.

Lack of SBIR would have slowed growth tremendously. In a highly competitive field, ViaSat requires a continuous significant stream of R&D to maintain and grow its share of the telecommunications manufacturing market. In the early years almost all R&D was either SBIR or contracts resulting from SBIR success. In 1994 the company began internal R&D (IR&D), which has amounted to as much as ten percent of its revenue. In spite of this large internal investment, SBIR remained vital in that it was used for higher risk, more innovative ideas. IR&D could then mature the proven idea. Forty-nine Phase I, 24 Phase II, and follow-on developmental contracts from DoD have provided ViaSat with a quarter billion dollars in R&D funding and a resulting wealth of products.

## INNOVATION AND IMPACT

ViaSat views Demand Assignment Multiple Access (DAMA) Networking as its most significant innovation. DAMA resulted from two Air Force Phase II awards in 1991 for 5 KHZ and 25 KHZ SATCOM DAMA modems. The Phase II awards provided credibility and money to exploit their key technology. For a period of time, ViaSat was recognized as the world expert. This development has resulted in an ongoing product line providing satellite terminal equipment for ships, aircraft, ground station terminals, and missiles. Using SBIR and additional R&D, ViaSat has advanced DAMA from its roots in Code Demand Multiple Access (CDMA) to two proprietary technologies, Paired Carrier Multiple Access (PCMA) and Code Reuse Multiple Access (CRMA).

ViaSat has established itself as a trusted provider of both equipment and technology development to DoD. DoD customers include the Army, Navy and Air Force. It is one of two prime contractors for Multifunctional Information Distribution System (MIDS) Link 16 systems, which provide the primary tactical data distribution system for DoD. In June 2004, ViaSat won a equipment delivery order valued at approximately $47 million for MIDS terminals from the Space and Naval Warfare Systems Command (SPAWAR), San Diego. MIDS provides secure, high capacity, jam resistant, digital data and voice communications capability for U.S. Navy, U.S. Air Force, and U.S. Army platforms. Soon after this production contract, in December 2004, ViaSat was awarded an Engineering Change Proposal modification and corresponding delivery order anticipated to be valued at approximately $60 million for development of a Joint Tactical Radio System (JTRS) compliant version of the MIDS terminal. JTRS is a programmable radio technology that contributes to the new "network-centric" vision of the military by enabling a variety of military wireless communications devices to easily communicate with each other.

In the commercial arena, ViaSat produces innovative satellite and other wireless communication products that enable fast, secure, and efficient communications to any location. Products include network security devices, and communication simulators. ViaSat also has a full line of VSAT products for data and voice applications, and is a market leader in Ka-band satellite systems, from user terminals to large gateways.

Just as technology developed under DoD SBIR has led to commercial products, ViaSat commercial satellite IP networking products are finding a number of applications for the military. For example ViaSat LinkStar® and LINKWAY® IP-based satellite networking products, widely used in commercial enterprise networking, are the core networking technology for the Coalition Military Network (CMN), recently fielded by Lockheed Martin for U.S. Central Command (USCENTCOM). Rather than multiple tactical Satcom units, the new commercial technology, under the Kuwait Iraq Command, Control, Communications and Computers (C4) Commercialization (KICC) project, is creating a permanent communications infrastructure.

## COMMERCIALIZATION STRATEGY

The focus of every ViaSat R&D effort, whether under contract or IR&D, is development of a product that it can manufacture and sell. Its expertise is in design, development, assembly and test. They contract out lower cost components like Chasses and cables, while retaining in house high value such as integration and test. One of its acquisitions, U.S. Monolithics (USM) is exceptionally adept at packaging RF transceivers in high performance low-cost MMIC modules, which are designed into ViaSat military and commercial products.

In the early years ViaSat subcontracted to larger DoD Primes. Now, ViaSat, due to its proprietary technology and Phase III noncompetitive awards, is the often prime and many of these larger firms subcontract to it.

## PRIVATE RETURNS AND SPILLOVER EFFECTS

ViaSat-developed technology provides increased capability at a lower cost. Communications Satellites have an inherent capacity using their as built technology. ViaSat software allows increasing that capacity without putting up a new satellite. Its research has driven the market to keep up. In other cases it has allowed them to keep up. It has increased efficiency allowing more users at improved quality of service. These improvements provide an increased number of messages/calls at any instant in time and over any period of time. They allow improved use and allocation of the spectrum. The net result is the same system of satellites can handle nine to ten times as many users, messages or calls.

ViaSat views its major competitors to be Rockwell, BAE, Harris and Raytheon. In the area of other SBIR success metrics (besides sales and growth), neither publications nor patents would provide much evidence of success. They make presentations at military Communications sessions and chair sessions but this is a relatively small effort. Presentation to the military user and RDTE community has value. Sharing with their competitors does not. As of 1999, ViaSat reported only one patent resulting from SBIR. Instead of patents, they rely on data rights from SBIR and rapid innovation and fielding to stay ahead of market. They do more patenting of the research funded by their IR&D program. The commercial side lacks the protection of DoD funded research.

## VIASAT VIEWS CONCERNING SBIR APPLICATION AND AWARD PROCESSES

The interviewee was not sure how ViaSat learned of SBIR, but during the timeframe of the first SBIR, the firm was actively seeking new sources of funding, and the founders were well connected with other leaders of small innovative companies in the communication technology rich San Diego area.

Geographical location was important to the firm's opportunity for proposing and receiving SBIR. San Diego in the 1980s and 1990s did quite well in communications technology. The University of California in San Diego was at the cutting edge in telecommunications. This impacted where the firm was founded. The Navy's SPAWAR, also in San Diego, had an active SBIR program in telecommunications.

Topics, technology, and prior experience have determined the agency to which the company proposed. The founders were experienced in working with DoD. DoD was in constant pursuit of improved communications. DoD provided the best opportunities. In addition to satellite communication, about half of the awards have dealt with line of sight, terrestrial communications.

All of their SBIR has been with DoD. ViaSat felt that the minor differences among the application and award processes of DoD component SBIR programs gave no perceived enduring advantage to one agency over another.

The firm's SBIR proposal strategy was to stay within its core competencies. It has had several awards on broad topics, others on narrow topics. It has worked with agencies to try to influence future topics. The number of proposals it submitted for any solicitation depended more on what topics were requested than any other factor. However, when business looked slow, more proposals may have occurred.

Most proposal work was done at night and on weekends. Its average investment on a Phase I was about $2000 to $3000. They once put in 22 proposals on a single solicitation. "Once you are doing a few, you often can raise that by a factor of two to three without that much additional work by taking advantage of similarities." The real work in obtaining a Phase II was in finding follow on sponsorship. This required finding and convincing other program managers to go to the SBIR sponsor and say that they wanted the result of the Phase II. SPAWAR, Hanscom AFB, MA; Rome AFB, NY; and Ft. Monmouth, NJ, were mentioned as locations that they visited

ViaSat has experience in applying for and receiving awards from other government R&D (non-SBIR) programs. In comparison to SBIR, they tend to partner more on other government R&D programs, which tend to be bigger and require much more complicated proposals. SBIR provides natural access and a much easier proposal process.

They would recommend reducing some of the bureaucratic requirements of SBIR application process. They pointed out that commercialization data requires more work. *(It should be noted that they had to enter data for many awards, and that they only participated in one solicitation that required that data before outgrowing the program. Once entered, updating for subsequent solicitations requires only a small fraction of the effort.)*

ViaSat believes some topics are well thought out, but some are not. Some are more rigorous, and validated. It is easier to propose if the topic is clear. It does believe that having some catch-all broad topics is a good thing. Topics should allow Phase I to focus on innovation.

From their perspective, the DoD practice of two SBIR solicitations per year was frequent enough.

They believe that the selection process appeared fair. The money pool seems to allow multiple awards on the same topic when appropriate.

## VIEWS ON FUNDING AMOUNTS AND TIMING

They often experienced delay between Phase I and II. Bridge funding, when available, was never enough. Delay used to require taking people off the effort. In the later years, once they had a healthy cash flow, it became an inconvenience, but in the early years delay was critical. Reduction in delay, and covering the gap would improve the program.

ViaSat no longer qualifies for Phase I but believes continuing to make small awards is better than increasing the size and giving less awards. However, the award needs to be large enough for the firms to demonstrate that they can make something of it at reasonable cost.

## OVERALL PROGRAM VIEWS

ViaSat identified dedicated government sponsors and noncompetitive Phase III as real strengths of the program. A sponsor who never gives up and advocates company efforts can be key to success. After Congress tried to clarify that the government could award Phase III noncompetitively, ViaSat still had difficulty convincing contracting officers that noncompetitive Phase III awards could be made. That gradually improved. ViaSat learned how to prove its case, but it may still be a problem for some firms and contracting officers. Noncompetitive Phase III gives small firms some leverage with primes. For ViaSat, outgrowing participation in Phases I and II does not prevent the award of Phase III. Continued eligibility for noncompetitive Phase III contracts is important to the continued positive impact of SBIR on firms that grow or are acquired.

A weakness that ViaSat perceives in the SBIR program is the disconnect between the SBIR firms and the primes. The primes have no incentives to use SBIR firms. Primes often see no advantage.

*Recommended Change* The government Planning and Programming process for R&D and Procurement makes it difficult to transition from Phase II. Every PE is programmed far in advance to be spent in a particular way. The successful Phase II becomes a spoiler. To get funded, you have to get support from an established program. They would like to see a change in the Planning and Programming process to make funding available for Phase III at conclusion of Phase II.

## RESEARCHER INSIGHTS

Each of the Armed Services are involved in what Secretary of Defense Rumsfeld refers to as Transformation—transformation in how they organize and how they fight. Central to all other DoD Transformation Initiatives is the concept of Network Centric Warfare (NCW). In simplest terms, NWC is waging war in the information age.

NCW means that information is acquired, processed into intelligence and provided to everyone who needs it seamlessly. Thus an aircraft would know exactly where all the threat acquisition and air defense weapons were on its path. A platoon would know the path to take to avoid observation and fires until it could flank and attack a threat force from an unexpected direction. A general would know where all the threat forces were and what they were doing, and everyone on the friendly side would be able to distinguish between threat combatants and friendly forces and noncombatants. Air, land and sea forces would be completely interoperable and mutually supporting.

There are many technical challenges to making the concept of NCW a reality, challenges in sensor systems, in information architectures, protocols and hardware, in understanding individual and group behavior and in the communications hardware required. ViaSat is a critical player in providing the interoperable communication pipes that will enable the Network. It has that capability because of the stimulus of SBIR.

The act establishing the SBIR program identified four goals for the program: technological innovation, commercialization, the use of small businesses to meet agencies' research and development needs, and participation by minorities and disadvantaged persons.

How do the SBIR awards at ViaSat measure up to these goals? ViaSat innovations spurred by SBIR have changed the industry; their $350 million in annual revenue (from a standing start) and their involvement in meeting not just the research and development needs of the agency, and in fact some of the most vital needs, gives evidence that they embody what the Congress was trying to foster. (They are not, however, minority or disadvantaged.)

Two points need to be made in light of current consideration by elements of the administration and Congress of Venture Capital and possible limits on the number of Phase II awards.

- ViaSat had Venture Capital before its first SBIR. Although the founding individuals never relinquished control of the company, this enormous program success and their significant contributions to National Defense might not have occurred under some interpretations of the presence of VC.
- ViaSat used 24 Phase II awards to develop its most innovative technology. It was a frequent winner. Some current initiatives under discussion for the SBIR Program would have eliminated it from SBIR, treating them as an "SBIR Mill." Such initiatives would limit SBIR eligibility to relatively few Phase II

awards per firm. The loss of the last nine or sixteen (choose the cutoff) of the ViaSat Phase II awards and the resulting Phase III would have materially reduced the contributions of ViaSat. We know that very few awards and very few SBIR companies succeed in achieving significant innovation, significant impact on meeting agency needs, and large-scale commercialization. Is SBIR best spent nurturing proven winners or in spreading it thinly with no focus on successful commercialization?

# Appendix E

# Bibliography


Acs, Z., and D. Audretsch. 1988. "Innovation in Large and Small Firms: An Empirical Analysis." *The American Economic Review* 78(4):678-690.

Acs, Z., and D. Audretsch. 1990. *Innovation and Small Firms.* Cambridge, MA: MIT Press.

Advanced Technology Program. 2001. *Performance of 50 Completed ATP Projects, Status Report 2.* National Institute of Standards and Technology Special Publication 950-2. Washington, DC: Advanced Technology Program/National Institute of Standards and Technology/U.S. Department of Commerce.

Alic, John A., Lewis Branscomb, Harvey Brooks, Ashton B. Carter, and Gerald L. Epstein. 1992. *Beyond Spinoff: Military and Commercial Technologies in a Changing World.* Boston, MA: Harvard Business School Press.

American Association for the Advancement of Science. "R&D Funding Update on NSF in the FY2007." Available online at <*http://www.aaas.org/spp/rd/nsf07hf1.pdf*>.

American Psychological Association. 2002. "Criteria for Evaluating Treatment Guidelines." *American Psychologist* 57(12):1052-1059.

Archibald, R., and D. Finifter. 2000. "Evaluation of the Department of Defense Small Business Innovation Research Program and the Fast Track Initiative: A Balanced Approach." In National Research Council. *The Small Business Innovation Research Program: An Assessment of the Department of Defense Fast Track Initiative.* Charles W. Wessner, ed. Washington, DC: National Academy Press.

Archibald, Robert, and David Finifter. 2003. "Evaluating the NASA Small Business Innovation Research Program: Preliminary Evidence of a Tradeoff Between Commercialization and Basic Research." *Research Policy* 32:605-619.

Arrow, Kenneth. 1962. "Economic welfare and the allocation of resources for invention." Pp. 609-625 in *The Rate and Direction of Inventive Activity: Economic and Social Factors.* Princeton, NJ: Princeton University Press.

Arrow, Kenneth. 1973. "The theory of discrimination." Pp. 3-31 in *Discrimination in Labor Market.* Orley Ashenfelter and Albert Rees, eds. Princeton, NJ: Princeton University Press.

Audretsch, David B. 1995. *Innovation and Industry Evolution.* Cambridge, MA: MIT Press.

Audretsch, David B., and Maryann P. Feldman. 1996. "R&D spillovers and the geography of innovation and production." *American Economic Review* 86(3):630-640.

Audretsch, David B., and Paula E. Stephan. 1996. "Company-scientist locational links: The case of biotechnology." *American Economic Review* 86(3):641-642.

Audretsch, D., J. Weigand, and C. Weigand. 2000. "Does the Small Business Innovation Research Program Foster Entrepreneurial Behavior." In National Research Council. *The Small Business Innovation Research Program: An Assessment of the Department of Defense Fast Track Initiative.* Charles W. Wessner, ed. Washington, DC: The National Academies Press.

Audretsch, D., and R. Thurik. 1999. *Innovation, Industry Evolution, and Employment.* Cambridge, MA: MIT Press.

Baker, Alan. No date. "Commercialization Support at NSF." Draft.

Barfield, C., and W. Schambra, eds. 1986. *The Politics of Industrial Policy.* Washington, DC: American Enterprise Institute for Public Policy Research.

Baron, Jonathan. 1998. "DoD SBIR/STTR Program Manager." Comments at the Methodology Workshop on the Assessment of Current SBIR Program Initiatives, Washington, DC, October.

Barry, C. B. 1994. "New directions in research on venture capital finance." *Financial Management* 23 (Autumn):3-15.

Bator, Francis. 1958. "The anatomy of market failure." *Quarterly Journal of Economics* 72: 351-379.

Bingham, R. 1998. *Industrial Policy American Style: From Hamilton to HDTV.* New York: M.E. Sharpe.

Birch, D. 1981. "Who Creates Jobs." *The Public Interest* 65 (Fall):3-14.

Branscomb, Lewis M., Kennth P. Morse, Michael J. Roberts, and Darin Boville. 2000. *Managing Technical Risk: Understanding Private Sector Decision Making on Early Stage Technology Based Projects.* Washington, DC: Department of Commerce/National Institute of Standards and Technology.

Branscomb, Lewis M., and Philip E. Auerswald. 2001. *Taking Technical Risks: How Innovators, Managers, and Investors Manage Risk in High-Tech Innovations*, Cambridge, MA: MIT Press.

Branscomb, L. M., and P. E. Auerswald. 2002. *Between Invention and Innovation: An Analysis of Funding for Early-Stage Technology Development.* Gaithersburg, MD: National Institute of Standards and Technology.

Branscomb, L. M., and P. E. Auerswald. 2003. "Valleys of Death and Darwinian Seas: Financing the Invention to Innovation Transition in the United States." *The Journal of Technology Transfer* 28(3-4).

Branscomb, Lewis M., and J. Keller. 1998. *Investing in Innovation: Creating a Research and Innovation Policy.* Cambridge, MA: MIT Press.

Brav, A., and P. A. Gompers. 1997. "Myth or reality?: Long-run underperformance of initial public offerings; Evidence from venture capital and nonventure capital-backed IPOs." *Journal of Finance* 52:1791-1821.

Brodd, R. J. 2005. *Factors Affecting U.S. Production Decisions: Why Are There No Volume Lithium-Ion Battery Manufacturers in the United States?* ATP Working Paper No. 05-01, June 2005.

Brown, G., and Turner J. 1999. "Reworking the Federal Role in Small Business Research." *Issues in Science and Technology* XV, no. 4 (Summer).

Bush, Vannevar. 1946. *Science—the Endless Frontier.* Republished in 1960 by U.S. National Science Foundation, Washington, DC.

Carden, S. D., and O. Darragh. 2004. "A Halo for Angel Investors." *The McKinsey Quarterly* 1.

Cassell, G. 2004. "Setting Realistic Expectations for Success." In National Research Council. *SBIR: Program Diversity and Assessment Challenges.* Charles W. Wessner, ed. Washington, DC: The National Academies Press.

Caves, Richard E. 1998. "Industrial organization and new findings on the turnover and mobility of firms." *Journal of Economic Literature* 36(4):1947-1982.

Christensen, C. 1997. *The Innovator's Dilemma.* Boston, MA: Harvard Business School Press.

Christensen, C. and M. Raynor. 2003. *Innovator's Solution*, Boston, MA: Harvard Business School.

Clinton, William Jefferson. 1994. *Economic Report of the President.* Washington, DC: U.S. Government Printing Office.

Clinton, William Jefferson. 1994. *The State of Small Business.* Washington, DC: U.S. Government Printing Office.

Coburn, C., and D. Bergland. 1995. *Partnerships: A Compendium of State and Federal Cooperative Technology Programs.* Columbus, OH: Battelle.

Cochrane, J. H. 2005. "The Risk and Return of Venture Capital." *Journal of Financial Economics* 75(1):3-52.

Cohen, L. R., and R. G. Noll. 1991. *The Technology Pork Barrel.* Washington, DC: The Brookings Institution.

Congressional Commission on the Advancement of Women and Minorities in Science, Engineering, and Technology Development. 2000. *Land of Plenty: Diversity as America's Competitive Edge in Science, Engineering and Technology.* Washington, DC: National Science Foundation/U.S. Government Printing Office.

Cooper, R. G. 2001. *Winning at New Products: Accelerating the process from idea to launch.* In Dawnbreaker, Inc. 2005. "The Phase III Challenge: Commercialization Assistance Programs 1990–2005." White paper. July 15.

Council of Economic Advisers. 1995. *Supporting Research and Development to Promote Economic Growth: The Federal Government's Role.* Washington, DC.

Council on Competitiveness. 2005. *Innovate America: Thriving in a World of Challenge and Change.* Washington, DC: Council on Competitiveness.

Cramer, Reid. 2000. "Patterns of Firm Participation in the Small Business Innovation Research Program in Southwestern and Mountain States." In National Research Council. 2000. *The Small Business Innovation Research Program: An Assessment of the Department of Defense Fast Track Initiative.* Charles W. Wessner, ed. Washington, DC: National Academy Press.

Cutler, D. 2005. *Your Money or Your Life.* New York: Oxford University Press.

David, P. A., B. H. Hall, and A. A. Tool. 1999. "Is Public R&D a Complement or Substitute for Private R&D? A Review of the Econometric Evidence." NBER Working Paper 7373. October.

Davidsson, P. 1996. "Methodological Concerns in the Estimation of Job Creation in Different Firm Size Classes." Working Paper. Jönköping International Business School.

Davis, S. J., J. Haltiwanger, and S. Schuh. 1994. "Small Business and Job Creation: Dissecting the Myth and Reassessing the Facts," *Business Economics* 29(3):113-122.

Dawnbreaker, Inc. 2005. "The Phase III Challenge: Commercialization Assistance Programs 1990–2005." White paper. July 15.

Dertouzos, M. L. 1989. *Made in America: The MIT Commission on Industrial Productivity.* Cambridge, MA: MIT Press.

Dess, G. G., and D. W. Beard. 1984. "Dimensions of Organizational Task Environments." *Administrative Science Quarterly* 29:52-73.

Devenow, A., and I. Welch. 1996. "Rational Herding in Financial Economics. *European Economic Review* 40(April):603-615.

DoE Opportunity Forum. 2005. "Partnering and Investment Opportunities for the Future." Tysons Corner, VA. October 24-25.

Ernst and Young. 2007. "U.S. Venture Capital Investment Increases to 8 percent to $6.96 Billion in First Quarter of 2007." April 23.

Eckstein, Otto. 1984. *DRI Report on U.S. Manufacturing Industries.* New York: McGraw Hill.

Eisinger, P. K. 1988. *The Rise of the Entrepreneurial State: State and Local Economic Development Policy in the United State.* Madison, WI: University of Wisconsin Press.

Evenson, R., P. Waggoner, and P. Ruttan. 1979. "Economic Benefits from Research: An Example from Agriculture," *Science*, 205(14 September):1101-1107.

Feldman, Maryann P. 1994. *The Geography of Knowledge.* Boston, MA: Kluwer Academic.

Feldman, Maryann P. 1994. "Knowledge complementarity and innovation." *Small Business Economics* 6(5):363-372.

Feldman, M. P. 2000. "Role of the Department of Defense in Building Biotech Expertise." In National Research Council. *The Small Business Innovation Research Program: An Assessment of the Department of Defense Fast Track Initiative.* Charles W. Wessner, ed. Washington, DC: The National Academies Press.

Feldman, M. P. 2001. "Assessing the ATP: Halo Effects and Added Value." In National Research Council, *The Advanced Technology Program: Assessing Outcomes.* Washington, DC: National Academy Press.

Feldman, M. P., and M. R. Kelley. 2001. "Leveraging Research and Development: The Impact of the Advanced Technology Program." In National Research Council. *The Advanced Technology Program.* Charles W. Wessner, ed. Washington, DC: National Academy Press.

Feldman, M. P., and M. R. Kelley. 2001. *Winning an Award from the Advanced Technology Program: Pursuing R&D Strategies in the Public Interest and Benefiting from a Halo Effect.* NISTIR 6577. Washington, DC: Advanced Technology Program/National Institute of Standards and Technology/U.S. Department of Commerce.

Fenn, G. W., N. Liang, and S. Prowse. 1995. *The Economics of the Private Equity Market.* Washington, DC: Board of Governors of the Federal Reserve System.

*Financial Times.* 2004. "Qinetiq set to make its first US acquisition," September 8.

Flamm, K. 1988. *Creating the Computer.* Washington, DC: The Brookings Institution.

Flender, J. O., and R. S. Morse. 1975. *The Role of New Technical Enterprise in the U.S. Economy.* Cambridge, MA: MIT Development Foundation.

Freear, J., and W. E. Wetzel Jr. 1990. "Who bankrolls high-tech entrepreneurs?" *Journal of Business Venturing* 5:77-89.

Freeman, Chris, and Luc Soete. 1997. *The Economics of Industrial Innovation.* Cambridge, MA: MIT Press.

Galbraith, J. K. 1957. *The New Industrial State.* Boston: Houghton Mifflin.

Geroski, Paul A. 1995. "What do we know about entry?" *International Journal of Industrial Organization* 13(4):421-440.

Geshwiler, J., J. May, and M. Hudson. 2006. "State of Angel Groups." Kansas City, MO: Kauffman Foundation.

Gompers, P. A., and J. Lerner. 1977. "Risk and Reward in Private Equity Investments: The Challenge of Performance Assessment." *Journal of Private Equity* 1:5-12.

Gompers, P. A. 1995. "Optimal investment, monitoring, and the staging of venture capital." *Journal of Finance* 50:1461-1489.

Gompers, P. A., and J. Lerner. 1996. "The use of covenants: An empirical analysis of venture partnership agreements." *Journal of Law and Economics* 39:463-498.

Gompers, P. A., and J. Lerner. 1998. "Capital formation and investment in venture markets: A report to the NBER and the Advanced Technology Program." Unpublished working paper. Harvard University.

Gompers, P. A., and J. Lerner. 1998. "What drives venture capital fund-raising?" Unpublished working paper. Harvard University.

Gompers, P. A., and J. Lerner. 1999. "An analysis of compensation in the U.S. venture capital partnership." *Journal of Financial Economics* 51(1):3-7.

Gompers, P. A., and J. Lerner. 1999. *The Venture Cycle.* Cambridge, MA: MIT Press.

Good, M. L. 1995. Prepared testimony before the Senate Commerce, Science, and Transportation Committee, Subcommittee on Science, Technology, and Space (photocopy, U.S. Department of Commerce).

Goodnight, J. 2003. Presentation at National Research Council Symposium. "The Small Business Innovation Research Program: Identifying Best Practice." Washington, DC May 28.

Graham, O. L. 1992. *Losing Time: The Industrial Policy Debate.* Cambridge, MA: Harvard University Press.

Greenwald, B. C., J. E. Stiglitz, and A. Weiss. 1984. "Information imperfections in the capital market and macroeconomic fluctuations." *American Economic Review Papers and Proceedings* 74:194-199.

Griliches, Z. 1990. *The Search for R&D Spillovers.* Cambridge, MA: Harvard University Press.

Groves, R. M., F. J. Fowler, Jr., M. P. Couper, J. M. Lepkowski, E. Singer, and R. Tourangeau. 2004. *Survey Methodology.* Hoboken, NJ: John Wiley & Sons, Inc.

Hall, Bronwyn H. 1992. "Investment and research and development: Does the source of financing matter?" Working Paper No. 92-194, Department of Economics/University of California at Berkeley.

Hall, Bronwyn H. 1993. "Industrial research during the 1980s: Did the rate of return fall?" Brookings Papers: *Microeconomics* 2:289-343.

Haltiwanger, J., and C. J. Krizan. 1999. "Small Businesses and Job Creation in the United States: The Role of New and Young Businesses" in *Are Small Firms Important? Their Role and Impact,* Zoltan J. Acs, ed., Dordrecht: Kluwer.

Hamberg, Dan. 1963. "Invention in the industrial research laboratory." *Journal of Political Economy* (April):95-115.

Hao, K. Y., and A. B. Jaffe. 1993. "Effect of liquidity on firms' R&D spending." *Economics of Innovation and New Technology* 2:275-282.

Hebert, Robert F., and Albert N. Link. 1989. "In search of the meaning of entrepreneurship." *Small Business Economics* 1(1):39-49.

Heilman, C. 2005. "Partnering for Vaccines: The NIAID Perspective" in Charles W. Wessner, ed. *Partnering Against Terrorism: Summary of a Workshop.* Washington, DC: The National Academies Press.

Held, B., T. Edison, S. L. Pfleeger, P. Anton, and J. Clancy. 2006. *Evaluation and Recommendations for Improvement of the Department of Defense Small Business Innovation Research (SBIR) Program.* Arlington, VA: RAND National Defense Research Institute.

Holland, C. 2007. "Meeting Mission Needs." In National Research Council. *SBIR and the Phase III Challenge of Commercialization.* Charles W. Wessner, ed. Washington, DC: The National Academies Press.

Himmelberg, C. P., and B. C. Petersen. 1994. "R&D and internal finance: A panel study of small firms in high-tech industries." *Review of Economics and Statistics* 76:38-51.

Hubbard, R. G. 1998. "Capital-market imperfections and investment." *Journal of Economic Literature* 36:193-225.

Huntsman, B., and J. P. Hoban Jr. 1980. "Investment in new enterprise: Some empirical observations on risk, return, and market structure." *Financial Management* 9 (Summer):44-51.

Institute of Medicine. 1998. "The Urgent Need to Improve Health Care Quality." National Roundtable on Health Care Quality. *Journal of the American Medical Association* 280(11):1003, September 16.

Jacobs, T. 2002. "Biotech Follows Dot.com Boom and Bust." *Nature* 20(10):973.

Jaffe, A. B. 1996. "Economic Analysis of Research Spillovers: Implications for the Advanced Technology Program." Washington, DC: Advanced Technology Program/National Institute of Standards and Technology/U.S. Department of Commerce).

Jaffe, A. B. 1998. "Economic Analysis of Research Spillovers: Implications for the Advanced Technology Program." Washington, DC: Advanced Technology Program/National Institute of Standards and Technology/U.S. Department of Commerce.

Jaffe, A. B. 1998. "The importance of 'spillovers' in the policy mission of the Advanced Technology Program." *Journal of Technology Transfer* (Summer).

Jewkes, J., D. Sawers, and R. Stillerman. 1958. *The Sources of Invention.* New York: St. Martin's Press.

Jarboe, K. P., and R. D. Atkinson. 1998. "The Case for Technology in the Knowledge Economy: R&D, Economic Growth and the Role of Government." Washington, DC: Progressive Policy Institute. Available online at <*http://www.ppionline.org/documents/CaseforTech.pdf*>.

Johnson, M. 2004. "SBIR at the Department of Energy: Achievements, Opportunities, and Challenges." In National Research Council. *SBIR: Program Diversity and Assessment Challenges.* Charles W. Wessner, ed. Washington, DC: The National Academies Press.

Kauffman Foundation. About the Foundation. Available online at *<http://www.kauffman.org/foundation.cfm>*.

Kleinman, D. L. 1995. *Politics on the Endless Frontier: Postwar Research Policy in the United States.* Durham, NC: Duke University Press.

Kortum, Samuel, and Josh Lerner. 1998. "Does Venture Capital Spur Innovation?" NBER Working Paper No. 6846, National Bureau of Economic Research.

Krugman, P. 1990. *Rethinking International Trade.* Cambridge, MA: MIT Press.

Krugman, P. 1991. *Geography and Trade.* Cambridge, MA: MIT Press.

Langlois, Richard N., and Paul L. Robertson. 1996. "Stop Crying over Spilt Knowledge: A Critical Look at the Theory of Spillovers and Technical Change." Paper prepared for the MERIT Conference on Innovation, Evolution, and Technology. Maastricht, Netherlands, August 25-27.

Langlois, R. N. 2001. "Knowledge, Consumption, and Endogenous Growth." *Journal of Evolutionary Economics* 11:77-93.

Lebow, I. 1995. *Information Highways and Byways: From the Telegraph to the 21st Century.* New York: Institute of Electrical and Electronic Engineering.

Lerner, J. 1994. "The syndication of venture capital investments." *Financial Management* 23 (Autumn):16-27.

Lerner, J. 1995. "Venture capital and the oversight of private firms." *Journal of Finance* 50: 301-318.

Lerner, J. 1996. "The government as venture capitalist: The long-run effects of the SBIR program." Working Paper No. 5753, National Bureau of Economic Research.

Lerner, J. 1998. "Angel financing and public policy: An overview." *Journal of Banking and Finance* 22(6-8):773-784.

Lerner, J. 1999. "The government as venture capitalist: The long-run effects of the SBIR program." *Journal of Business* 72(3):285-297.

Lerner, J. 1999. "Public venture capital: Rationales and evaluation." In *The SBIR Program: Challenges and Opportunities*. Washington, DC: National Academy Press.

Levy, D. M., and N. Terleckyk. 1983. "Effects of government R&D on private R&D investment and productivity: A macroeconomic analysis." *Bell Journal of Economics* 14:551-561.

Liles, P. 1977. *Sustaining the Venture Capital Firm.* Cambridge, MA: Management Analysis Center.

Link, Albert N. 1998. "Public/Private Partnerships as a Tool to Support Industrial R&D: Experiences in the United States." Paper prepared for the working group on Innovation and Technology Policy of the OECD Committee for Science and Technology Policy, Paris.

Link, Albert N., and John Rees. 1990. "Firm size, university based research and the returns to R&D." *Small Business Economics* 2(1):25-32.

Link, Albert N., and John T. Scott. 1998. "Assessing the infrastructural needs of a technology-based service sector: A new approach to technology policy planning." *STI Review* 22:171-207.

Link, Albert N., and John T. Scott. 1998. *Overcoming Market Failure: A Case Study of the ATP Focused Program on Technologies for the Integration of Manufacturing Applications (TIMA).* Draft final report submitted to the Advanced Technology Program. Gaithersburg, MD: National Institute of Technology. October.

Link, Albert N., and John T. Scott. 1998. *Public Accountability: Evaluating Technology-Based Institutions.* Norwell, MA: Kluwer Academic.

Link, Albert N., and John T. Scott. 2005. *Evaluating Public Research Institutions: The U.S. Advanced Technology Program's Intramural Research Initiative.* London: Routledge.

Longini, P. 2003. "Hot buttons for NSF SBIR Research Funds," Pittsburgh Technology Council, *TechyVent*. November 27.

Malone, T. 1995. *The Microprocessor: A Biography.* Hamburg, Germany: Springer Verlag/Telos.

Mankins, John C. 1995. *Technology Readiness Levels: A White Paper.* Washington, DC: NASA Office of Space Access and Technology, Advanced Concepts Office.

Mansfield, E. 1985. "How Fast Does New Industrial Technology Leak Out?" *Journal of Industrial Economics* 34(2).

Mansfield, E. 1996. *Estimating Social and Private Returns from Innovations Based on the Advanced Technology Program: Problems and Opportunities.* Unpublished report.

Mansfield, E., J. Rapoport, A. Romeo, S. Wagner, and G. Beardsley. 1977. "Social and private rates of return from industrial innovations." *Quarterly Journal of Economics* 91:221-240.

Martin, Justin. 2002. "David Birch." *Fortune Small Business* (December 1).

McCraw, T. 1986. "Mercantilism and the Market: Antecedents of American Industrial Policy." In C. Barfield and W. Schambra, eds. *The Politics of Industrial Policy.* Washington, DC: American Enterprise Institute for Public Policy Research.

Mervis, Jeffrey D. 1996. "A $1 Billion 'Tax' on R&D Funds." *Science* 272:942–944.

Moore, D. 2004. "Turning Failure into Success." In National Research Council. *The Small Business Innovation Research Program: Program Diversity and Assessment Challenges.* Charles W. Wessner, ed. Washington, DC: The National Academies Press.

Morgenthaler, D. 2000. "Assessing Technical Risk," in L. M. Branscomb, K. P. Morse, and M. J. Roberts, eds. *Managing Technical Risk: Understanding Private Sector Decision Making on Early Stage Technology-Based Project.* Gaithersburg, MD: National Institute of Standards and Technology.

Mowery, D. 1998. "Collaborative R&D: how effective is it?" *Issues in Science and Technology* (Fall):37-44.

Mowery, D., ed. 1999. *U.S. Industry in 2000: Studies in Competitive Performance.* Washington, DC: National Academy Press.

Mowery, D., and N. Rosenberg. 1989. *Technology and the Pursuit of Economic Growth.* New York: Cambridge University Press.

Mowery, D., and N. Rosenberg. 1998. *Paths of Innovation: Technological Change in 20th Century America.* New York: Cambridge University Press.

Murphy, L. M., and P. L. Edwards. 2003. *Bridging the Valley of Death—Transitioning from Public to Private Sector Financing.* Golden, CO: National Renewable Energy Laboratory. May.

Myers, S., R. L. Stern, and M. L. Rorke. 1983. *A Study of the Small Business Innovation Research Program.* Lake Forest, IL: Mohawk Research Corporation.

Myers, S. C., and N. Majluf. 1984. "Corporate financing and investment decisions when firms have information that investors do not have." *Journal of Financial Economics* 13:187-221.

National Aeronautics and Space Administration. 2002. "Small Business/SBIR: NICMOS Cryo-cooler—Reactivating a Hubble Instrument." *Aerospace Technology Innovation* 10(4):19-21.

National Aeronautics and Space Administration. 2005. "The NASA SBIR and STTR Programs Participation Guide." Available online at <*http://sbir.gsfc.nasa.gov/SBIR/zips/guide.pdf*>

National Institutes of Health. 2003. Road Map for Medical Research. Available online at <*http://nihroadmap.nih.gov/*>.

National Institutes of Health. 2005. *Report on the Second of the 2005 Measures Updates: NIH SBIR Performance Outcomes Data System (PODS).*

National Research Council. 1986. *The Positive Sum Strategy: Harnessing Technology for Economic Growth.* Washington, DC: National Academy Press.

National Research Council. 1987. *Semiconductor Industry and the National Laboratories: Part of a National Strategy.* Washington, DC: National Academy Press.

National Research Council. 1991. *Mathematical Sciences, Technology, and Economic Competitiveness.* James G. Glimm, ed. Washington, DC: National Academy Press.

National Research Council. 1992. *The Government Role in Civilian Technology: Building a New Alliance.* Washington, DC: National Academy Press.

National Research Council. 1995. *Allocating Federal Funds for R&D.* Washington, DC: National Academy Press.

National Research Council. 1996. *Conflict and Cooperation in National Competition for High-Technology Industry.* Washington, DC: National Academy Press.

National Research Council. 1997. *Review of the Research Program of the Partnership for a New Generation of Vehicles: Third Report.* Washington, DC: National Academy Press.

National Research Council. 1999. *The Advanced Technology Program: Challenges and Opportunities.* Charles W. Wessner, ed. Washington, DC: National Academy Press.

National Research Council. 1999. *Funding a Revolution: Government Support for Computing Research.* Washington, DC: National Academy Press.

National Research Council. 1999. *Industry-Laboratory Partnerships: A Review of the Sandia Science and Technology Park Initiative.* Charles W. Wessner, ed. Washington, DC: National Academy Press.

National Research Council. 1999. *New Vistas in Transatlantic Science and Technology Cooperation.* Charles W. Wessner, ed. Washington, DC: National Academy Press.

National Research Council. 1999. *The Small Business Innovation Research Program: Challenges and Opportunities.* Charles W. Wessner, ed. Washington, DC: National Academy Press.

National Research Council. 2000. *The Small Business Innovation Research Program: An Assessment of the Department of Defense Fast Track Initiative.* Charles W. Wessner, ed. Washington, DC: National Academy Press.

National Research Council. 2000. *U.S. Industry in 2000: Studies in Competitive Performance.* Washington, DC: National Academy Press.

National Research Council. 2001. *The Advanced Technology Program: Assessing Outcomes.* Charles W. Wessner, ed. Washington, DC: National Academy Press.

National Research Council. 2001. *Attracting Science and Mathematics Ph.Ds to Secondary School Education.* Washington, DC: National Academy Press.

National Research Council. 2001. *Building a Workforce for the Information Economy.* Washington, DC: National Academy Press.

National Research Council. 2001. *Capitalizing on New Needs and New Opportunities: Government-Industry Partnerships in Biotechnology and Information Technologies.* Charles W. Wessner, ed. Washington, DC: National Academy Press.

National Research Council. 2001. *A Review of the New Initiatives at the NASA Ames Research Center.* Charles W. Wessner, ed. Washington, DC: National Academy Press.

National Research Council. 2001. *Trends in Federal Support of Research and Graduate Education.* Washington, DC: National Academy Press.

National Research Council. 2002. *Government-Industry Partnerships for the Development of New Technologies: Summary Report.* Charles W. Wessner, ed. Washington, DC: The National Academies Press.

National Research Council. 2002. *Making the Nation Safer: The Role of Science and Technology in Countering Terrorism.* Washington, DC: The National Academies Press.

National Research Council. 2002. *Measuring and Sustaining the New Economy.* Dale W. Jorgenson and Charles W. Wessner, eds. Washington, DC: National Academy Press.

National Research Council. 2002. *Partnerships for Solid-State Lighting.* Charles W. Wessner, ed. Washington, DC: The National Academies Press.

National Research Council. 2004. *An Assessment of the Small Business Innovation Research Program: Project Methodology.* Washington, DC: The National Academies Press.

National Research Council. 2004. Capitalizing on Science, Technology, and Innovation: An Assessment of the Small Business Innovation Research Program/Program Manager Survey. Completed by Dr. Joseph Hennessey.

National Research Council. 2004. *Productivity and Cyclicality in Semiconductors: Trends, Implications, and Questions.* Dale W. Jorgenson and Charles W. Wessner, eds. Washington, DC: The National Academies Press.

National Research Council. 2004. *The Small Business Innovation Research Program: Program Diversity and Assessment Challenges.* Charles W. Wessner, ed. Washington, DC: The National Academies Press.

National Research Council. 2006. *Beyond Bias and Barriers: Fulfilling the Potential of Women in Academic Science and Engineering.* Washington, DC: The National Academies Press.

National Research Council. 2006. *Deconstructing the Computer.* Dale W. Jorgenson and Charles W. Wessner, eds. Washington, DC: The National Academies Press.

National Research Council. 2006. *Software, Growth, and the Future of the U.S. Economy.* Dale W. Jorgenson and Charles W. Wessner, eds. Washington, DC: The National Academies Press.

National Research Council. 2006. *The Telecommunications Challenge: Changing Technologies and Evolving Policies.* Dale W. Jorgenson and Charles W. Wessner, eds. Washington, DC: The National Academies Press.

National Research Council. 2007. *Enhancing Productivity Growth in the Information Age: Measuring and Sustaining the New Economy.* Dale W. Jorgenson and Charles W. Wessner, eds. Washington, DC: The National Academies Press.

National Research Council. 2007. *India's Changing Innovation System: Achievements, Challenges, and Opportunities for Cooperation.* Charles W. Wessner and Sujai J. Shivakumar, eds. Washington, DC: The National Academies Press.

National Research Council. 2007. *Innovation Policies for the 21st Century.* Charles W. Wessner, ed. Washington, DC: The National Academies Press.

National Research Council. 2007. *SBIR and the Phase III Challenge of Commercialization.* Charles W. Wessner, ed. Washington, DC: The National Academies Press.

National Research Council. 2008. *An Assessment of the SBIR Program at the Department of Defense.* Charles W. Wessner, ed. Washington, DC: The National Academies Press.

National Research Council. 2008. *An Assessment of the SBIR Program at the Department of Energy.* Charles W. Wessner, ed. Washington, DC: The National Academies Press.

National Research Council. 2008. *An Assessment of the SBIR Program at the National Science Foundation.* Charles W. Wessner, ed. Washington, DC: The National Academies Press.

National Research Council. 2009. *An Assessment of the SBIR Program at the National Aeronautics and Space Administration.* Charles W. Wessner, ed. Washington, DC: The National Academies Press.

National Research Council. 2009. *An Assessment of the SBIR Program at the National Institutes of Health.* Charles W. Wessner, ed. Washington, DC: The National Academies Press.

National Science Board. 2005. *Science and Engineering Indicators 2005.* Arlington, VA: National Science Foundation.

National Science Board. 2006. *Science and Engineering Indicators 2006.* Arlington, VA: National Science Foundation.

National Science Foundation. 2004. *Federal R&D Funding by Budget Function: Fiscal Years 2003-2005 (historical tables).* NSF 05-303. Arlington, VA: National Science Foundation.

National Science Foundation. 2006. "SBIR/STTR Phase II Grantee Conference, Book of Abstracts." Office of Industrial Innovation. May 18-20, 2006, Louisville, Kentucky.

National Science Foundation. Committee of Visitors Reports and Annual Updates. Available online at <*http://www.nsf.gov/eng/general/cov/*>.

National Science Foundation. Emerging Technologies. Available online at <*http://www.nsf.gov/eng/sbir/eo.jsp*>.

National Science Foundation. Guidance for Reviewers. Available online at <*http://www.eng.nsf.gov/sbir/peer_review.htm*>.

National Science Foundation. National Science Foundation at a Glance. Available online at <*http://www.nsf.gov/about*>.

National Science Foundation. National Science Foundation Manual 14, *NSF Conflicts of Interest and Standards of Ethical Conduct.* Available online at <*http://www.eng.nsf.gov/sbir/COI_Form.doc*>.

National Science Foundation. The Phase IIB Option. Available online at <*http://www.nsf.gov/eng/sbir/phase_IIB.jsp#ELIGIBILITY*>.

National Science Foundation. Proposal and Grant Manual. Available online at <*http://www.inside.nsf.gov/pubs/2002/pam/pamdec02.6html*>.

National Science Foundation. 2005. Synopsis of SBIR/STTR Program. Available online at <*http://www.nsf.gov/funding/pgm_summ.jsp?Phase Ims_id=13371&org=DMII*>.

National Science Foundation. 2006. "SBIR/STTR Phase II Grantee Conference, Book of Abstracts." Office of Industrial Innovation. May 18-20, 2006, Louisville, Kentucky.

National Science Foundation. 2006. "News items from the past year." Press Release. April 10.

National Science Foundation, Office of Industrial Innovation. Draft Strategic Plan, June 2, 2005.

National Science Foundation, Office of Legislative and Public Affairs. 2003. SBIR Success Story from News Tip. Web's "Best Meta-Search Engine," March 20.

National Science Foundation, Office of Legislative and Public Affairs. 2004. SBIR Success Story: GPRA Fiscal Year 2004 "Nugget." Retrospective Nugget–AuxiGro Crop Yield Enhancers.

Nelson, R. R. 1982. *Government and Technological Progress.* New York: Pergamon.

Nelson, R. R. 1986. "Institutions supporting technical advances in industry." *American Economic Review, Papers and Proceedings* 76(2):188.

Nelson, R. R., ed. 1993. *National Innovation System: A Comparative Study.* New York: Oxford University Press.

Office of Management and Budget. 1996. "Economic analysis of federal regulations under Executive Order 12866."

Office of Management and Budget. 2004. *"What Constitutes Strong Evidence of Program Effectiveness."* Available online at <*http://www.whitehouse.gov/omb/part/2004_program_eval.pdf*>.

Office of the President. 1990. *U.S. Technology Policy.* Washington, DC: Executive Office of the President.

Organization for Economic Cooperation and Development. 1982. *Innovation in Small and Medium Firms.* Paris: Organization for Economic Cooperation and Development.

Organization for Economic Cooperation and Development. 1995. *Venture Capital in OECD Countries.* Paris: Organization for Economic Cooperation and Development.

Organization for Economic Cooperation and Development. 1997. *Small Business Job Creation and Growth: Facts, Obstacles, and Best Practices.* Paris: Organization for Economic Cooperation and Development.

Organization for Economic Cooperation and Development. 1998. *Technology, Productivity and Job Creation: Toward Best Policy Practice.* Paris: Organization for Economic Cooperation and Development.

Organization for Economic Cooperation and Development. 2006. "Evaluation of SME Policies and Programs: Draft OECD Handbook." *OECD Handbook.* CFE/SME 17. Paris: Organization for Economic Cooperation and Development.

Pacific Northwest National Laboratory. SBIR Alerting Service. Available online at <*http://www.pnl.gov/edo/sbir*>.

Perko, J. S., and F. Narin. 1997. "The Transfer of Public Science to Patented Technology: A Case Study in Agricultural Science." *Journal of Technology Transfer* 22(3):65-72.

Perret, G. 1989. *A Country Made by War: From the Revolution to Vietnam—The Story of America's Rise to Power.* New York: Random House.

Porter, Michael E. 1998. "Clusters and Competition: New Agendas for Government and Institutions." In Michael E. Porter, ed. *On Competition.* Boston, MA: Harvard Business School Press.

Powell, J. W. 1999. *Business Planning and Progress of Small Firms Engaged in Technology Development through the Advanced Technology Program.* NISTIR 6375. National Institute of Standards and Technology/U.S. Department of Commerce.

Powell, Walter W., and Peter Brantley. 1992. "Competitive cooperation in biotechnology: Learning through networks?" In N. Nohria and R. G. Eccles, eds. *Networks and Organizations: Structure, Form and Action.* Boston, MA: Harvard Business School Press. Pp. 366-394.

Price Waterhouse. 1985. *Survey of Small High-tech Businesses Shows Federal SBIR Awards Spurring Job Growth, Commercial Sales.* Washington, DC: Small Business High Technology Institute.

Roberts, Edward B. 1968. "Entrepreneurship and technology." *Research Management* (July): 249-266.

Romer, P. 1990. "Endogenous technological change." *Journal of Political Economy* 98:71-102.

Rosa, Peter, and Allison Dawson. 2006. "Gender and the commercialization of university science: Academic founders of spinout companies." *Entrepreneurship & Regional Development* 18(4):341-366. July.

Rosenberg, N. 1969. "The Direction of Technological Change: Inducement Mechanisms and Focusing Devices." *Economic Development and Cultural Change*, 18:1-24.

Rosenbloom, R., and Spencer, W. 1996. *Engines of Innovation: U.S. Industrial Research at the End of an Era.* Boston, MA: Harvard Business School Press.

Rubenstein, A. H. 1958. *Problems Financing New Research-Based Enterprises in New England.* Boston, MA: Federal Reserve Bank.

Ruegg, Rosalie, and Irwin Feller. 2003. *A Toolkit for Evaluating Public R&D Investment Models, Methods, and Findings from ATP's First Decade.* NIST GCR 03-857.

Ruegg, Rosalie, and Patrick Thomas. 2007. *Linkages from DoE's Vehicle Technologies R&D in Advanced Energy Storage to Hybrid Electric Vehicles, Plug-in Hybrid Electric Vehicles, and Electric Vehicles.* Washington, DC: U.S. Department of Energy/Office of Energy Efficiency and Renewable Energy.

Sahlman, W. A. 1990. "The structure and governance of venture capital organizations." *Journal of Financial Economics* 27:473-521.

Saxenian, Annalee. 1994. *Regional Advantage: Culture and Competition in Silicon Valley and Route 128.* Cambridge, MA: Harvard University Press.

SBIR World. SBIR World: A World of Opportunities. Available online at <*http://www.sbirworld.com*>.

Scherer, F. M. 1970. *Industrial Market Structure and Economic Performance.* New York: Rand McNally College Publishing.

Schumpeter, J. 1950. *Capitalism, Socialism, and Democracy.* New York: Harper and Row.

Scotchmer, S. 2004. *Innovation and Incentives.* Cambridge MA: MIT Press.

Scott, J. T. 1998. "Financing and leveraging public/private partnerships: The hurdle-lowering auction." *STI Review* 23:67-84.

Scott, J. T. 2000. "An Assessment of the Small Business Innovation Research Program in New England: Fast Track Compared with Non-Fast Track." In National Research Council. *The Small Business Innovation Research Program: An Assessment of the Department of Defense Fast Track Initiative.* Charles W. Wessner, ed. Washington, DC: National Academy Press.

Siegel, D., D. Waldman, and A. Link. 2004. "Toward a Model of the Effective Transfer of Scientific Knowledge from Academicians to Practitioners: Qualitative Evidence from the Commercialization of University Technologies." *Journal of Engineering and Technology Management* 21(1-2).

Silverstein, S. C., H. H. Garrison, and S. J. Heinig. 1995. "A Few Basic Economic Facts about Research in the Medical and Related Life Sciences." FASEB 9:833-840.

Society for Prevention Research. 2004. *Standards of Evidence: Criteria for Efficacy, Effectiveness and Dissemination.* Available online at <*http://www.preventionresearch.org/softext.php*>.

Sohl, Jeffrey. 1999. *Venture Capital* 1(2).

Sohl, Jeffery, John Freear, and W.E. Wetzel Jr. 2002. "Angles on Angels: Financing Technology-Based Ventures—An Historical Perspective." *Venture Capital: An International Journal of Entrepreneurial Finance* 4(4).

Solow, R. S. 1957. "Technical Change and the Aggregate Production Function." *Review of Economics and Statistics* 39:312-320.

Stiglitz, J. E., and A. Weiss. 1981. "Credit rationing in markets with incomplete information." *American Economic Review* 71:393-409.

Stokes, Donald E. 1997. *Pasteur's Quadrant: Basic Science and Technological Innovation.* Washington, DC: The Brookings Institution.

Stowsky, J. 1996. "Politics and Policy: The Technology Reinvestment Program and the Dilemmas of Dual Use." Mimeo. University of California.

Tassey, Gregory. 1997. *The Economics of R&D Policy.* Westport, CT: Quorum Books.

Tibbetts, R. 1997. "The Role of Small Firms in Developing and Commercializing New Scientific Instrumentation: Lessons from the U.S. Small Business Innovation Research Program," in J. Irvine, B. Martin, D. Griffiths, and R. Gathier, eds. *Equipping Science for the 21st Century.* Cheltenham UK: Edward Elgar Press.

Tirman, John. 1984. *The Militarization of High Technology.* Cambridge, MA: Ballinger.

Tyson, Laura, Tea Petrin, and Halsey Rogers. 1994. "Promoting entrepreneurship in Eastern Europe." *Small Business Economics* 6:165-184.

University of New Hampshire Center for Venture Research. 2007. *The Angel Market in 2006.* Available online at *<http://wsbe2.unh.edu/files/Full%20Year%202006%20Analysis%20Report%20-%20March%202007.pdf>.*

U.S. Congress, House Committee on Science, Space, and Technology. 1992. *SBIR and Commercialization: Hearing Before the Subcommittee on Technology and Competitiveness of the House Committee on Science, Space, and Technology, on the Small Business Innovation Research [SBIR] Program.* Testimony of James A. Block, President of Creare, Inc. Pp. 356-361.

U.S. Congress. House Committee on Science, Space, and Technology. 1992. *The Small Business Research and Development Enhancement Act of 1992.* House Report (REPT. 102-554) Part I (Committee on Small Business).

U.S. Congress. House Committee on Science, Space, and Technology. 1998. *Unlocking Our Future: Toward a New National Science Policy: A Report to Congress by the House Committee on Science, Space, and Technology.* Washington, DC: Government Printing Office. Available online at *<http://www.access.gpo.gov/congress/house/science/cp105-b/science105b.pdf>.*

U.S. Congress. House Committee on Small Business. Subcommittee on Workforce, Empowerment, and Government Programs. 2005. *The Small Business Innovation Research Program: Opening Doors to New Technology.* Testimony by Joseph Hennessey. 109th Cong., 1st sess., November 8.

U.S. Congress. House Committee on Science, Space, and Technology. Subcommittee on Technology and Innovation. 2007. Hearing on "Small Business Innovation Research Authorization on the 25th Program Anniversary." Testimony by Robert Schmidt. April 26.

U.S. Congress. Senate Committee on Small Business. 1999. Senate Report 106-330. Small Business Innovation Research (SBIR) Program. August 4, 1999. Washington, DC: U.S. Government Printing Office.

U.S. Congress. Senate Committee on Small Business. 1981. Small Business Research Act of 1981. S.R. 194, 97th Congress.

U.S. Congress. Senate Committee on Small Business. 1999. Senate Report 106-330. *Small Business Innovation Research (SBIR) Program.* August 4. Washington, DC: U.S. Government Printing Office.

U.S. Congress. Senate Committee on Small Business. 2006. *Strengthening the Participation of Small Businesses in Federal Contracting and Innovation Research Programs.* Testimony by Michael Squillante. 109th Cong., 2nd sess., July 12.

U.S. Congressional Budget Office. 1985. *Federal financial support for high-technology industries.* Washington, DC: U.S. Congressional Budget Office.

U.S. Department of Education. 2005. "Scientifically-Based Evaluation Methods: Notice of Final Priority." *Federal Register* 70(15):3586-3589.

U.S. Food and Drug Administration. 1981. Protecting Human Subjects: Untrue Statements in Application. 21 C.F.R. §314.12

U.S. Food and Drug Administration. *Critical Path Initiative.* Available online at *<http://www.fda.gov/oc/initiatives/criticalpath/>.*

U.S. General Accounting Office. 1987. *Federal research: Small Business Innovation Research participants give program high marks.* Washington, DC: U.S. General Accounting Office.

U.S. General Accounting Office. 1989. *Federal Research: Assessment of Small Business Innovation Research Program.* Washington, DC: U.S. General Accounting Office.

U.S. General Accounting Office. 1992. *Federal Research: Small Business Innovation Research Program Shows Success but Can Be Strengthened.* RCED–92–32. Washington, DC: U.S. General Accounting Office.

U.S. General Accounting Office. 1997. *Federal Research: DoD's Small Business Innovation Research Program.* RCED–97–122, Washington, DC: U.S. General Accounting Office.

U. S. General Accounting Office. 1998. *Federal Research: Observations on the Small Business Innovation Research Program.* RCED–98–132. Washington, DC: U.S. General Accounting Office.

U.S. General Accounting Office. 1999. *Federal Research: Evaluations of Small Business Innovation Research Can Be Strengthened.* RCED–99–198, Washington, DC: U.S. General Accounting Office.

U.S. Government Accountability Office. 2006. *Small Business Innovation Research: Agencies Need to Strengthen Efforts to Improve the Completeness, Consistency, and Accuracy of Awards Data*, GAO-07-38, Washington, DC: U.S. Government Accountability Office.

U.S. Government Accountability Office. 2006. *Small Business Innovation Research: Information on Awards made by NIH and DoD in Fiscal years 2001-2004.* GAO-06-565. Washington, DC: U.S. Government Accountability Office.

U.S. Public Law 106-554, Appendix I–H.R. 5667—Section 108.

U.S. Small Business Administration. 1992. *Results of Three-Year Commercialization Study of the SBIR Program.* Washington, DC: U.S. Government Printing Office.

U.S. Small Business Administration. 1994. *Small Business Innovation Development Act: Tenth-Year Results.* Washington, DC: U.S. Government Printing Office.

U.S. Small Business Administration. 1998. "An Analysis of the Distribution of SBIR Awards by States, 1983-1996." Washington, DC: Small Business Administration.

U.S. Small Business Administration. 2003. "Small Business by the Numbers." SBA Office of Advocacy. May.

U.S. Small Business Administration. 2006. *Frequently Asked Questions*, June 2006. Available online at *<http://www.sba.gov/advo/stats/sbfaq.pdf>*.

U.S. Small Business Administration. 2006. "Small Business by the Numbers." SBA Office of Advocacy. May.

Venture Economics. 1988. *Exiting Venture Capital Investments.* Wellesley, MA: Venture Economics.

Venture Economics. 1996. "Special Report: Rose-colored asset class." *Venture Capital Journal* 36 (July):32-34.

VentureOne. 1997. National Venture Capital Association 1996 annual report. San Francisco: VentureOne.

Wallsten, S. J. 1996. The Small Business Innovation Research Program: Encouraging Technological Innovation and Commercialization in Small Firms. Unpublished working paper. Stanford University.

Wallsten, S. J. 1998. "Rethinking the Small Business Innovation Research Program," in *Investing In Innovation.* L. M. Branscomb and J. Keller, eds., Cambridge, MA: MIT Press.

*Washington Technology.* 2007. "Top 100 Federal Prime Contractors: 2004." May 14.

Weiss, S. 2006. "The Private Equity Continuum." Presentation at the Executive Seminar on Angel Funding, University of California at Riverside, December 8-9, Palm Springs, CA.

Wessner, Charles W. 2004. *Partnering Against Terrorism.* Washington, DC: The National Academies Press.